British Fortifications, 1485–1945

ALSO BY JEAN-DENIS G.G. LEPAGE
AND FROM McFARLAND

Dutch Fortifications: An Illustrated History from the Roman Era to the Cold War (2021)

Military Trains and Railways: An Illustrated History (2017)

Hitler's Armed Forces Auxiliaries: An Illustrated History of the Wehrmachtsgefolge, 1933–1945 (2015)

An Illustrated Dictionary of the Third Reich (2014)

British Fortifications Through the Reign of Richard III: An Illustrated History (2012)

Vauban and the French Military Under Louis XIV: An Illustrated History of Fortifications and Strategies (2010)

French Fortifications, 1715–1815: An Illustrated History (2010)

Hitler Youth, 1922–1945: An Illustrated History (2009)

Aircraft of the Luftwaffe, 1935–1945: An Illustrated Guide (2009)

The French Foreign Legion: An Illustrated History (2008)

German Military Vehicles of World War II: An Illustrated Guide to Cars, Trucks, Half-Tracks, Motorcycles, Amphibious Vehicles and Others (2007)

The Fortifications of Paris: An Illustrated History (2006; paperback 2010)

Medieval Armies and Weapons in Western Europe: An Illustrated History (2005; paperback 2014)

Castles and Fortified Cities of Medieval Europe: An Illustrated History (2002; paperback 2011)

British Fortifications, 1485–1945
An Illustrated History

Jean-Denis G.G. LEPAGE

McFarland & Company, Inc., Publishers
Jefferson, North Carolina

Library of Congress Cataloguing-in-Publication Data

Names: Lepage, Jean-Denis G. G., 1952– author.
Title: British fortifications, 1485-1945 : an illustrated history / Jean-Denis G.G. Lepage.
Description: Jefferson, North Carolina : McFarland & Company, Inc., Publishers, 2023 | Includes bibliographical references and index.
Identifiers: LCCN 2023027137 | ISBN 9781476689715 (paperback : acid free paper) ∞
ISBN 9781476647913 (ebook)
Subjects: LCSH: Fortification—Great Britain—History. | Great Britain—History, Military. | Great Britain—History—1485–
Classification: LCC UG429.G7 L468 2023 | DDC 623.1941—dc23/eng/20230616
LC record available at https://lccn.loc.gov/2023027137

British Library cataloguing data are available

**ISBN (print) 978-1-4766-8971-5
ISBN (ebook) 978-1-4766-4791-3**

© 2023 Jean-Denis G.G. Lepage. All rights reserved

No part of this book may be reproduced or transmitted in any form or by any means, electronic or mechanical, including photocopying or recording, or by any information storage and retrieval system, without permission in writing from the publisher.

Front cover: Walmer Castle in the sunset (Shutterstock/Oszibusz)

Printed in the United States of America

*McFarland & Company, Inc., Publishers
Box 611, Jefferson, North Carolina 28640
www.mcfarlandpub.com*

Acknowledgments

The author wishes to express his gratitude to Jeannette à Stuling, Cyril Hamiaux, Véronique Janty, Antoinette-Anna Genessey and Nicole-Juliette Lapaux, Hervé François, Jan à Stuling, Siepje Kroonenberg, Eltjo de Lang and Ben Marcato, and Simone and Bernard Lepage.

Table of Contents

Acknowledgments   v
Introduction   1

Part 1: Tudor Fortifications   3
   Historical Background, 1485–1603   3
   Residential Palaces   4
   Henry VIII's Coastal Forts   6
   The Elizabethan Era (1558–1603)   24
   Bastion Fortifications   25
   Early Bastioned Fortifications in England   28

Part 2: Bastioned Fortifications in the 17th and 18th Centuries   36
   The English Civil War   36
   Fortifications During the English Civil Wars   36
   Anglo-Dutch Wars   54
   Wars Against Louis XIV's France   61
   Jacobite Rebellion   63
   British Fortifications in North America   67
   American War of Independence   77
   British Fortifications in the West Indies   81
   British Fortifications in India and Asia   84

Part 3: British Fortifications During the Napoleonic Era   89
   Napoleonic Wars   89
   Martello Towers   90
   Coastal Forts   97
   Dover Western Heights   100
   Chatham Dockyard Defenses   105
   The Royal Military Canal   108
   The Lines of Torres Vedras   109

Part 4: Fortifications in Britain in the 19th Century — 114
Britain in the 19th Century — 114
Polygonal Fortification — 123
Palmerston Forts — 128
Sea Forts — 140
Defense of Dover — 143
Ferro-Concrete Fortification — 144
Defense of London — 153

Part 5: Fortifications in the British Empire in the 19th Century — 156
Colonialism — 156
The British Empire — 157
Colonial Fortifications — 158
British Fortifications in Canada — 163
British Fortifications in Gibraltar — 167
British Fortifications in Malta — 167
British Fortifications in India — 172
British Fortifications in New Zealand and Australia — 173
British Fortifications in South Africa — 175
British Fortifications in the Boer Wars — 176

Part 6: British Fortifications in World War I — 178
The First Phase of the War — 178
Trench Warfare — 179
Tactics and Weapons — 186
Allied Victory — 190

Part 7: British Fortifications in World War II — 192
Evolution of Fortifications in the Interwar — 192
Operation Sea Lion — 193
Anti–Invasion Preparations — 195
Coastal Defenses — 197
Inland Fortifications — 201
FW3 Pillboxes — 208
Obstacles — 231
Sea Forts — 238
Churchill's Command Bunker — 240
Evaluation — 241
German Fortifications in the Channel Islands — 242
Other Theaters of Operation — 243

Conclusion	249
British Fortifications Today	251
Appendices	255
1. British Kings and Queens, 1485–2022	255
2. British Prime Ministers, 1721–1945	256
Bibliography	259
Index	261

Introduction

This book is the continuation of *British Fortifications Through the Reign of Richard III*, previously published by McFarland in 2012. This second part covers the British fortifications from the Tudor period until the end of the Cold War. The Tudor period opened a new chapter in the history of Britain and also in the history of fortification, as, since then, warfare was dominated by the use of firearms.

By 1500 CE, Europe began to enter into a new phase of its existence, which was marked by the foundation of colonial empires beyond seas and oceans and by the gradual spread of its influence and domination throughout the inhabitable globe.

By that time, the British Isles ceased to occupy a remote position at the western edge of the known world. Instead, its location widely open to the Atlantic Ocean gave the British people a tremendous advantage they were quick to seize in the great days of exploration. During the next four centuries, British sailors, ruthless adventurers, fighting men, and greedy merchants laid the foundations for the most widespread and most prosperous colonial empire the world had ever seen. An important contribution made by Great Britain to the modern world lies in the vigorous democratic spirit, which shaped its constitution and became a model for Western democracies. More than a hundred years before the French, the British people made it plain that no king could expect to enjoy divine rights over them. At the same time and until today, they favor traditions, and the British reserve, combined with the genius they have manifested in so many fields of human activities, are admired by the rest of the world and are part of what makes Britain unique, fascinating, and great.

The word *fortification* (the military scientific art of strengthening positions against attack) comes from the Latin *fortis*, meaning "strong," and *facere*, meaning "to make." It thus implies the creation of man-made defense works often reinforcing natural obstacles.

The purposes of fortification are various and have always depended on the development of attacking weapons and siege techniques. A vast enceinte around a town was intended to give shelter to the population and their wealth. The same idea may be seen in medieval feudal lords who built their personal castles for security against attacks by their neighbors and, for example, also to watch over towns, bridges, or fords from which they drew toll revenue. Vertical fortifications (featuring high walls, towers, and gatehouses) governed the art of fortification since the beginning of mankind. However, vertical defenses lost a great deal of their efficiency when firearms were introduced, developed, and used by the end of the Middle Ages. Since then, siege warfare has been dominated by artillery fire and the use of explosives. Until the 20th century, fortresses have been constantly adapted and modernized in order to cope with the advances made by firearms and explosives in a never-ending race between offense and defense.

Introduction

In this book, the emphasis is on British coastal fortifications and on combinations of fortresses used for more general strategic purposes, namely, to protect points of utmost importance, such as capital cities, military depots, and dockyards or at strategic points both in Britain and in the British Empire.

Fortresses do not decide the issue of a campaign; they can only influence it or deter enemy attack. The actual problems that soldiers, military engineers, and statesmen have to consider are too complex to be dealt with in generalities, and no mere theorical treatise can take the place of knowledge, thought, good sense, experience, and practice.

Of course, this book reflects the author's rather anglophile inclination, personal predilections, biased opinions, and particular areas of interest. Largely based on the works of previous historians of British fortifications, this book is the author's "own version" of a well-known and already-told story including both well-known and more obscure fortifications.

Facing abundant sources, it was necessary to make a drastic subjective selection, to neglect or even omit important buildings and facts, in order to present and examine only examples displaying the development of British fortifications in a chronological manner, highlighting major changes in building, design, defenses, weapons, and warfare with drawings illustrating the major features discussed. The purpose of this illustrated book is to present in a simple and accessible form the rather complex issue of British fortifications to a wide and nonacademic public, to give a rather general overview of the subject, to provide a reliable source of information and reference for the professional and amateur military historian, the serious student, and the interested and curious lay reader alike. It is hoped that it will stimulate interest and point to new areas for research.

PART 1

Tudor Fortifications

Historical Background, 1485–1603

Henry VII

England, finally defeated in the long struggle with the French Valois kings (the so-called Hundred Years' War from 1337 to 1453), fell into a bitter civil war—known as the Wars of the Roses, which lasted from 1455 to 1485, fought between the Houses of York and Lancaster. After many vicissitudes, battles, skirmishes, sieges, battles, political machinations, and assassinations, Henry Tudor, a powerful Welsh lord, Earl of Richmont, and descendant of John of Gaunt, Duke of Lancaster, took the English crown by force of arms and held it by strength and skill. So Henry the Welshman became King Henry VII, and England settled down to a new age of orderly government under the royal House of Tudor. Henry VII's government was prudent, managerial, efficient, and unostentatious. His foreign policy was studiedly unadventurous, giving peace, prosperity, and security to the country for 24 years. Henry VII died in 1509 and was succeeded by his son Henry VIII, born in 1491.

Henry VIII

The wise and rather prudent astuteness of the father strangely contrasted with the extravagant flamboyance of the son. Henry VIII slammed the door of medieval history and brought England into the Renaissance. He flung away his father's carefully hoarded treasures in wild pursuit of ostentatious magnificence, expensive glamor, as well as foreign adventures and wars that placed England in great danger of invasion. In the late 1530s, Henry was desperate for a male heir, and when his wife, Catherine of Aragon, failed to produce a living son, he decided to divorce her. An annulment of the marriage could come from only the pope, who was under the influence of Catherine's nephew, the king of Spain. So began a serious conflict with the mighty Hapsburg dynasty (which effectively controlled Spain, the Holy German Empire, and the Low Countries) and with Rome in 1534. The quarrel about the divorce led to a serious crisis, followed by a far-reaching clash: Henry VIII ordered the dissolution of all Catholic monasteries in England in 1536–39, the confiscation of their wealth, and most important, the establishment of an independent national state Anglican Church. Almost in one fell swoop, Henry VIII shattered the whole medieval ecclesiastical culture and prepared the way for renewed religious, social, and economic activity. After Catherine of Aragon, Henry VIII married five more times, and after his death in 1547, he was succeeded by his young son,

Edward VI (born in 1537). The sick and fragile Edward reigned from 1547 until his death in 1553, and was only a pawn in the hands of his uncle the Duke of Somerset and later the Duke of Northumberland. Edward was succeeded by his half-sister Mary (the daughter of Catherine of Aragon). Although the fanatical Catholic queen Mary Tudor (reigned from 1553 to 1558) attempted by intrigue, betrayal, terror, force, and violence to reverse her father's Reformation and bring the country back into the Catholic orbit, England moved even further into the sphere of the northern European Protestant world. A direct confrontation with the rampant forces of the Spanish-Austrian Hapsburgs, the champion of Catholic Europe, was soon inevitable.

Residential Palaces

During the Renaissance, changes in society gradually led to the decline of the traditional castle. Where the castle had once served an important defensive, administrative, and residential role, these functions were now being better served by other buildings. Nobles then looked for more comfortable and airy homes while forts manned by professional soldiers took over the defensive duties. Some old medieval castles remained centers for local administration, and many were used as prisons long after they had ceased to serve a residential role. Some existing castles were turned into luxurious palaces, and new residences were constructed.

Monarchs of the Renaissance regarded militarily strong private castles as an inherent threat. Legal and other measures were taken to eliminate them. In Britain, the Tudors were particularly effective in eliminating great noble castles as part of a well-designed program to establish the state's monopoly on war and violence.

Tudor residential buildings were wholly Gothic in form and were nearly all secular. The confiscation of the Catholic properties provided large funds and enabled Henry VIII to launch an ambitious building and modernizing program with great, elegant, and luxurious residential palaces such as Hampton Court, Windsor Castle, Whitehall Palace, and St. James's Palace.

Located upstream on the banks of the river Thames about 12 miles from Central London in Middlesex, Hampton Court is certainly one of England's most fascinating and impressive royal palaces. Originally a 14th-century manor house, Thomas Wolsey, then-archbishop of York and chief minister to Henry VIII, spent lavishly to turn it into the palace it is today, only to have to hand it over to his king as he fell from favor in 1529. Construction continued until 1694.

As the oldest royal residence in Britain to have remained in constant use by successive monarchs, Windsor Castle in Berkshire represents an amazing transformation originating from a simple and basic Norman timber motte-and-bailey castle constructed c. 1080 to a modern and large royal palatial residence.

Part 1. Tudor Fortifications

Windsor Castle.

Henry VIII's Coastal Forts

Transitional Fortifications

The loss of French territories under Henry VI and years of internal strife had reduced England to a lesser power on the international stage, while France had emerged from the turmoil of the Hundred Years' War as a powerful nation. When the proud and bellicose Henry VIII acceded to the throne in 1509, he was determined to put England back on to the European map. However, through his policies and aggression, as well as his complicated conjugal life, Henry brought England to the brink of disaster. In 1539, England was threatened by a simultaneous triple invasion from Scotland, France, and Spain. Preparations to repel the imminent attacks were hastily set in motion. Musters were levied in each county, and the navy was mobilized to counter the Spanish fleet that was gathering at Antwerp, Belgium. Fortunately, the invasion did not materialize, but the fear and hysteria generated by the threat prompted Henry VIII to build in a hurry new coastal defensive works on the English southern shores.

By the reign of Henry VIII, offensive fire weaponry was universally adopted. Firearms were slowly developed, but gradually since the middle of the 15th century, artillery had made tremendous advances, cannons were capable of breaching any fortification, and handguns (pistol, arquebus, and musket) were rather common infantry weapons. Traditional castles, whose strength was based on high walls and high towers, now only offered superb targets for enemy gunners. The design of the older castles meant they could no longer stand up to attack by cannon fire. Besides, their essentially vertical nature was not intended and not suited to integrate defensive artillery arrangements. Guns being placed high on the walls were made ineffective by bad mountings, which did not allow for proper depression. With the existing vertical system of defense becoming obsolete, new ideas came to the fore. The latter part of the 15th century and the early 16th century saw the development of fortresses designed not only with a view to resisting attack by artillery but also to integrating artillery in the defense. Roughly speaking to simplify a complicated matter, two main movements to fortification design emerged in Europe in the wake of the crisis brought about by the development of siege warfare with artillery.

First, in Italy the approach consisted of low, compact, and geometrical designs focusing on flanking. Advocated by such Renaissance artists, thinkers, architects, designers, and engineers as Michelangelo, Niccolò Machiavelli, Leonardo da Vinci, Francesco di Giorgio Martini, Jacomo Castriotto, Girolame Maggi, Francesco de Marchi, Antonio and Guiliano da San Gallo, and many others, geometrical angled fortification ultimately led to the emergence of a sound and complete system of defense known as *tracé à l'italienne* or *bastioned fortification*.

Second, the northern European approach to artillery fortification came up with a logical product of medieval thought in the form of mighty, roundish, projecting stone-built gun towers. Advocated by German military architects, notably by the artist Albrecht Dürer in his treaty on fortifications titled *Etliche underricht zu befestigung der Stett, Schlosz, und flecken* from 1527, this school of thought emphasized frontal firing with a multitude of guns placed in casemates (vaulted chambers with gunports allowing the gun muzzle to protrude) and behind embrasures cut into thick sloping parapets in open platforms.

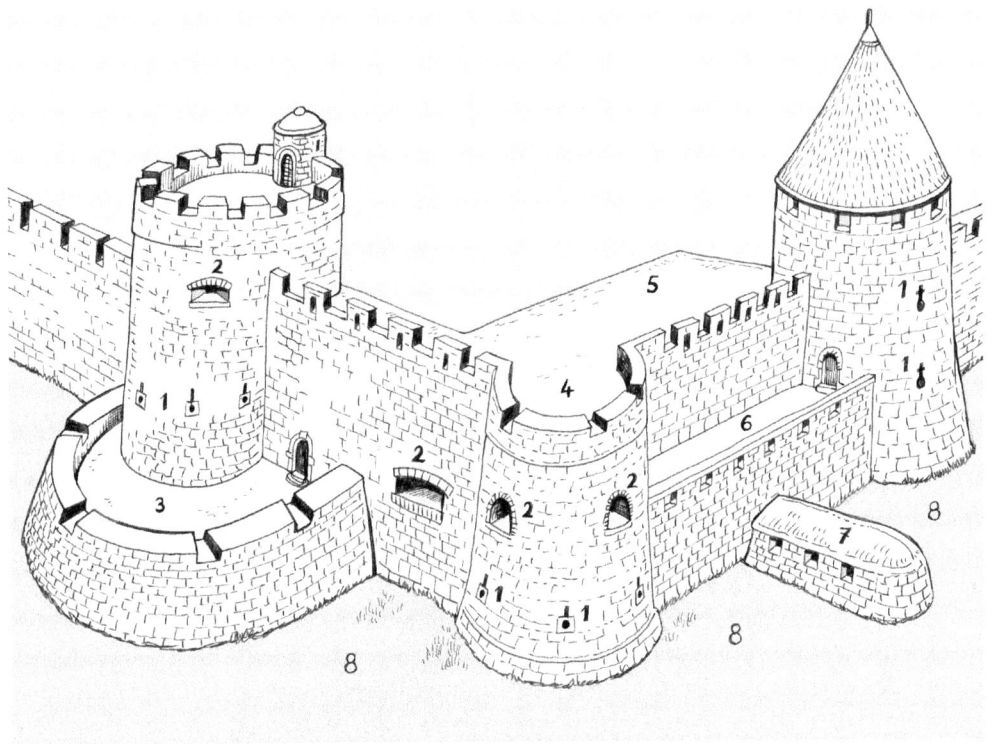

Transitional fortifications. The main characteristics of the so-called transitional fortifications are (1) arrow loops adapted to the use of portable firearms; (2) embrasure in casemates especially made for defensive cannon; (3) low gun tower called bulwark; (4) medieval tower lowered to the level of rampart; (5) rampart—a thick layer of earth added behind an existing medieval wall intended to absorb enemy fire; (6) faussebraye—a kind of artillery bulwark often placed between two existing medieval towers; (7) caponier—a flanking emplacement placed across the ditch; and (8) deep ditch.

Design of Henry VIII's Coastal Forts

For many years, the practice of fortification in Britain had dwindled into insignificance, and the revolutionary system designed by the Italians seemed no more than a distant rumor. It is therefore quite understandable that when Henry VIII decided to start his program of fortification, the angled Italian system was not considered simply because it was not known. Instead, it was the German school of thought that was adopted, probably because of the involvement between 1539 and 1543 of the German-born military engineer Stefan von Haschenperg (1511–1571). Although much is known about the organization and functioning of the workforces on Henry VIII's projects, very little can be glimpsed of Haschenperg's origins, background, and previous career except that he was a "gentleman from Moravia and a competent surveyor." Other designers involved in the Henrician coastal scheme were as follows. Thomas Cromwell (1485–1540) was the chief organizer of the fortification scheme. He was also responsible for the dissolution of the Catholic monasteries. He was created Earl of Essex but later fell out of grace and was executed by order of the king.

Sir Christopher Morris was Master of the Ordnance, and together with master and deviser of the works James Needham, was involved in the surveying of various projects. Master Mason John Rogers and Surveyor of Works Richard Lee worked at Calais and Guisnes in northern France and at Hull. Rogers and Lee introduced Italian elements in the design of late-Henrician forts. The king himself took a great interest in the latest developments in the military science of the day including warship and artillery. Henry VIII's coastal fortifications were to provide the backbone of the land defense against an invasion, but the first and main weapon against aggression was to be the new, apparently mighty but totally untested naval force. Henry is traditionally cited as one of the founders of the Royal Navy. His reign featured some naval warfare and, more significantly, large royal investment in shipbuilding (including spectacular great ships such as the *Mary Rose* and *Henry Grâce à Dieu*), dockyards (such as the naval base at Portsmouth), and naval innovations (such as the pioneering use of cannons placed in tiers on lower decks firing "broadsides" through gunports protected by hinged shutters). Henry VIII was also greatly involved in fortress design. After the Reformation of the 1530s, Henry VIII's exchequer had been considerably buttressed from the spoils gained through the confiscation of Roman Catholic Church properties, so he had a comfortable budget enabling him to undertake the most expensive scheme of coastal defense that England had ever witnessed so far.

Henry VIII's coastal forts not only took account of the new firearms, but they also represented a new concept of military defense. They were designed with a view to resisting attack by artillery and with the view to using artillery. First of all, it must be said that although Henry VIII's coastal defenses were, and still are, termed "castle," they were not castles but actually forts. A fort is quite different from a traditional medieval castle. It is a stronghold designed, built, and maintained by the state. It is not a private home or a baronial residence built and occupied by a feudal lord, his family, his guards, and his retinue. A fort was (and still is) a surveillance and combat place manned by a professional, exclusively military garrison headed by a governor—an officer appointed and paid by the state. The fort's function could include ruling and administrating a territory, but its primary role consisted of defending a strategic point purely for the purpose of national defense.

The majority of the Henrician forts were built along the south coast from Kent to Cornwall, and virtually all were intended to protect havens, ports, anchorages, estuaries, islands, inlets, bays, and beaches where an enemy invasion force could disembark troops. Built on high ground, their purpose was to make difficult for a force to land safely and speedily. If the forts could not prevent a landing, they were to disrupt and delay it until the local militias, navy, and royal army could be gathered and massed for counteroffensive. Every hour the fort held out gained time for Henry VIII's field army and warships to come into action. Completely new forts were built notably at Sandown, Deal, Walmer, and Sandgate in Kent, at Calshot and Hurst on the Solent (the strait separating the Isle of Wight from the mainland of England), and at Portland in Dorset. In addition, blockhouses—simpler fortified structures often mounting only a few guns and musketeers—were erected, namely at Tilbury and Gravesend to protect the Thames and London, at Cowes on the Solent, and at Sandsfoot opposite Portland. Temporary batteries made of earth were added to existing strongholds, e.g., at Dover and Camber in Kent, Dale and Angle in Wales, and at Calais in northern France. Henry VIII's coastal castles are not to be seen in isolation but in the context of a complete fortification system. Henry VIII's coastal forts, built between 1539 and 1543, represented an astonishing achievement, as they were the first integrated designs of military work in England, which took the various functions and requirements of artillery into full account. They broke completely fresh ground even though the precise construction varied from fort to fort. The variations in design suggest that different military architects were involved, but nonetheless, all forts have a very strong family resemblance, drawing heavily on the north European large gun tower approach to artillery fortification. Indeed, circular and symmetrical multistory gun towers had already been built notably at Modon in Greece in 1480, at Prague in 1484, at Kufstein in North Tyrol, Austria, between 1518 and 1522, and at the entrance of the harbor of Rhodes by the Knights of St. John between 1480 and 1520.

Basically, Henry VIII's multitiered forts were solid, compact, and cohesive gun platforms. In design, curves dominated and the forts were often symmetrical with a clover-leaf plan. Many consisted of a central cylindrical tower topped by a cupola providing daytime illumination and fitted with a spiral staircase giving access to all floors. The central core was surrounded by low, D-shaped, half-round or circular towers often with rounded battlements to present a small deflective target to enemy cannonballs. They were sunk in the ground, thus offering no easy target to enemy gunners, and all were surrounded by a deep, artificial ditch. Most of them were self-defensive—capable of all-round defense. Heavy guns were placed in casemates and mounted on open platforms on top of the towers. The theory of their operation was based on the superiority of shore armament over contemporary ship weaponry. The heaviest gun commonly found on 16th-century sailing ships was the 18-pounder culverin, while the forts (where weight and stability were no issues) were provided on their sea faces with the 32-pounder demicannon, which was more powerful and had a longer range. Besides, coastal artillery occasionally made use of hot shots, which were incendiary cannonballs heated in a furnace. As can easily be imagined, a hit with a red-hot shot caused serious damage to a wooden sailing ship.

So the coastal fort was the master of the ship, and the only way a ship could approach the shore was to do so with the assistance of a land force, which of course had to land elsewhere, out of reach and sight of the fort and marching overland to attack it in the back. In order to thwart this possible tactical maneuver, Henry VIII's forts had all-round fire capability using 18-pounder culverins and 9-pounder demiculverins. In addition, there were ample provision of smaller gunports for light handguns for close-range firing to the vicinity and flanking the ditch. All forts had a single entrance with a fixed bridge and a movable drawbridge crossing the deep ditch, as well as a gatehouse fitted with portcullis, a trapdoor covering a pit in the entrance, murder holes in the ceiling, and angled passage covered by firing ports for handguns.

The internal arrangements were as well designed as the external, and each fort obviously included quarters for the men and their commanding officers, magazines for ammunitions, powder, and weapons, and food store while water was obtained from wells and cisterns. The forts were intended to be garrisoned by professional soldiers. In peacetime, the garrison was small (only 24 men headed by a captain at Deal, 15 at Camber, for example). In wartime, this select cadre would train additional volunteers for the use of heavy cannons and handguns.

Henry VIII's forts had none of the splendors and graces of a medieval castle, but they presented a remarkable unity of style. Although they were erected to cope with an emergency, the workmanship throughout is remarkable. They were beautifully built, highly utilitarian, and with their fine, spare, decorative detailing, powerful and beautiful architectural realizations.

Disadvantages

In 1539, invasion did not come, and but for a few small, isolated raids, Henry VIII's forts were not put to the test of war. This may have been quite fortunate because they were not without functional faults, clearly showing that their designers were not acquainted with the most up-to-date development in south European fortification design. Henry VIII's forts indeed presented a number of disadvantages and shortcomings.

First of all, their stone walls were not ramparted—meaning not reinforced with a thick mass of earth at the back that would have given them a greater capacity to absorb the impact of enemy cannon shot. Also, their roofs and upper gun platforms were not vaulted in masonry but made of timber and lead, and thus highly vulnerable to plunging projectiles lobbed by mortars. Although Henrician coastal forts had formidable firepower (e.g., no fewer than 145 gun and handgun locations at Deal Castle), the curvilinear towers festooned with numerous apertures were structurally weak, and their general curved form left wide and dangerous "dead ground" around their bases—blind spots that could not be seen and not swept by the fire of their guns. Closed casemates gave excellent protection to guns, gunners, and ammunitions, but their embrasures considerably reduced the observation possibility and limited the traverse of the weapons. Operating firearms generates toxic gases, thus the dispersion of fumes in a confined casemate was a serious issue for the gunners after a few shots had been fired. In spite of vents and chimney flues, the casemate was an unhealthy place full of choking and acrid smoke in combat and a dark, humid, and cold cave in peacetime. Finally, living conditions were rather primitive and poorly suited in the event of a long siege.

Haschenperg was sacked in 1543, officially for bad behavior and extravagance, but perhaps Henry VIII had a shrewd idea that the German engineer's concept was already obsolete. In the end, the Germanic approach looked backward instead of forward and rapidly proved a short-lived dead-end street, only a temporary transitional form between the high medieval castle that had dominated European fortifications for 500 years and the low, solid bastioned fortification described below. Obviously, the future belonged to the geometrical Italian angled bastioned system, which became the universal standard of European fortification for almost three centuries.

A quick and remarkable transformation took place in English fortress design in the second half of the 1540s. In those years, war with France broke out again, and under the pressure of a French invasion in 1545, and the brilliant endeavor of the king, the clover-leaf design, embodied by the beautiful symmetry of Deal Castle, was abandoned almost as soon as it had come into being. In fact, the first Henrician forts were outdated even before they were completed.

Henry VIII's acceptance of the mainstream of the Italian angled-fortress design is quite evident, e.g., at Southsea and Yarmouth Forts built in the late 1540s.

By adopting the revolutionary Italian bastioned system under the reigns of Edward VI (1547–1553), "Bloody" Mary Tudor (1553–1558), and Elizabeth I (1558–1603), England hoisted itself in the art of fortification with amazing rapidity from a quasi-medieval situation to a position where it could at least claim some parity with other European powers.

In spite of their limitations and shortcomings, some of the Henrician forts played a military role during the Spanish War (1588–1598), the British civil wars (1642–1651), and the Napoleonic Wars (1790–1815). Some were later refurbished and incorporated within modern coastal defenses or served as bases for policing the coastline against smuggling in the 18th and 19th centuries. As late as 1940, during the Battle of Britain, some were used as observatories. Today, several forts have suffered from coastal erosion, but fortunately some have survived. They represent a rich historical and architectural legacy spaced along the coast of southern England, and many are open to the public.

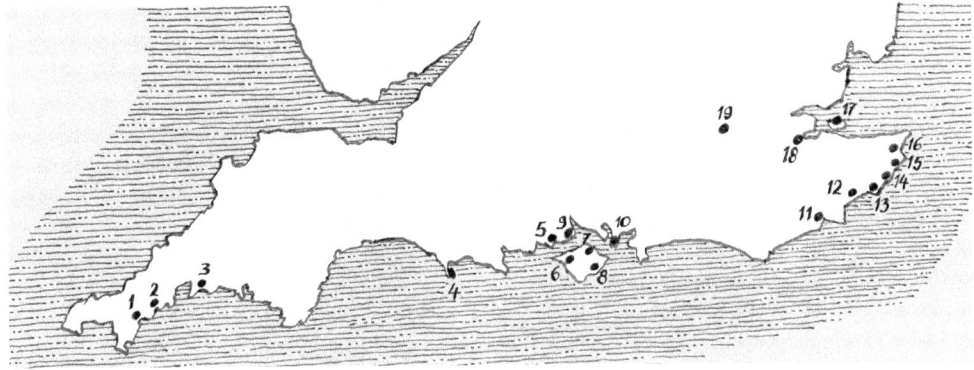

Map of South England with Henry VIII's main coastal forts. (1) Pendennis, (2) St. Mawes, (3) St. Catherine's, (4) Portland, (5), Hurst, (6) Yarmouth, (7) Cowes, (8) Sandown, (9) Calshot, (10) Southsea, (11) Camber, (12) Sandgate, (13) Dover, (14) Walmer, (15) Deal, (16) Sandwich, (17) Queensborough, (18) Upnor, (19) Gravesand.

Deal Castle

Deal Castle is the largest, most important and best preserved of Henry VIII's coastal forts. Despite later alterations, it remains a remarkable fortress. Deal was linked to neighboring forts Walmer and Sandown by a continuous ditch, an earth wall, and circular interval gun batteries. The three forts clearly formed a distinctive group along the shore for a common purpose: the defense of the Downs—a stretch of water between the Kent coast and the Goodwin Sands. The Downs provided a safe anchorage and was a key position in the days of sail, particularly in times of war when hostile ships were frequently spotted. The three forts and the interval batteries worked in concert with one another to provide overlapping arcs of fire. Although Deal and all other Henrician coastal forts were built to cope with an emergency, the workmanship throughout is superb. Deal Castle, with its double-clover-leaf plan, embodies the main ideas underlying the design of Henry VIII's forts. It is interesting to note that the fundamental principles of the concentric castle were revived here together with modern features. The entire complex was sunk into the earth and surrounded by a stone-lined flat-bottom dry ditch about 50 feet wide and 16 feet deep, whose plan echoes the six-lobbed shape of the main structure. The fort is composed of six low projecting semicircular towers whose flat roofs were arranged as gun platforms with parapets and embrasures. From a continuous gallery in the basement, light handheld guns could sweep the moat from 53 gunports should intruders penetrate that far. The central keep had a remarkable structure based on a large hollow pillar containing a well in the basement, and spiral stairs leading to the first floor and the gun terraces. Around and within the central keep, there are six inner, semicircular gun towers. Here, again, numerous hand-gun ports in the walls of these towers could command the outer areas if need be, while heavy guns were mounted on the flat terraced roofs. The fort had a single entrance on the west side via a stone-built causeway with originally a drawbridge. The tower facing due west formed the gatehouse with a porter's lodge, an entrance hall, and a passageway including a portcullis and five murder holes.

Deal Castle.

Part 1. Tudor Fortifications

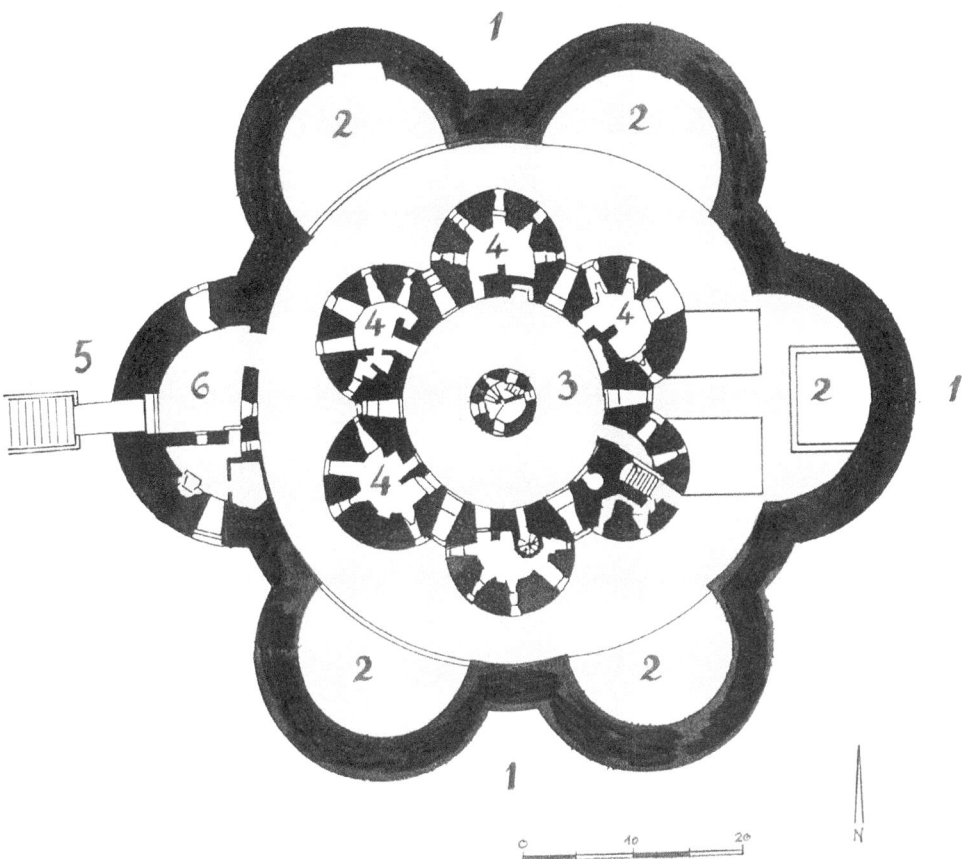

Plan of Deal Castle. Deal Castle, with its double clover-leaf plan, embodies the main ideas underlying the design of Henry VIII's forts. It is interesting to notice that the fundamental principles of the concentric castle were revived here with modern features. The entire complex was sunk into the earth and surrounded by a stone-lined, flat-bottomed dry ditch (1) about 50 feet wide and 16 feet deep, whose plan echoes the six-lobbed shape of the main structure. The fort is composed of six low, projecting, semi-circular towers (2) whose flat roofs were arranged as gun platforms with parapets and embrasures. From a continuous gallery in the basement, light hand-held guns could sweep the moat from 53 gunports, should intruders penetrate that far. The central keep (3) had a remarkable structure, based on a central hollow pillar containing a well in the basement and spiral stairs leading to the first floor and the gun terraces. Around and within the central keep, there are six inner, semi-circular gun towers (4). Here again, numerous hand-gun ports in the walls of these towers could command the outer areas if need be, while heavy guns were mounted on their flat terraced roofs. The fort had a single entrance on the west side via a stone-built causeway originally with a drawbridge (5). The tower facing due west formed the gatehouse (6) with a porter's lodge, an entrance hall, and a passageway including a portcullis and five murder holes.

Fort Walmer

Fort Walmer displays the same characteristics as Deal but on a smaller and simpler scale. It has a dry revetted moat crossed by a causeway, but instead of the six towers, it has only four (a quatrefoil plan), and the central feature is a single round tower. Its arcs of fire were designed to overlap with those of nearby earth batteries and Deal Castle to the north. Walmer is substantially intact but has been much added to by reason of its function as the official residence of the lords warden of the Cinque Ports.

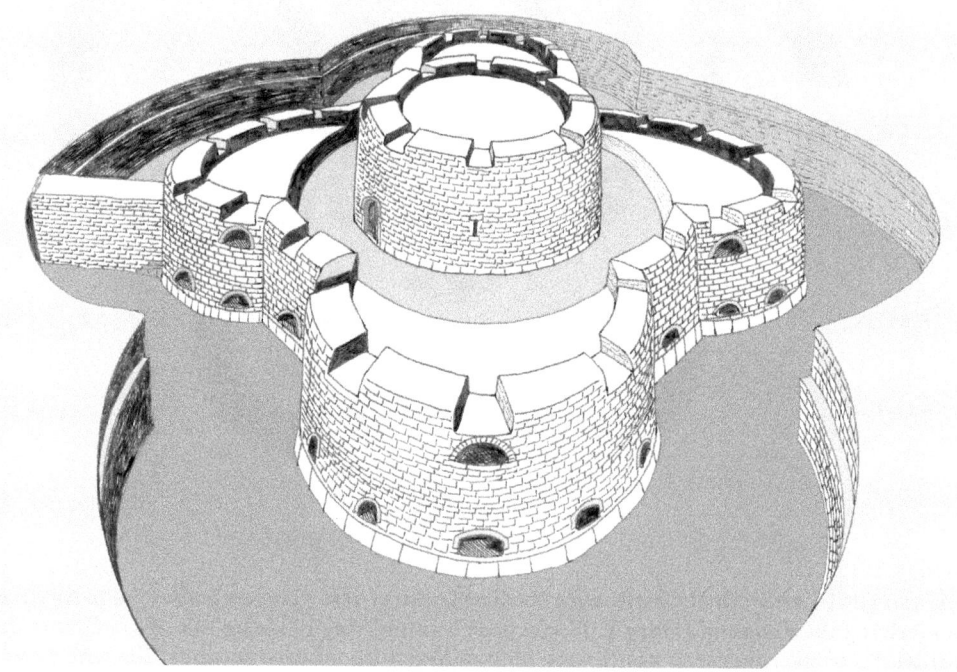

Fort Walmer.

Sandgate

Unlike other Henrician coastal defenses, the fort of Sandgate, located only 25 miles from the French coast, was not designed to protect a harbor or an anchorage but was intended to repulse a landing on the vulnerable beach between Dover and Folkestone. The fort, built in 1539–40, was centrally planned to provide three lines of fire including a central three-story high keep, an oval inner curtain flanked by three low artillery towers, and an outer curtain whose outline echoed the main inner structure. The entrance was defended by an outwork, aka barbican. Today, the top tower has been washed away by coastal erosion.

Part 1. Tudor Fortifications

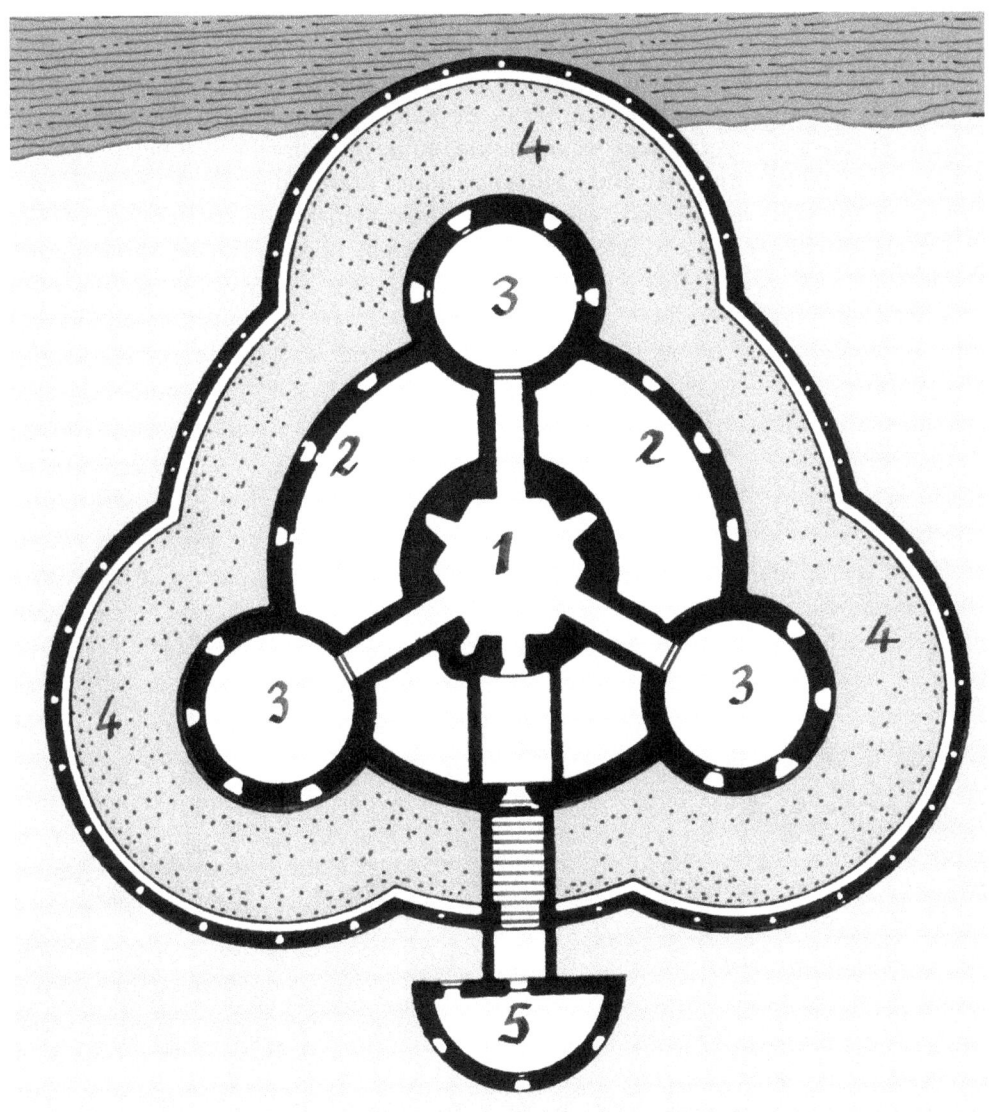

Sandgate Fort plan. Unlike other Henrician coastal defences, the fort of Sandgate, located only 25 miles from the French coast, was not designed to protect a harbor or an anchorage but intended to repulse a landing along the vulnerable beach between Dover and Folkestone. The fort, built in 1539–40, was centrally planned to provide three lines of fire including a central three-story-high keep (1), an oval inner curtain (2) flanked by three low artillery towers (3), and an outer curtain (4) whose outline echoed the main inner structure. The entrance was defended by an outwork also known as a barbican (5). Today the top tower has been washed away by coastal erosion.

Calshot Castle

This Henrician fort, located on Calshot Spit at the Solent near Fawley, was intended to guard the entrance to Southampton Water. Built in 1540, it featured a low circular gun battery, a three-story central keep using recycled stone from nearby Beaulieu (Catholic) Abbey, a ditch filled with water, and a gatehouse with a drawbridge. The outer walls were lowered in 1774, and the gatehouse was rebuilt in order to provide more living space. The castle was in military use until 1956. It is now owned by English Heritage and open to the public.

Calshot Castle.

St. Mawes Castle

St. Mawes Castle in Cornwall, together with the fort of Pendennis, was built between 1540 and 1543 in order to defend the important Falmouth River and its anchorage. St. Mawes has a remarkable unity of design and perfection of workmanship. It presents a regular geometric clover-leaf design, and its outer walls were adorned with fine royal coats of arms. The fort was, however, sited with extraordinary incompetence—namely, on the slope of a hill—so that it was utterly undefendable if attacked from the dominating landward side.

Part 1. Tudor Fortifications

St. Mawes Castle.

Plan of Fort St. Mawes. St. Mawes included a symmetrical single clover-leaf plan with a guardhouse (1) defending the drawbridge, a rock-cut dry ditch (2) surrounding three half-round artillery towers (3), and a 44-foot-high central circular keep (4).

Pendennis Castle

Built between 1540 and 1545, Pendennis, together with the fort of St. Mawes, was intended to defend Falmouth in Cornwall. The fort comprises a simple round tower and gate enclosed by a low wall fitted with artillery embrasures. It is now in the care of English Heritage.

Pendennis Castle.

Portland

Portland Castle, situated in Dorset, was built between 1539 and 1540 to protect the vital anchorage of Portland Roads. The fort appears as a segment of circle including a central two-story inner keep and a terraced gun battery flanked by wings on each side, as well as casemates (firing chambers) and embrasures at ground level to accommodate heavy cannons. In fact, Portland was more a coastal battery facing the sea than a proper fort as it lacked all-round defenses. Nonetheless, its roundish design places it within the general Henrician style observed at Deal, Walmer, and St. Mawes.

Portland. Top left: Plan.

Southsea Castle

Southsea Castle was built in 1544 in order to defend the deep-water channel into Portsmouth harbor from mid–Solent. One of the late-Henrician castles, this fort was quite different from the other Tudor coastal works with their rounded gun towers. Southsea reflected new concepts in the art of fortification developed in Italy and France in which circularity was replaced with angular outline. The angled fort showed a quick and brilliant appreciation of the new problems of fortification: the abrupt change from the curved shape to the square and triangular forms. Artillery fortifications with angled walls had been in existence in Italy and the eastern Mediterranean since the end of the 15th century, and this principle was adopted at Southsea. Indeed, straight angled walls fitted with gun ports did not leave "dead spots" and offered a better flanking. Southsea's design was actually a foretaste of the bastioned fortification that was to become universal standard.

Portsmouth

Portsmouth on the English Channel coast of Hampshire had always been an important harbor since Roman times. From the Tudor period to the 20th century, Portsmouth's fortifications were always subject to continuous modernization. Under King Henry VIII with funds obtained from the confiscation of Roman Catholic properties, the round tower was rebuilt in stone, the square tower was raised, and Italian-styled bastions were added.

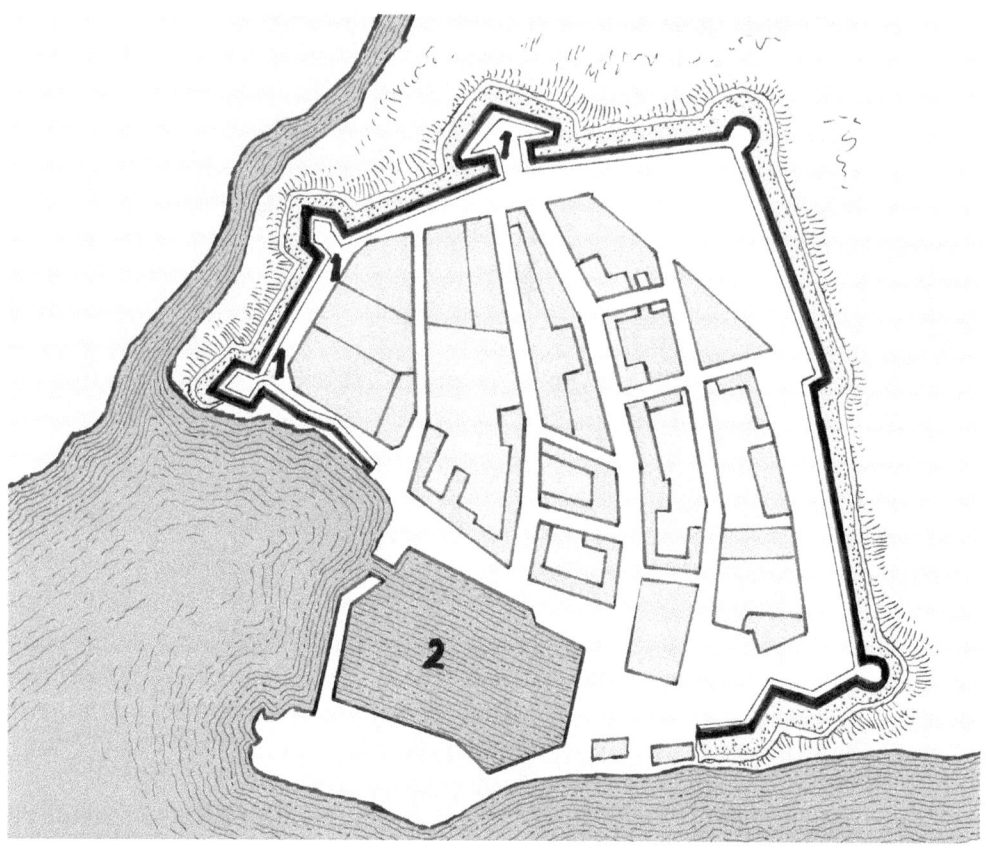

Portsmouth, c. 1540. Portsmouth on the English Channel coast of Hampshire has always been an important harbor since Roman times. Through the Tudor period, Portsmouth's fortifications were subject to almost continuous modernization. Under King Henry VIII the Round Tower was rebuilt out of stone, the Square Tower was raised, and Italian-style bastions (1) were added. It was at this time that Robert Brygandine and Sir Reginald Bray, with the support of the king, commenced the establishment of the country's first dry dock (2). In 1527 with some of the money obtained from the dissolution of the Catholic properties and monasteries, Henry VIII built the fort which became known as Southsea Castle.

Yarmouth

Yarmouth Castle, located on the Isle of Wight, was built in 1547 after Henry VIII's death. Concentric forts with clover-leaf curved walls had become obsolete by the mid–1540s, and the Italian bastioned style with angled bastions, covered way, ravelins, and other refined features imposed itself as the fortification of the future. In stark contrast to its Henrician predecessors of the 1539–40 period, Yarmouth has no keep and displays a simple diamond-shaped enclosure with thick parapets fitted with embrasures facing the sea, a small early-styled Italian "arrowhead" two-story bastion, and a 30-feet-wide dry ditch defending the landward side and the entrance at the rear. However, the lack of earth backing behind the wall (thus unable to absorb the impact of enemy artillery), the timber platform, and the unprotected faces of the bastion revealed the ignorance and uncertainty of English designers who had not yet fully understood the potential and effectiveness of the Italian bastioned system. Yarmouth Castle was intended to reinforce and work in conjunction with another of Henry VIII's coastal fort: Sandham Castle. Sandham, intended to protect the only feasible landing place on the eastern coast of the Isle of Wight, was quadrangular with a bastion, in the shape of an arrow, on the landside to give extra protection to the land walls. Today, Yarmouth Castle is absorbed into the city's construction, but views over the Solent are still magnificent. The fort is open to the public.

Yarmouth.

Cornet Castle

Cornet Castle is located on Cornet Rock outside the safe anchorage of St. Peter Port in Guernsey (British Channel Islands). From the Neolithic to the Bronze Age, the site was used by traders. The Channel Islands and Guernsey were linked to the English Crown by the conquest of the Norman duke William I in 1066. In 1204, King Philippe Auguste of France took back the mainland of Normandy from King John, but Guernsey remained an English possession. Castle Cornet was built as an English stronghold to help protect the island's safe anchorage at St. Peter Port. The first castle, begun in the mid–13th century, consisted of a half round tower and a square tower defended by walls, a ditch, and a drawbridge, and later a donjon and a barbican (outer defensive passageway) were added. There were constant attacks from French raiders during the 14th and 15th centuries, and the castle suffered great damage during these troubled times. By the end of the medieval period, parts of the castle had been rebuilt and made stronger. Between 1545 and 1548, Henry VIII ordered Cornet Castle to be adapted to the use of artillery weapons by the addition of bastioned fortifications.

Castle Cornet.

Hull

King Henry VIII visited the town and port of Hull in October 1540 and made an inspection of the existing medieval fortifications. Soon, the king gave orders for the construction of new defenses for which he himself took a particular interest in the details of the work. The Tudor defenses of Hull, carried out c. 1540–1542, consisted of Henry VIII's updating of the medieval enceinte and an additional defensive line, designed by the military engineer John Rogers. This was constructed north of the port beyond the river Hull with the purpose of protection from an overland attack from the east or a seaborne attack from the adjacent river Hull. Nearly half a mile in length, the line included a thick wall with a ditch connecting a central stone squarish fort and two roundish bulwarks (or blockhouses) at each end, all capable of mounting artillery. The blockhouses were planned as a trefoil and resembled a club on a playing card. The entrance was arranged in the square or flat part of the structure and communicated with an open courtyard 37 feet square giving access to three projecting slightly pointed towers. The blockhouses featured two stories, with walls 15 feet in thickness. Circular staircases of stone built within the walls provided means of communication between the upper and lower floors of both the central castle and blockhouses. The south section of this linear work was later incorporated within the 17th-century bastioned citadel.

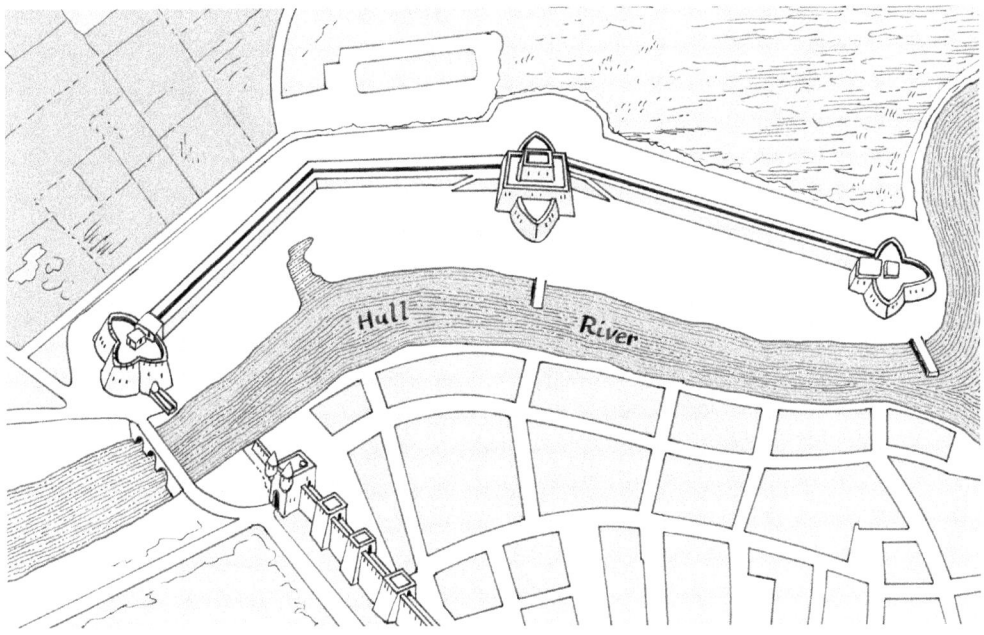

Henrician fortifications at Hull (c. 1542).

Ambleteuse

Ambleteuse, situated north of Boulogne and Wimereux, in the mouth of the river Slack, as well as Calais in northern France had been ceded to England according to the Treaty of Brétigny in 1360 during the Hundred Years' War. For two centuries, Ambleteuse and Calais remained integral parts of England. Eventually adapted to the use of artillery during the reign of Henry VIII to maintain a show of power toward the French, those English advanced bridgeheads in France were fortified in Italian style with angled bastioned fortifications designed by the English military engineer John Rogers.

Ambleteuse was reconquered in 1549 by the French king Henri II. Eventually in 1558, Calais was also ultimately retaken by the French.

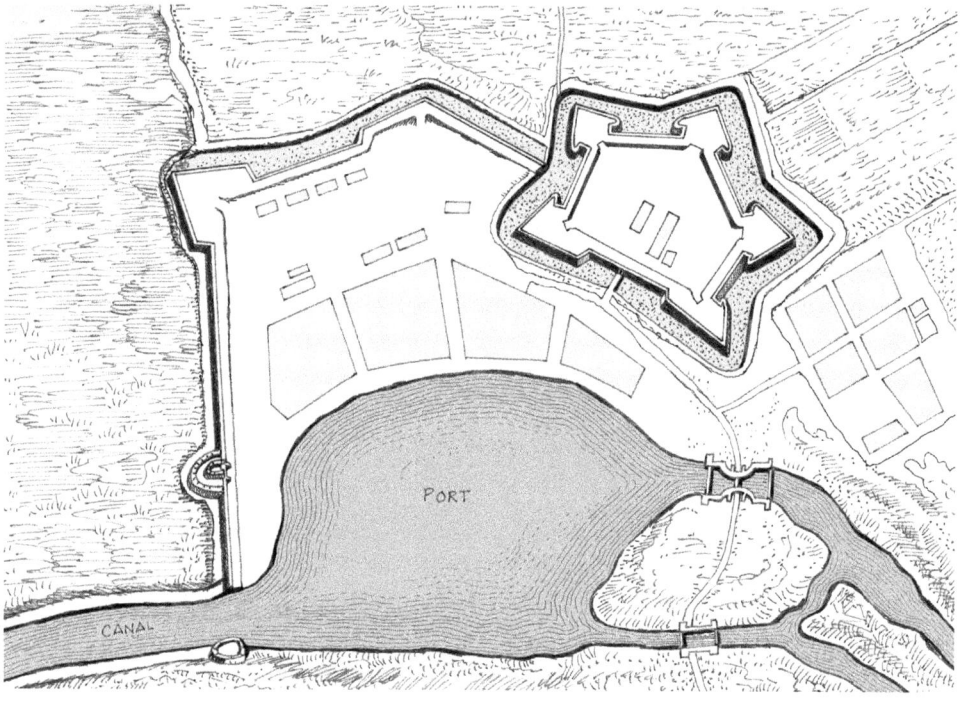

Ambleteuse.

The Elizabethan Era (1558–1603)

Historical Background

The long reign of Elizabeth I (1558–1603) marked a new period in English history. Elizabeth, born in 1533, was the daughter of the executed Anne Boleyn, and the fifth and last Tudor monarch. From her father, Henry VIII, she had inherited strength, good sense, determined will, pride, ferocity, a dazzling energy, lust for life, and passion for pleasure and splendor. Elizabeth adopted a general policy of avoiding involvement in major continental conflicts, although she intrigued constantly in European affairs. Elizabeth's main enemy was Philip II, king of Spain. The Reformation and the establishment of the national Anglican Church brought about a strategic shift in European alliances. England's traditional ally was Spain helping to fight the traditional nemesis, France. But following Henry VIII's break with Rome, relations between Catholic Spain and Anglican England greatly deteriorated, and hostility soon reached a climax. Raids and provocations on Philip's overseas empire by daring English privateers goaded him into open war. In 1588, he launched a formidable fleet, known as the Great Armada, with the intention to invade England, overthrow Elizabeth, and restore Catholicism. The Henrician sea forts were hastily put on a war foot, and measures were taken to repulse an invasion, but no Spanish troop ever landed in England. The "invincible" Armada was scattered by storm and successfully attacked by the English navy, which had been reconstituted as a powerful force—built most effectively on the sound practices and principles established during Henry VIII's reign. The ultimate defeat of the Armada resulted in fervent patriotism in England. The danger from Spain was temporary removed, the Anglican/Protestant faith was saved in England and northern Europe, and the nicknamed "Virgin Queen" or "Good Queen Bess," now at the summit of her glory, continued to reign. Ever since, Elizabeth I has been one of the best-loved sovereigns in English history.

The culture and spirit of the Elizabethan age is reflected by the work of the great poet and playwright William Shakespeare (1564–1616) and by adventurous privateers (aka sea dogs) and explorers like John Hawkins, Francis Drake, Walter Raleigh, and others, who mixed aspects of pirates, explorers, naval commanders, black slave traders, and businessmen. Mainly operating between 1560 and 1605, the sea dogs initiated the era of explorations, maritime adventures, and naval wars.

After the death of Henry VIII in 1547, less building schemes were undertaken, for the king's spendthrift policy had left the country practically bankrupt. However, at the accession of Elizabeth, the religious turmoil was largely quelled, and the prosperity of England began to revive.

Bastion Fortifications

On the matter of military architecture, the last year of the reign of Mary I and during the reign of Elizabeth I saw the general adoption of the Italian bastioned fortification. The structure of defenses changed radically by the start of the 16th century as cannons gained more widespread adoption as siege weapons. As already mentioned, cannons made the previous vertical defensive structures, such as those surrounding medieval castles and walled cities, practically useless. A new defensive system, known as bastioned fortification, was designed in Italy and was developed and adopted with astonishing speed all over Europe.

Ditch, Rampart, and Bastion

In order to counteract the power of the new weapons, defensive walls were made lower and thicker. A ditch was dug in front of them, and the earth used from the excavation was piled behind the walls to create a rampart—a thick and solid structure able to absorb the shock of artillery shots. To improve the defense of the fortress, covering and flanking fire was provided by a new structure called a bastion jutting out at regular interval along the walls.

A bastion had two main characteristics. First, in cross-section, it was a solid structure made of a rampart (piled earth held by masoned revetments) strong enough to resist enemy fire. Second, in plan, it had an angular wedged-shaped outline enabling the siting of defenders and weapons on a roomy dominating firing platform. Bastions projected the defense forward and thus subjected attackers to defensive flanking cross fire, thereby eliminating all dead ground in the ditch.

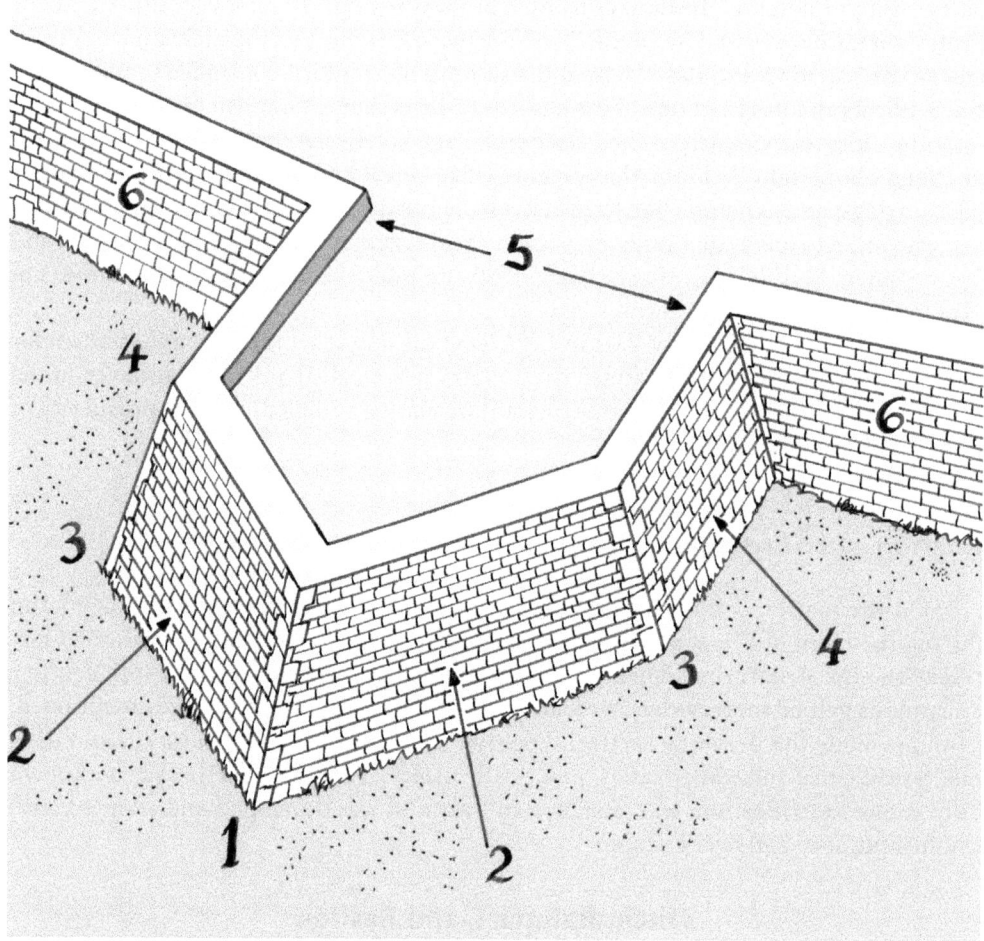

Bastion. The basic features of a bastion are (1) salient, (2) face, (3) shoulder, (4) flank, (5) gorge, and (6) adjacent curtain (wall).

Bastioned Front

The basis of the bastioned fortification was the bastioned front, an ensemble of elements related by rules and geometrical ratios. A bastioned front included two half bastions and a section of curtain (wall), a deep ditch, very often a triangular outwork called a demilune (aka ravelin placed in the ditch), a covered way (a continuous patrol lane and advanced combat emplacement established on the counterscarp), and a flat and bare glacis (denying attackers any cover). Bastioned fronts could be attached and repeated at will with endless variety of lengths and angles in order to form a fort, a wall around a city, or establish a straight continuous line of defense. The bastioned system permitted maximum defense with a relatively small garrison, it protected with efficiency from enemy fire and at the same time provided an excellent defensive platform. The bastions allowed for cross fire directed against attackers to the front and flanking fire against attackers making direct assault on the walls. The space around a bastioned fortress was transformed into a highly dangerous killing ground as every

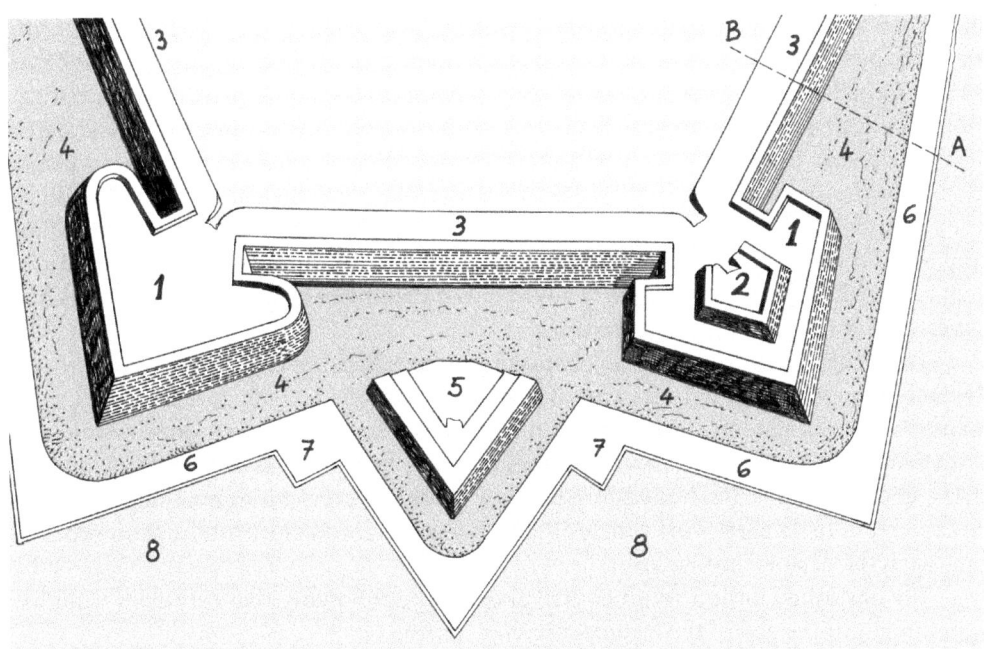

New Italian bastioned front. (1) bastion, (2) cavalier, (3) curtain (wall), (4) ditch, (5) ravelin (demilune), (6) covered way, (7) place of arms, (8) glacis.

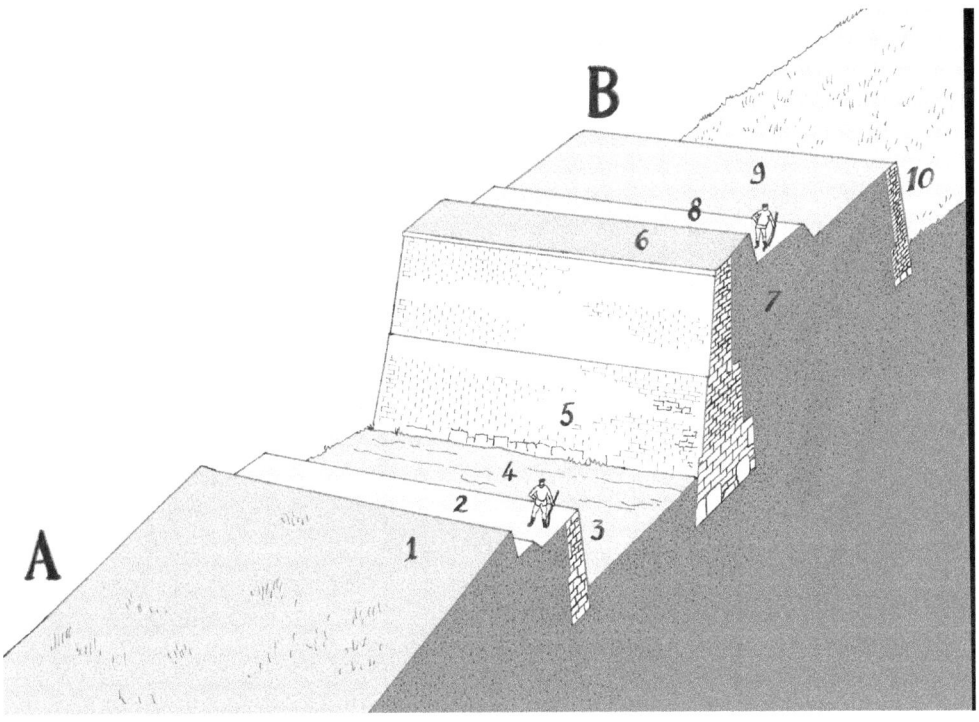

Cross-section of bastioned fortification. The cross-section shows the following: (1) glacis, (2) covered way, (3) counterscarp, (4) ditch, (5) scarp, (6) parapet or breastwork, (7) rampart (composed of masoned walls holding thick mass of piled earth), (8) firestop, (9) terreplein, (10) inner slope.

element was always covered by another. An all-round defense was achieved with minimum expenditure on a few cannons and light handguns. The disposition of elements in space and height allowed for a good command (a fundamental principle of fortification taking advantage of height). The places and distances of all works were calculated to offer defense in depth. Attackers would not only come under fire from many directions, but the deeper they infiltrated into the fortifications, the more intense would be the fire unleashed on them. The system thus eliminated all blind spots (zones below and beyond which the ground cannot be seen or fired at). In short, the bastioned system restored the balance of arms in favor of the defenders as quickly as cannons had reversed it at the end of the 15th century.

The general adoption of the bastioned system had far-reaching consequences. Notably, it marked the definite end of the private fortified castle and announced the monopoly of the state in matters of national defense. Indeed, the bastioned system was so expensive to design, build, and maintain that only rich cities, senior rulers, and monarchs could afford it. Bastioned fortification made the creation of new siege technics necessary, notably earthworks including gun batteries, trenches, and saps in order to approach the defenses undercover.

The bastioned design retained its value for the following two centuries, roughly speaking until the end of the Napoleonic Wars in 1815.

Early Bastioned Fortifications in England

The bastioned system spread from Italy in the 1530s and 1540s, and although the cost was astronomical, it was employed rapidly all over Europe. Occasionally, Italian engineers accepted commissions for military works for wealthy or royal foreign patrons, but due to religious turmoil, this practice was soon gradually discontinued—at

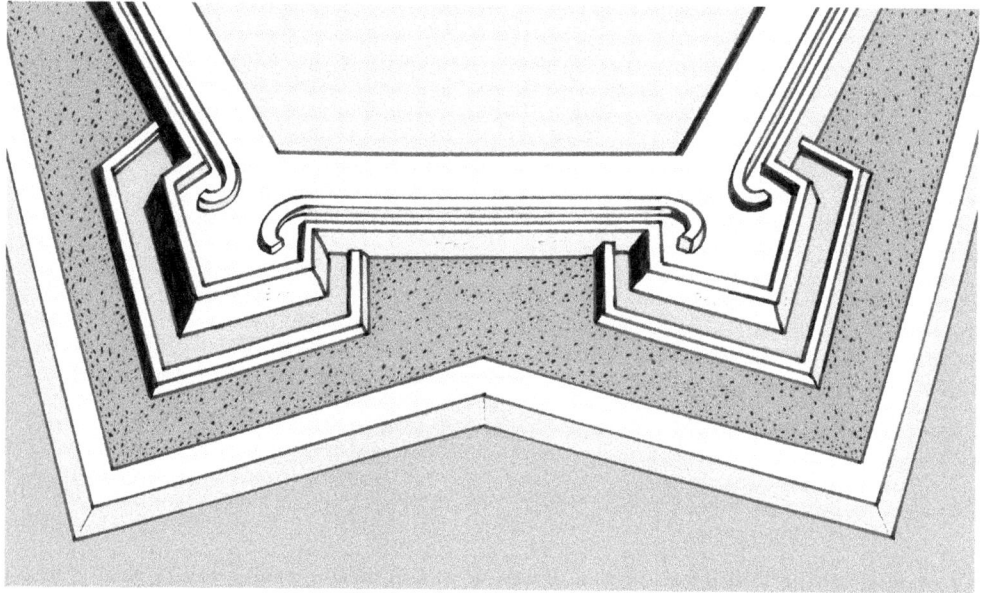

Bastioned front by Paul Ive, 1589.

least in England. Soon, a generation of British theorists, designers, and master builders emerged. Paul Ive (d. 1604), for example, was one of the most important figures in English military engineering history at the end of the 16th century. In 1589, he published a treaty—one of the first in English—titled *Practice of Fortification*.

Upnor Castle

Upnor Castle is a small Tudor fort situated on the west bank of the river Medway opposite Chatham dockyard. It was built in 1559 to protect warships anchoring in the river. It consisted of a large and low bastion facing the river and a residential towered block. The gatehouse and moat were added later. Around the turn of the 17th century, the palisade in front of the water bastion was constructed along with the moat around the landward side, with flankers to defend the curtain wall. The gatehouse and drawbridge were also built at this time. The fort was attacked by the Dutch during the Second Dutch War (1665–1667) as discussed in part 2. After the poor showing in 1667, Upnor Castle was reduced in importance and turned into stores and magazine. More modifications were made to the castle, and by 1698, the castle looked much as it does today. To the southwest of the castle, barracks were built. The castle remained in use as a magazine until 1827 and, after that date, was turned into an ordnance experimental ground and laboratory. In 1891, Upnor Castle was taken over by the Admiralty from the War Office.

Upnor Castle.

Since then, the castle has had various uses. In World War II, it was used as a depot again. After the war, some restoration work was carried out on the castle, and since 1961, it has been maintained as a national monument. Today, the castle and some of the grounds are open to the public.

Star Castle, Scilly Isles

By the 1590s, the existing defenses of Scilly Islands in Cornwall were clearly inadequate in the face of Spanish privateering and the threat of an invasion fleet. In 1593, an artillery fort, named Star Castle, was built on Queen Elizabeth's order by Francis Godolphin, the island's governor. The castle features a central keep and an enclosure in the shape of an eight-point star surrounded by a dry moat. A stone bridge over the moat leads to a defensive gatehouse, which could be closed by a portcullis. Star Castle, built in the so-called tenaille system, was not bastioned but consisted of salient walls. Each of the fort's eight points provided oblique fire flanking the ground immediately in front of the works so that a complete cover of deadly fire could be achieved. Although the tenaille system was revived in the 19th century, it never supplanted the bastioned method. Eventually, Star Castle served as a prison as well as a fortress, and it is now a privately owned hotel.

Star Castle, Scilly Isles.

Part 1. Tudor Fortifications

Corgarff Castle

Located in Aberdeenshire, Scotland, Corgarff Castle has always been of strategic importance, guarding the quickest way from Deeside to Speyside, a route later followed by the military road from Blairgowrie to Fort George. Its location contributed to Corgarff Castle's eventful and sometimes tragic history. The castle was built c. 1550 by John Forbes of Towie. It initially comprised a tower house set within a walled enclosure.

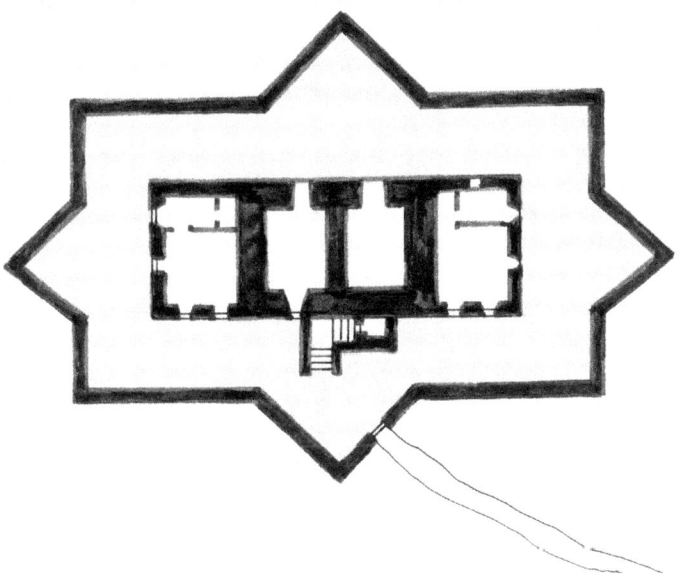

Plan and view from the southeast of Corgarff Castle, Scotland.

The tower house was similar to the structure we see today, and the surrounding wall (built in a later period) presented a tenaille (angled) plan with protruding redans (one on each side of the rectangular enceinte) fitted with musket firing holes allowing a good flanking.

Berwick-upon-Tweed

Berwick-upon-Tweed is a town in the county of Northumberland and is the northernmost town in England, on the east coast at the mouth of the river Tweed. It is situated 2.5 miles (4 km) south of the Scottish border. Strategically sited, the town had already been fortified, notably under Edward I (1272–1307). Under Henry VIII,

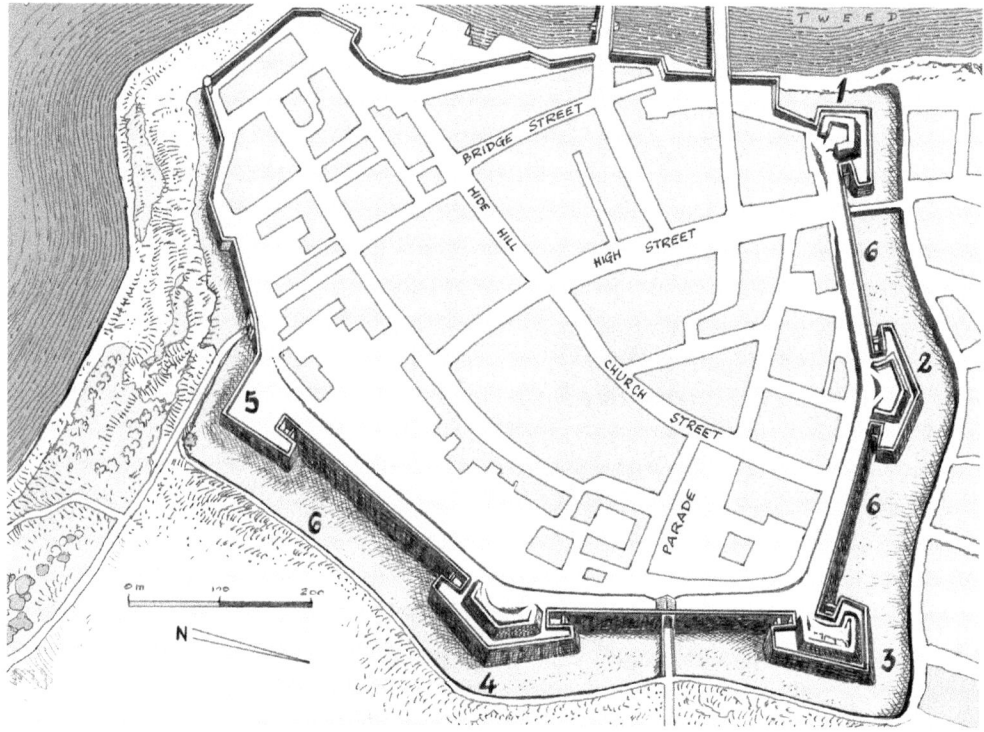

Berwick-upon-Tweed. The map shows (1) Meg's Mount, (2) Cumberland Bastion, (3) Brass Bastion, (4) Windmill Bastion, (5) King's Mount, and (6) Dry Ditch.

attempts were made to adapt the medieval defenses for artillery, and under Edward VI (1547–53), projecting bastions were added to the main walls. These structures were still unfinished in 1557, and during the reign of Queen Elizabeth I, vast sums were spent on fortifications in the new bastioned Italian style. The fortifications of Berwick, designed by Sir Richard Lee and built in the late 1550s, embody the concept of Italian bastioned fortification. Today well preserved, they have survived when so much else in the field of artillery defense has been destroyed. Therefore, they are historically very important because they display early classical arrowhead bastion type directly inspired by the Italian school.

Carisbrooke Castle

Carisbrooke Castle, located near Newport, Isle of Wight, originated from a Norman motte-and-bailey castle, later improved with stone walls, towers, a keep, and gatehouse. During the reign of Queen Elizabeth I in the period 1597–1600, large bastions and curtains, designed by the Italian military engineer Federigo Giambelli, were added. Today, the site is a historical museum.

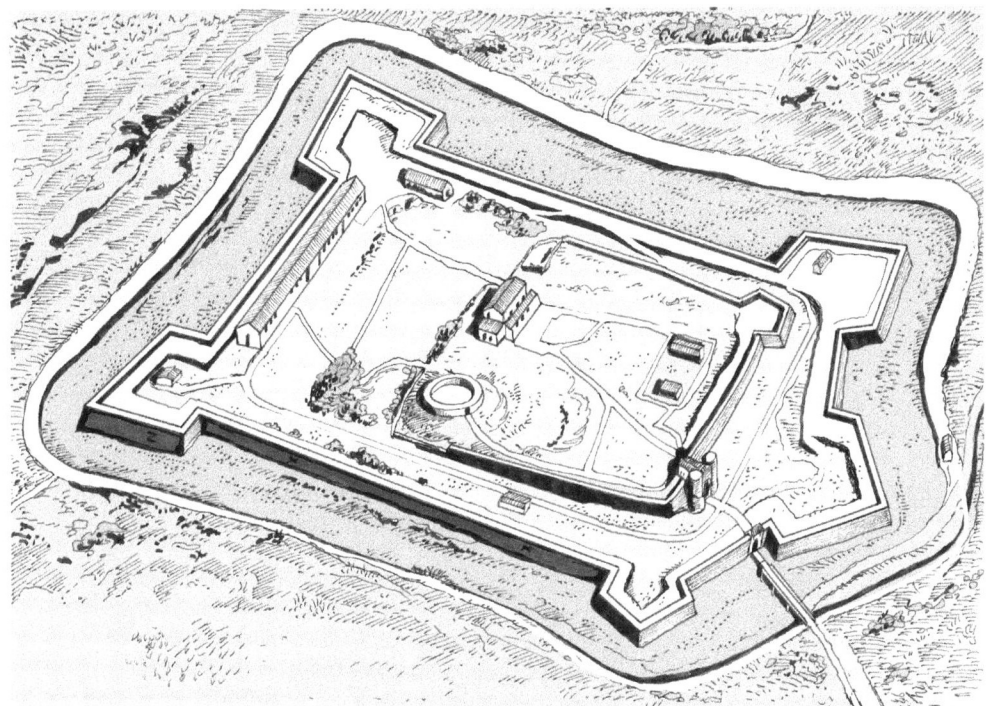

Carisbrooke.

Pendennis Castle

The coastal fortress of Pendennis and that of St. Mawes in Cornwall guarded the anchorage of Carrick Roads and the port of Falmouth for over 400 years. To Henry VIII's coastal circular fort built in 1539, a bastioned enclosure was added in the 1590s following the threat of Spanish invasion. Pendennis defenses were tested during a siege by a Parliamentary army in March–August 1646 during the English Civil War. The fortifications were periodically updated, notably during the Napoleonic Wars, later in the years 1880–1910, and during World War II. The site is intact and is now in the care of English Heritage.

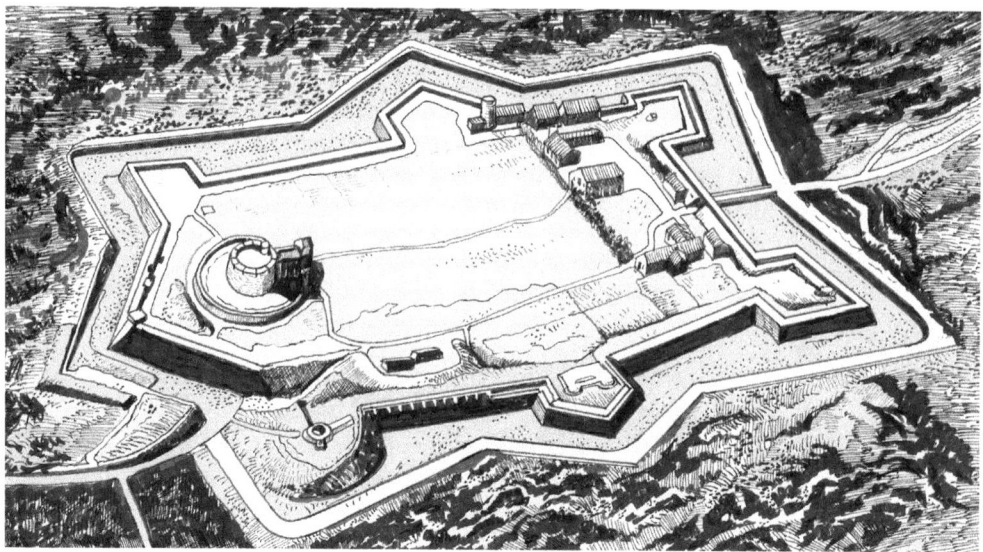

Pendennis Castle.

Roanoke

Primitive forms of bastioned fortifications were also used in the colonies, notably in the territories later known as the United States of America. The Roanoke Colony

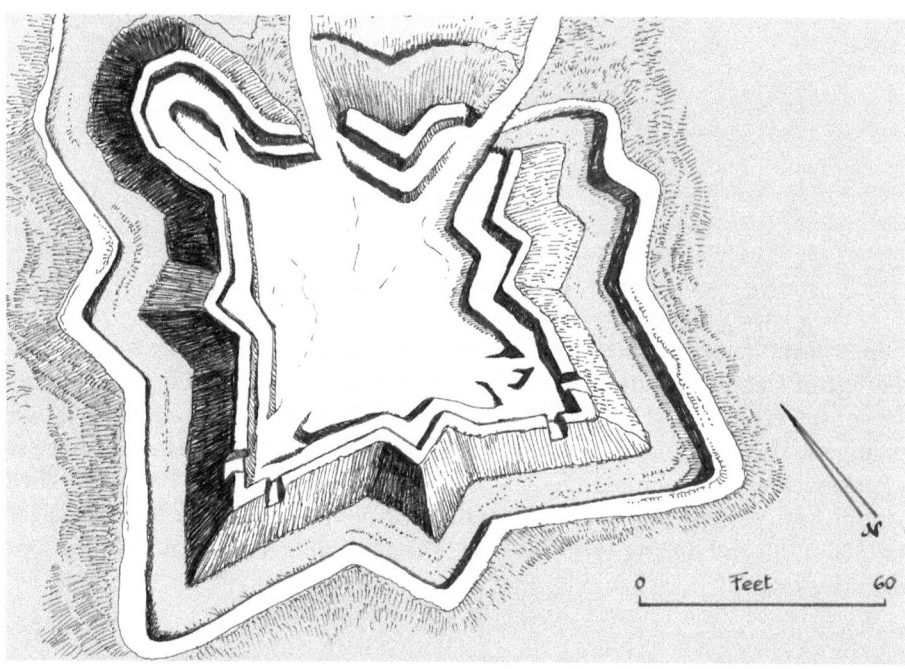

Fort Raleigh, Roanoke Island, c. 1585. The illustration is a conjectured reconstruction of the fort (consisting of a dry ditch, wooden stockade, earthen embankments with gun platforms) probably designed by Ralph Lane during 1585–86.

on Roanoke Island (in present-day Dare County in the Outer Banks of North Carolina) was an enterprise financed and organized by the English aristocrat, writer, poet, soldier, courtier, and explorer Sir Walter Raleigh (1552–1618). It was carried out by Ralph Lane and Richard Grenville (Raleigh's cousin) in the late 16th century in order to establish a permanent English settlement in the Virginia Colony. Raleigh and Queen Elizabeth I also expected that the venture should provide riches from the New World and a base from which to send privateers on raids against the treasure fleets of Spain. Between 1585 and 1587, several groups attempted to establish a colony but either abandoned the settlement or disappeared leading to the continuing unsolved mystery known as "The Lost Colony"—a question that has haunted historians and archaeologists for hundreds of years. The stronghold was also called "the new Fort in Virginia," which originally commanded a view of Roanoke Sound. Fort Raleigh was rather strange in plan, illustrating the primitive knowledge of military architecture of British sailors and early colonists.

Part 2

Bastioned Fortifications in the 17th and 18th Centuries

The English Civil War

Historical Background

The English Civil War (1642–1651) was a series of armed conflicts between Parliamentarians (Roundheads, later known as Whigs and Liberals) and Royalists (Cavaliers, later known as Tories and Conservatives). There were several causes for these wars, notably the fact that the autocratic and unwise King Charles I (born 1600, reigned 1625–1649) wanted to rule alone and his friendship toward the Catholic Church that increased Protestant fears. Bitter confrontations between the monarch and the Parliament, religious and constitutional issues mingled with economically based rivalry erupted into an armed conflict in 1642. The first (1642–46) and second (1648–49) civil wars pitted the supporters of King Charles I against the supporters of the Long Parliament, while the third war (1649–51) saw fighting between supporters of King Charles II and supporters of the Rump Parliament (so called because it was purged from Royalists). The Royalist controlled most of the north and west of the country, while Parliament controlled East Anglia and the southeast, including London. The Parliament party was supported by the navy, by most of the merchants, and by the population of London and therefore controlled the most important national and international source of wealth. Indeed, London played an important role during the English Civil War when its inhabitants took the side of the Parliament.

The civil war ended with the Parliamentary army's victory at the Battle of Worcester in September 1651. The civil war led to the trial and execution of Charles I in 1649, the exile to France of his son, Charles II (b. 1630), and replacement of the English monarchy with, first, the Commonwealth of England (1649–53) and then with a Protectorate (1653–59) under Oliver Cromwell's personal rule. After Cromwell's death in 1658, the monarchy was restored by the army in the person of Charles II in 1660.

Fortifications During the English Civil Wars

For years, military engineering had been neglected in Britain, whereas in continental Europe it had flourished and developed during the Thirty Years' War (1618–1648). The British were slow to form a corps of engineers capable of holding their own with their counterparts on the continent. When the civil war broke out in 1642, the country had only a few towns defended by permanent fortifications and only a few military engineers to design and build new modern ones. Sieges with the purpose of isolating and taking fortified places became the dominant instrument for prosecuting the war, and the conflict threw Britain into a frenzy of construction of forts, redoubts, sconces, fortlets, urban defenses, batteries, and other works to resist attack. In most cases, there was little time, not a lot of money, and not enough qualified engineers to design and build permanent masonry fortifications in the sophisticated Italian or French styles. So British civil war fortifications were marked by hasty compromises, ad hoc arrangements, and cost-cutting measures. As a result, temporary, cheap, and semipermanent earthworks were quickly erected.

Dutch Influence

The model chosen was largely inspired by the Dutch bastioned system developed during the Eighty Years' War (1568–1648) in which the independent Protestant Dutch people opposed Catholic Spain. The Dutch bastioned style was fully adapted to the peculiar conditions of the flat, low, and marshy land. Designed by the mathematician Simon Stevin and engineers Daniel Specklin, Adam Freitag, Samuel Marolois, and several others, it was developed in the practice and widely used by the engineers Adrian Anthonisz, Johan van den Kornput, Johan van Valckenburgh, Jacob Kemp, Johan van Rijswijk, and several others.

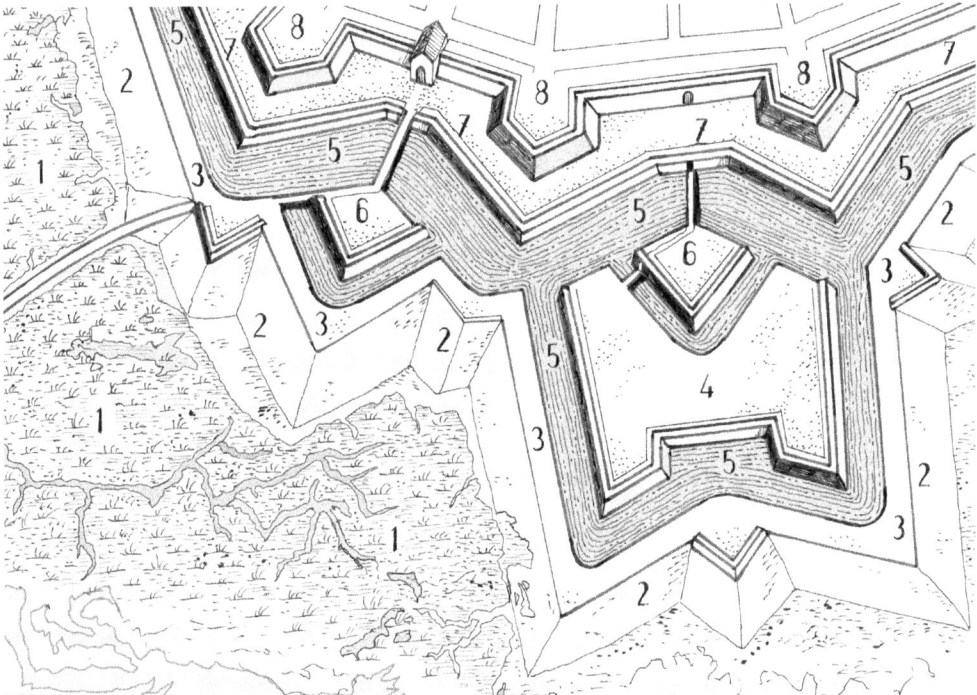

Dutch bastioned system, early 17th century. (1) Flooding, (2) glacis, (3) covered way with places of arms, (4) hornwork, (5) wet ditch, (6) ravelin or demilune, (7) faussebraye, (8) bastion.

The Dutch system was characterized by elements made of earth with very few masonry parts. It featured large floodings (intended to interdict the approach and construction of enemy siege approaching works and mining), a glacis (a wide and bare zone denying the enemy of any cover), a covered way with places of arms (first advanced defensive positions on the counterscarp), broad wet ditches (obstacles difficult to pass), hornwork and crownwork (advanced bastioned positions), ravelins or demilunes (other projecting outworks intended to defend the ditch), and often a faussebraye (a continuous low breastwork built on the scarp at the foot of bastions and curtains). The Dutch method, making use of earth banks and ditches filled with water, was rather easily and quickly built and relatively cheap—at least always cheaper than bastioned systems using expensive masonry. The Dutch system was largely exported to Scandinavia and northern Germany and was quickly adopted in Britain where attention had begun to be given to theories of fortifications.

Military Engineers

The British engineer Henry Hexham published in 1637 a book titled *The Principles of the Art Militare as Practised in the Warres of the United Netherlands*, which was the most exhaustive treatment of the Dutch military system of warfare and fortification. Hexman's book provided much-needed military instruction at the onset of the English Civil War.

The British engineer Robert Ward published in 1639 a treatise titled *Animadversions of Warre*, in many ways the most comprehensive book available to officers at the beginning of the English Civil War. Largely inspired by Dutch warfare, Ward's book included theory and design of fortification, with analysis on the attack and defense of places, and instructions on the use of mines and countermines.

The British engineer Richard Norwood published in 1640 a work titled *Fortification or Architecture Military* in which he provided detailed instructions on the design of Dutch-styled earthen fortification.

Several Dutch engineers were invited to England and hired for their skills. Johan Rosworme (c. 1630–1660), for example, was a Dutch soldier and military engineer who served the Parliamentarian cause during the English Civil War. He participated in the fortifications of Manchester in 1642, Preston in 1643, Liverpool in 1643, and Yarmouth in 1651. After being appointed engineer general of the army in 1659, there is no further record of him. It is thought that he died in exile following the Restoration.

Bernard de Gomme (1620–1685)

The most important military engineer in 17th-century Britain was a Dutchman named Bernard de Gomme. De Gomme was born in 1620 at Terneuzen near Antwerp, Belgium. After attending engineering school at Leiden University, the Netherlands, he served in the army of the Prince of Orange, Frederick Henry, notably in the siege of Breda in 1637 and in the Gennep campaign of 1641. Rapidly gaining experience, he was recruited by the colorful Prince Rupert (1619–1682—nephew of Charles I and leader of the Royalist cavalry) who took him to England. De Gomme served with conspicuous ability in the Royalist army as engineer and quartermaster general from August 1642 to May 1646, designing notably the fortifications of Liverpool, Oxford, and Newark. For

Part 2. Bastioned Fortifications in the 17th and 18th Centuries

his excellent service, de Gomme was knighted in 1645 but left England after the 1646 defeat of the first English Civil War. He then returned to the Netherlands, where he probably worked as a civil engineer in the construction of polders in Flanders. In June 1649, de Gomme received a commission from Charles II, then at Breda, to be quartermaster general of all forces to be raised in England and Wales. He took part in the Battle of the Dunes near Dunkirk in 1658, and after the English Restoration, he was appointed surveyor general of fortifications in 1660. Back in England, he was tasked with repairing the Dover pier and entrusted with designing fortifications at Dunkirk, France, as well as defense works at Tangier, Morocco (this place being temporary British possession). For political reasons, King Charles II was married to the Portuguese princess infanta Catarina Henriqueta de Bragança (1638–1705). The bride's dowry allowed Britain to secure the Seven Islands of Bombay in India and trading privileges in Brazil and the West Indies. Also, the city of Tangier in Morocco, then a Portuguese possession, was granted to Britain in 1661. The British gave the city a garrison and a charter which made it equal to other English towns. They also planned to improve the harbor by building a mole and refurbished the fortifications in modern bastioned style. A crippling blockade imposed by Sultan Moulay Ismael forced the British to abandon the port in 1684. Catarina of Bragança is often credited with introducing the habit of drinking tea to England, and the borough of Queens in New York City, United States, was named after her in 1683.

In 1665, de Gomme was commissioned to build new fortifications and a new citadel at Plymouth, and in 1673 and 1675, he made surveys at Dublin. The highly regarded Sir Bernard de Gomme died in November 1685 and was buried in the chapel of the Tower of London.

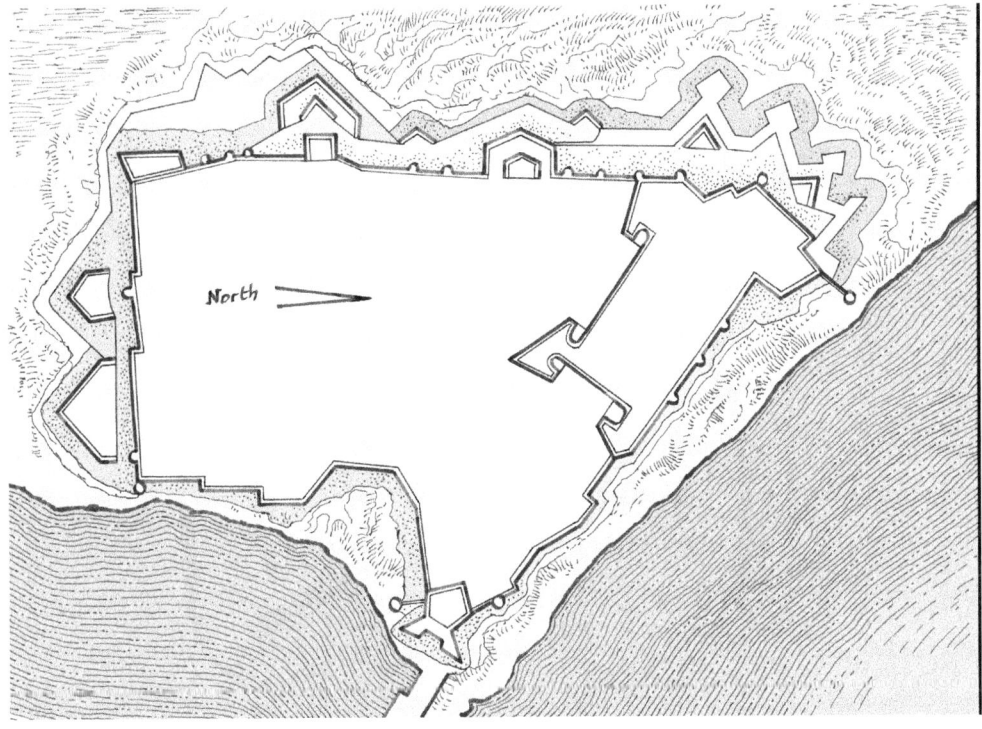

Tangier, c. 1662.

40 British Fortifications, 1485–1945

After the end of the civil war, many fortifications were dismantled, and the fate of many medieval castles was sealed by their role in the civil war.

Following their victory against the Royalists, the Parliamentary forces adopted the policy of slighting—partially or even totally demolishing castles to prevent their potential use in any future civil conflicts.

There is no point here in describing each and every numerous British fortifications influenced by the Dutch style and all designs carried out by de Gomme. The following examples illustrate what English Civil War fortifications were and underline the magnitude of de Gomme's work.

Horsey Hill

This sconce (small fort), probably built in October 1644, was an irregular pentagon. It stands near Peterborough in northwest Cambridgeshire and occupies the angle between the old course of the river Nene and an ancient watercourse known as King's

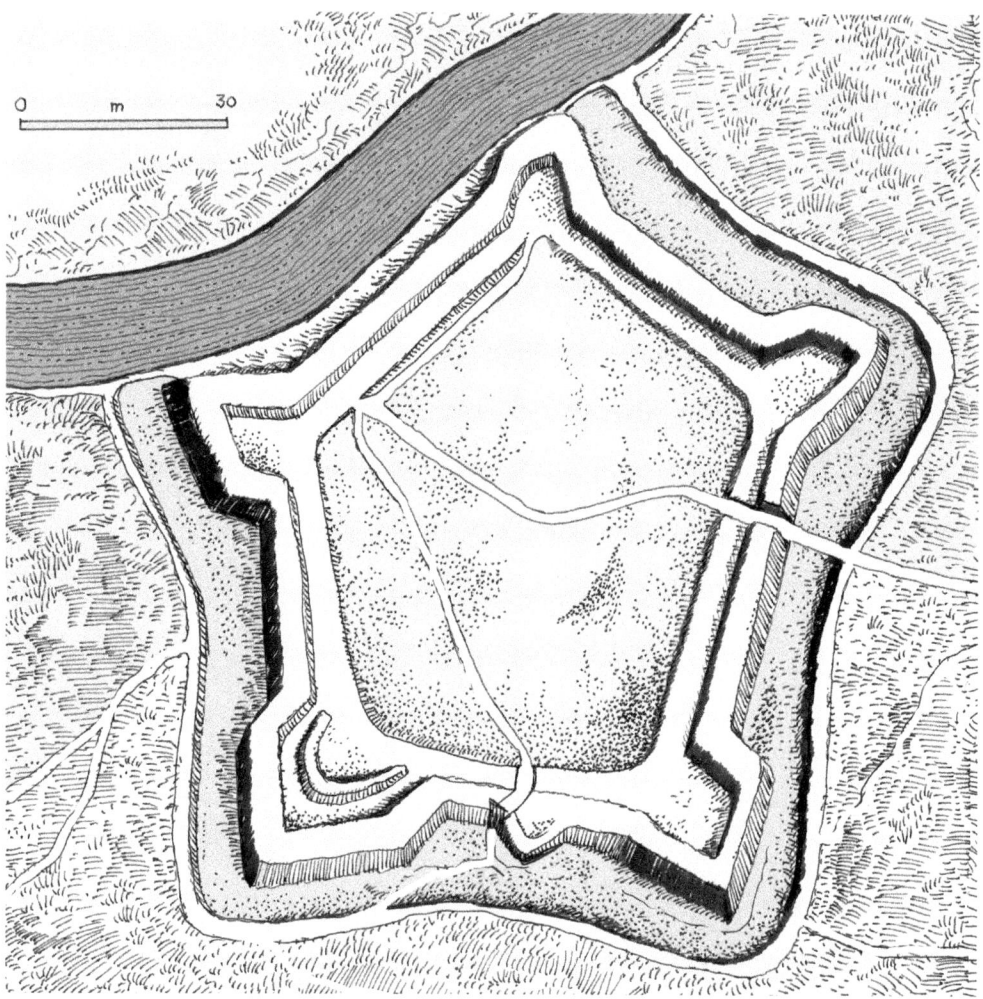

Horsey Hill.

Dyke, and is some 600 ft east of the bridge over the former stream called Horsey Bridge. The small fort with its earthen walls and bastions surrounded by a ditch is a good example of fortifications influenced by the Dutch style.

Donnington Castle

During the English Civil War, medieval castles, whether derelict or still occupied, were refortified and used as bases by the opposing forces. Donnington Castle, located near Newbury in the county of Berkshire, was seized by Royalist forces after the first Battle of Newbury. Colonel John Boys was put in charge of its defense. He ordered the construction of a series of bastioned earthworks in the shape of a star around the castle. The garrison successfully withstood a Parliamentary siege in July 1644; in October, King Charles marched to the relief of the castle, and the second Battle of Newbury was fought around it. This time, the Royalists were forced to withdraw, but Colonel John Boys refused to surrender the castle. After an 18-month siege, the garrison finally accepted terms for an honorable surrender and were allowed to march out of the castle and join Royalist forces in Wallingford.

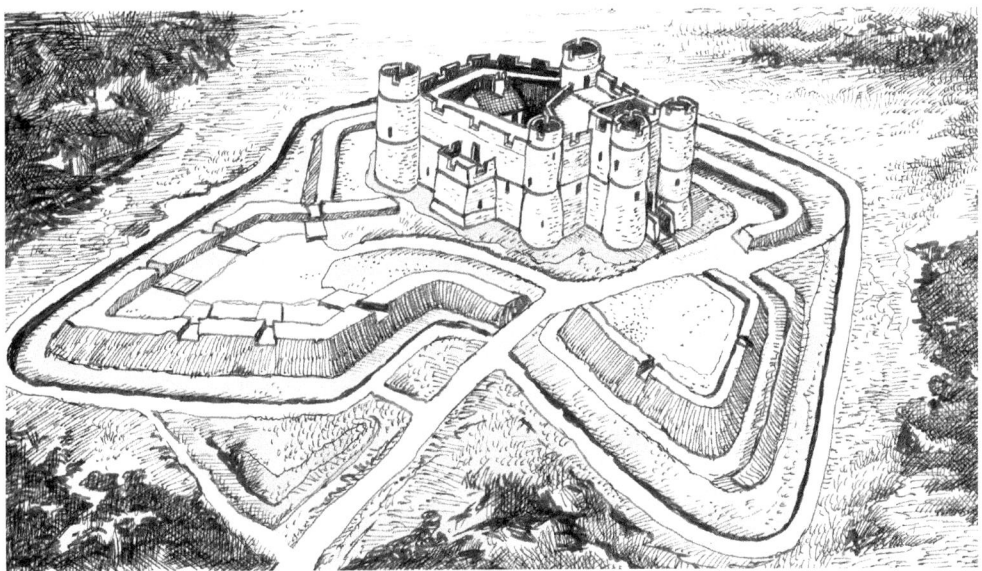

Donnington Castle, c. 1643.

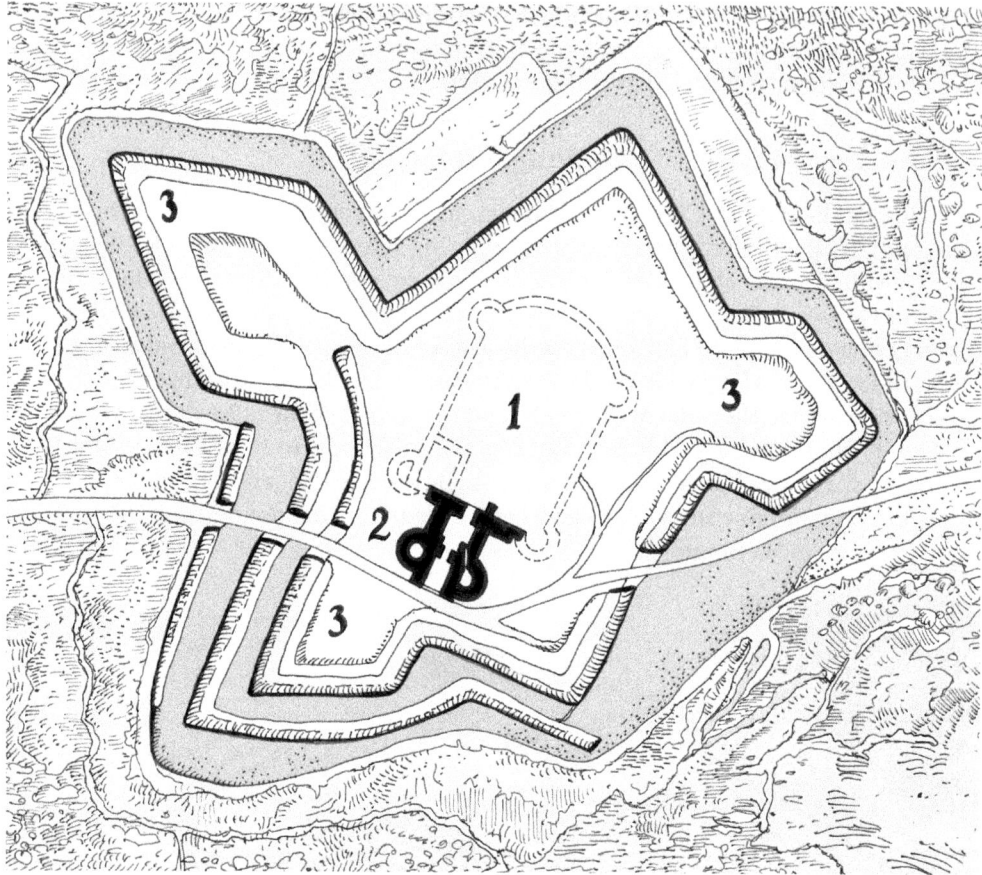

Plan of Donnington Castle. The ground plan shows the medieval castle (1) which was destroyed during the battles; the gatehouse (2) that still stands today; and the temporary earthworks, banks, bastions, wooden palisades, and ditches (3) erected during the Civil War.

Oxford

The city of Oxford, located in southeast England was founded in the 8th century. During the civil war, the military engineer Bernard de Gomme made an ambitious project to fortify the prestigious city that housed the court of Charles I in 1642 after the king was expelled from London. The town yielded to Parliamentarian forces under General Fairfax after the siege of 1646. De Gomme's project was not completed.

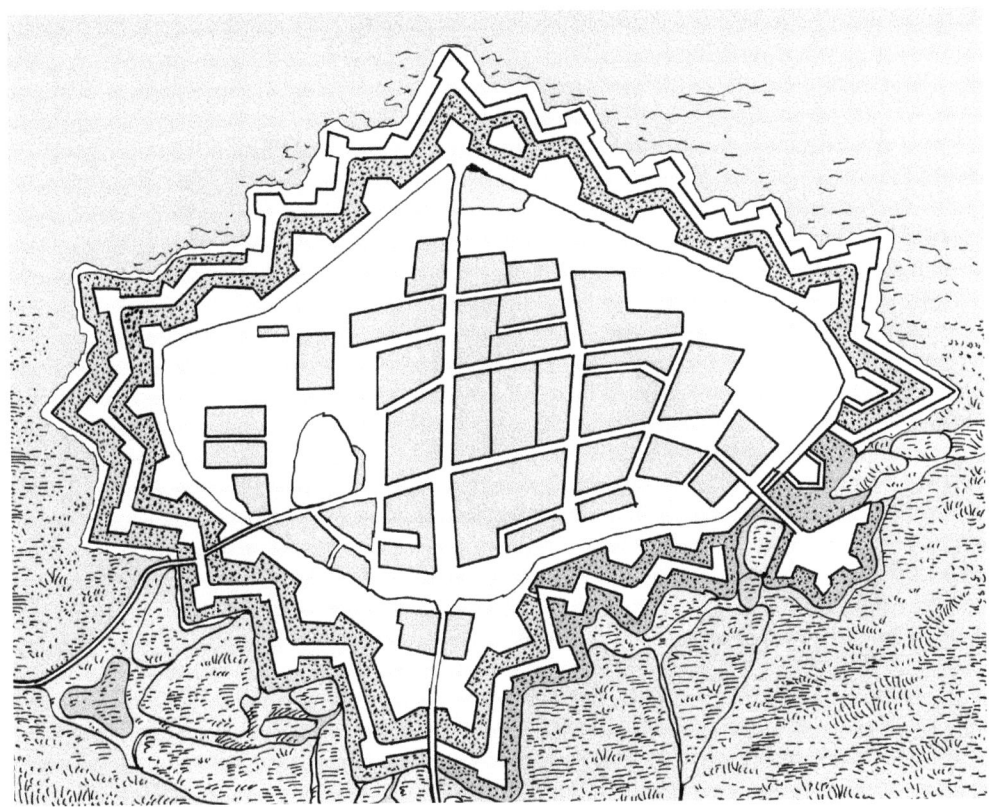

Oxford.

Gallants Bower

Gallants Bower, located near Dartmouth, was a civil war fort built between 1643 and 1645 by the Royalist garrison of Dartmouth. The fort was one of a pair; the second was at Mount Ridley on the eastern side of the Dart Estuary. It was besieged in January 1646 and capitulated. Today, Gallants Bower is a well-preserved earthwork and forms a good example of the English Civil War military engineering based on the Dutch style.

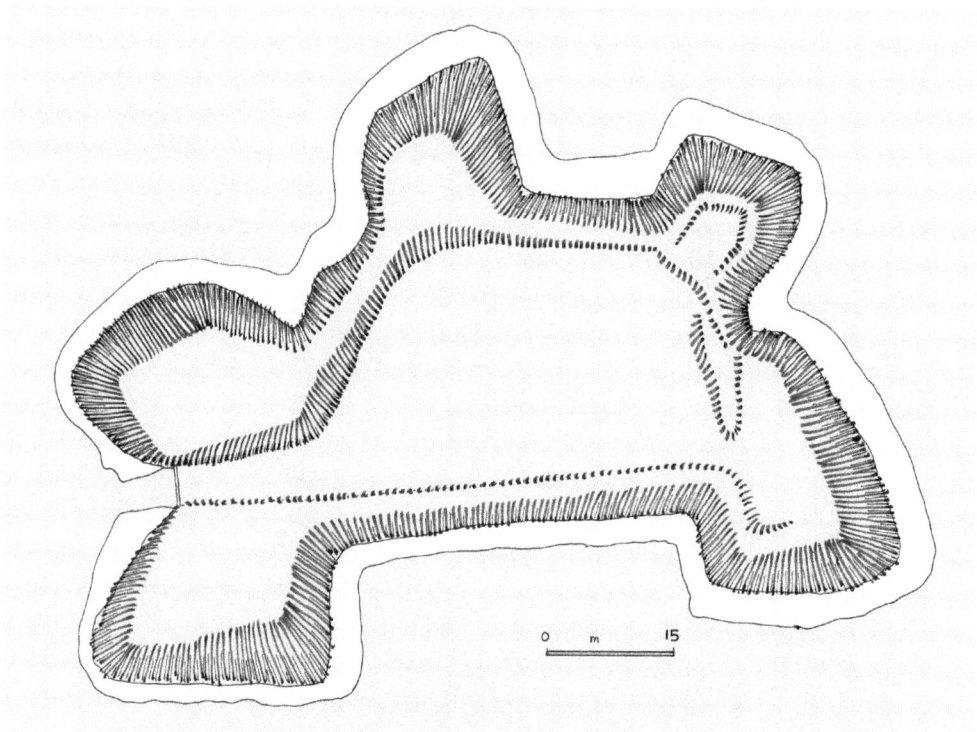

Gallants Bower.

Newark-on-Trent

Newark-on-Trent is a market town in Nottinghamshire in the East Midlands region of England. Its origins are possibly Roman as it lies on an important Roman road, the Fosse Way. The medieval town grew around Newark Castle, now ruined, and a large marketplace, now lined with historical buildings. It was a local center for wool and cloth trade. During the English Civil War, Newark was fortified by the Dutch-born engineer Bernard de Gomme. Newark was a mainstay of the Royalist cause, and it was besieged three times by Parliamentary forces, The first two sieges were relieved by Royalist armies. It was after the second siege that the Royalist forces decided to greatly strengthen their defensive positions around the town.

Part 2. Bastioned Fortifications in the 17th and 18th Centuries 45

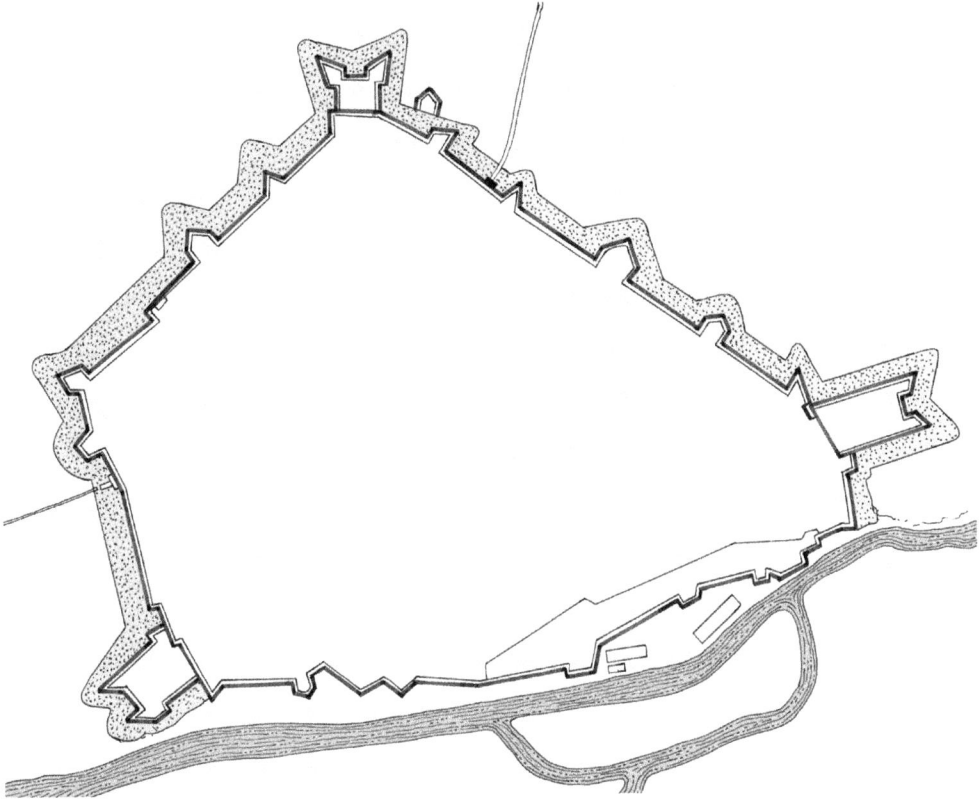

Newark-on-Trent, c. 1646.

Queen's Sconce

The term *sconce* is probably derived from the Dutch word "schans" meaning small fort or redoubt. The Queen's Sconce is located on elevated ground near the town of Newark-on-Trent in Nottinghamshire. It is one of Britain's best preserved earthwork fortifications of the English Civil War period. The Royalist stronghold was named after the wife of Charles I, Queen Henrietta. There was also a King's Sconce to the northeast of the town. The sconce served to control access to the town across the river Trent. It covers an area of 1.2 hectares; it is 9 m (29 ft 6 in) high and 92 m (302 ft) wide. The ditch (at its widest) is 23 m (75 ft 5 in) wide and 4.5 m (14 ft 9 in) deep. The sconce, in typical Dutch style, is a square with four bastions at the corners and is made of earth and gravel from the flood plain of the nearby river. Originally, the walls were probably strengthened by a palisade—a fence of wooden stakes fixed in the ground. The garrison was normally 20 soldiers but 150 when under threat serving five artillery pieces during the siege of 1646. Since 1957, the ditches and banks of the earthwork have been cleared of a dense jungle of small trees, bushes, shrubs, and brambles and is now the centerpiece of a pleasant public park.

Queen's Sconce (Newark).

Part 2. Bastioned Fortifications in the 17th and 18th Centuries 47

Bristol

The city of Bristol on the river Avon in southwest England received a royal charter in 1155. An important port on the Bristol Channel, the city was fortified during the English Civil War. It was occupied by Royalist military, who built the Royal Fort House (a pentagonal fort with five bastions designed by Bernard de Gomme) on the site of an earlier Parliamentarian stronghold. The fort was designed as the western headquarters of the Royalist army under Prince Rupert. Royalists retreated into the fort when the Parliamentarians broke through the lines in the siege of 1645, before eventually surrendering to Cromwell's forces.

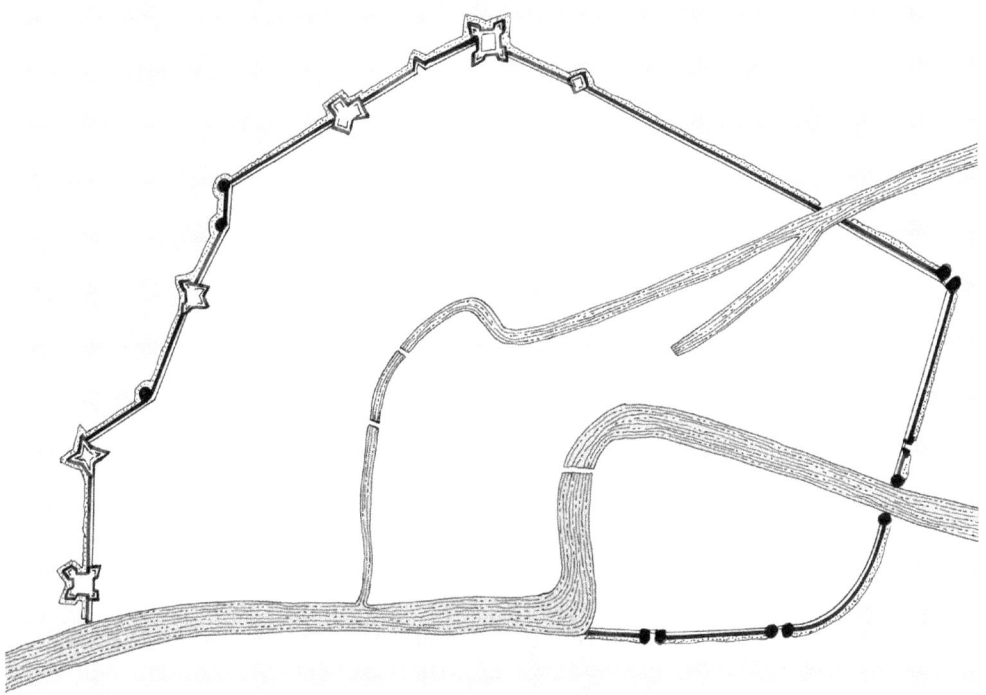

Bristol.

Worcester

The city of Worcester, located in the West Midlands in England on the river Severn, was founded by the Romans in the 1st century CE and became a thriving city in the

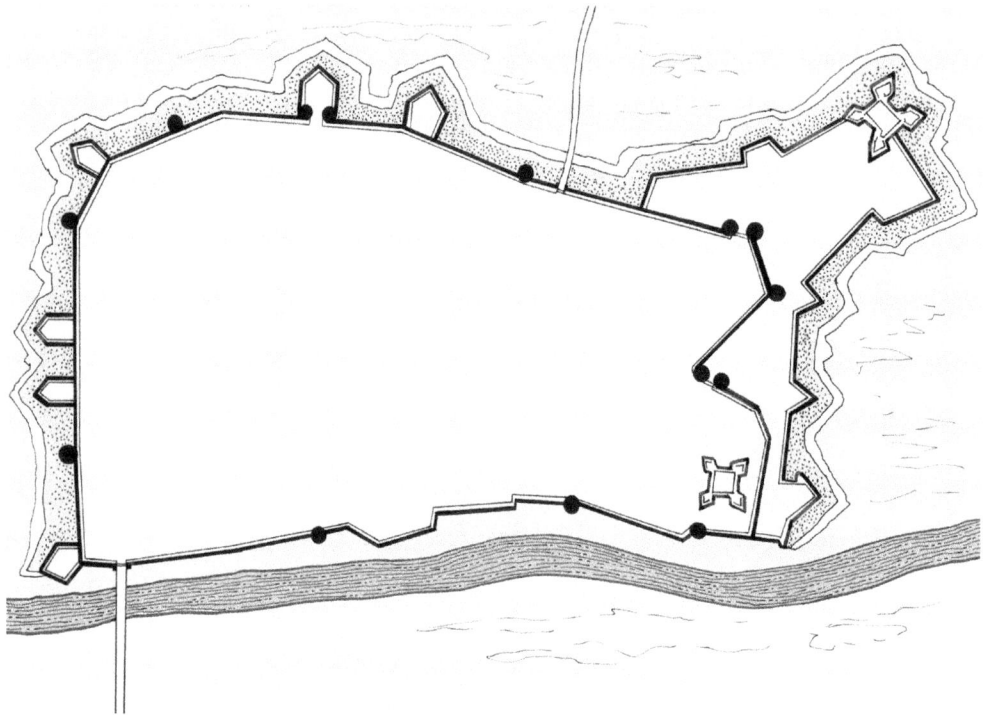

Worcester.

Middle Ages. Worcester was the site of the Battle of Worcester (September 3, 1651) in the fields to the west and south of the city, near the village of Powick, when Charles II attempted to forcefully regain the crown. Charles II was defeated and returned to his headquarters in what is now known as King Charles House in the Cornmarket, before fleeing in disguise to Boscobel House in Shropshire from where he eventually escaped to France. Worcester was one of the cities loyal to the king in that war, for which it was given the epithet "Fidelis Civitas" (The Faithful City).

London

London played an important role during the English Civil War when its inhabitants took the side of the Parliament. After the battle of Brentford in 1642, the city of Lon-

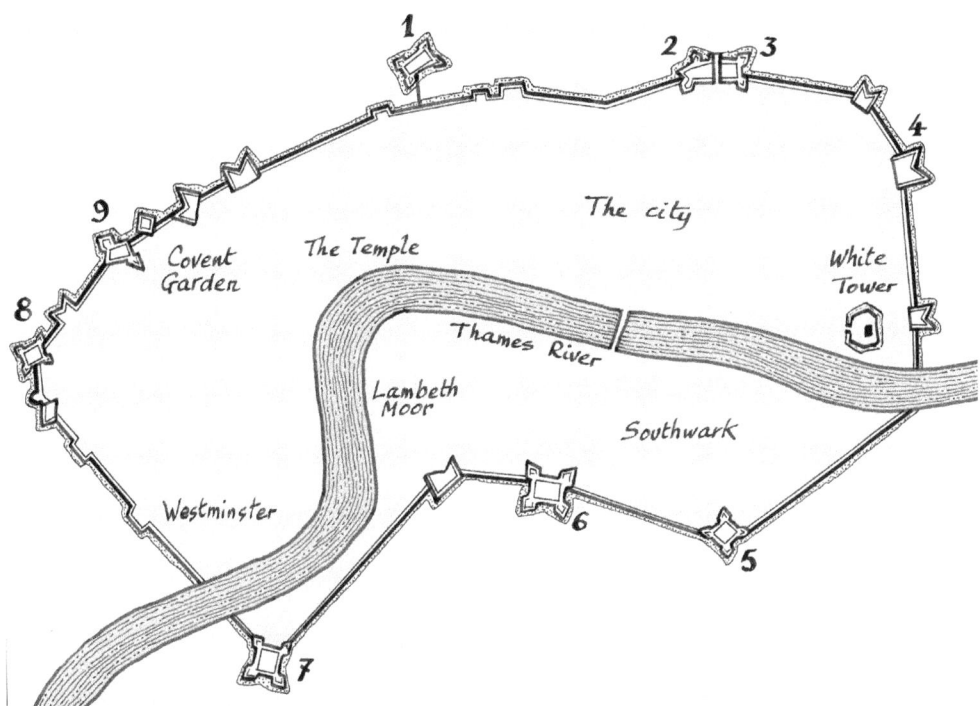

London. The illustration shows the fortifications of London c. 1646 with the most important strongholds: (1) New River Fort, (2) Kingsland's Road Redoubt, (3) Hackney Road Redoubt, (4) White Chapel's Hornwork, (5) Kent Street's Redoubt, (6) Blackman Street's Fort, (7) Vauxhall Fort, (8) Hyde Park Fort, (9) Wardour Street Fort.

don organized a new makeshift army, and King Charles hesitated and retreated. Subsequently, an extensive system of fortifications was built by thousands of men, women, and children to protect London from a renewed attack by the Royalists. These so-called Lines of Communication comprised strong earthen ramparts and ditches punctuated with forts and redoubts guarding the main accesses to the city. The temporary earth fortifications stretched for 11 miles and were well beyond the city's medieval walls as they encompassed the whole urban area, including Westminster and Southwark. London was not seriously threatened by the Royalists again, and the financial resources of the city made an important contribution to the Parliamentarian victory in the war.

Portsmouth

During the English Civil War, the city of Portsmouth was initially held by the Royalist faction before falling to Parliament after a siege in August–September 1642. In 1665, King Charles II ordered Bernard de Gomme to begin the reconstruction of Portsmouth's fortifications in bastioned style—a process that took many years.

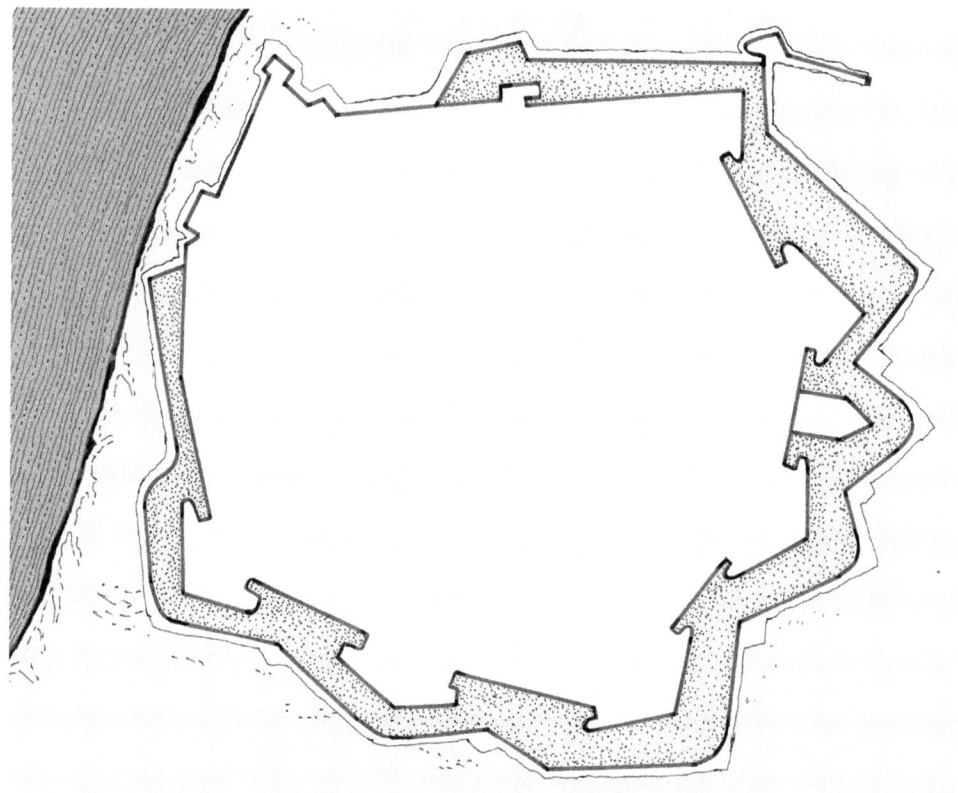

Portsmouth.

Leicester

Leicester in the East Midlands region of England was a Parliamentarian stronghold during the English Civil War. In 1645, Prince Rupert decided to attack the city to draw the New Model Army away from the Royalist headquarters of Oxford. Royalist guns were set up on Raw Dykes, and after an unsatisfactory response to a demand for surrender, the Newarke neighborhood was stormed and the city was sacked on May 30, 1645.

Part 2. Bastioned Fortifications in the 17th and 18th Centuries

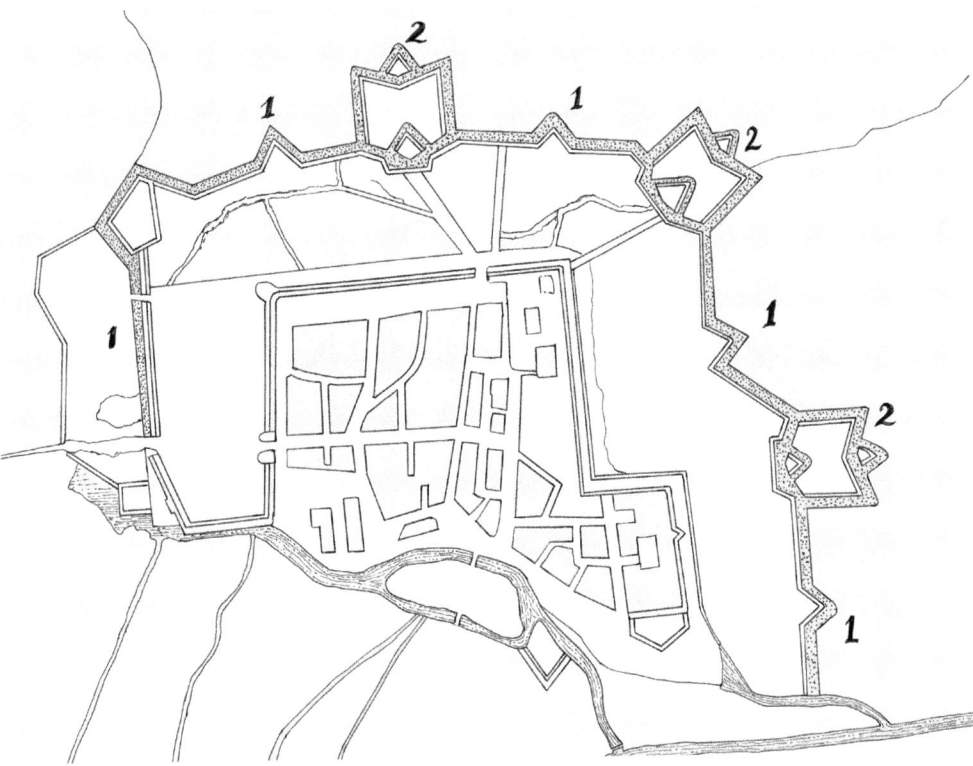

Leicester, 1645. The fortifications of Leicester c. 1645 included an earth wall with redans and a ditch (1), and three forts in the shape of projecting tenailles (2) with ravelins in the gorge and front.

Liverpool

In 1644, the Dutch-born engineer Bernard de Gomme made a plan for the fortifications of Liverpool. The design would have included a square bastioned citadel built around the medieval castle in the southwest and walls, bastions, and ditches around the town. De Gomme's design was not completed.

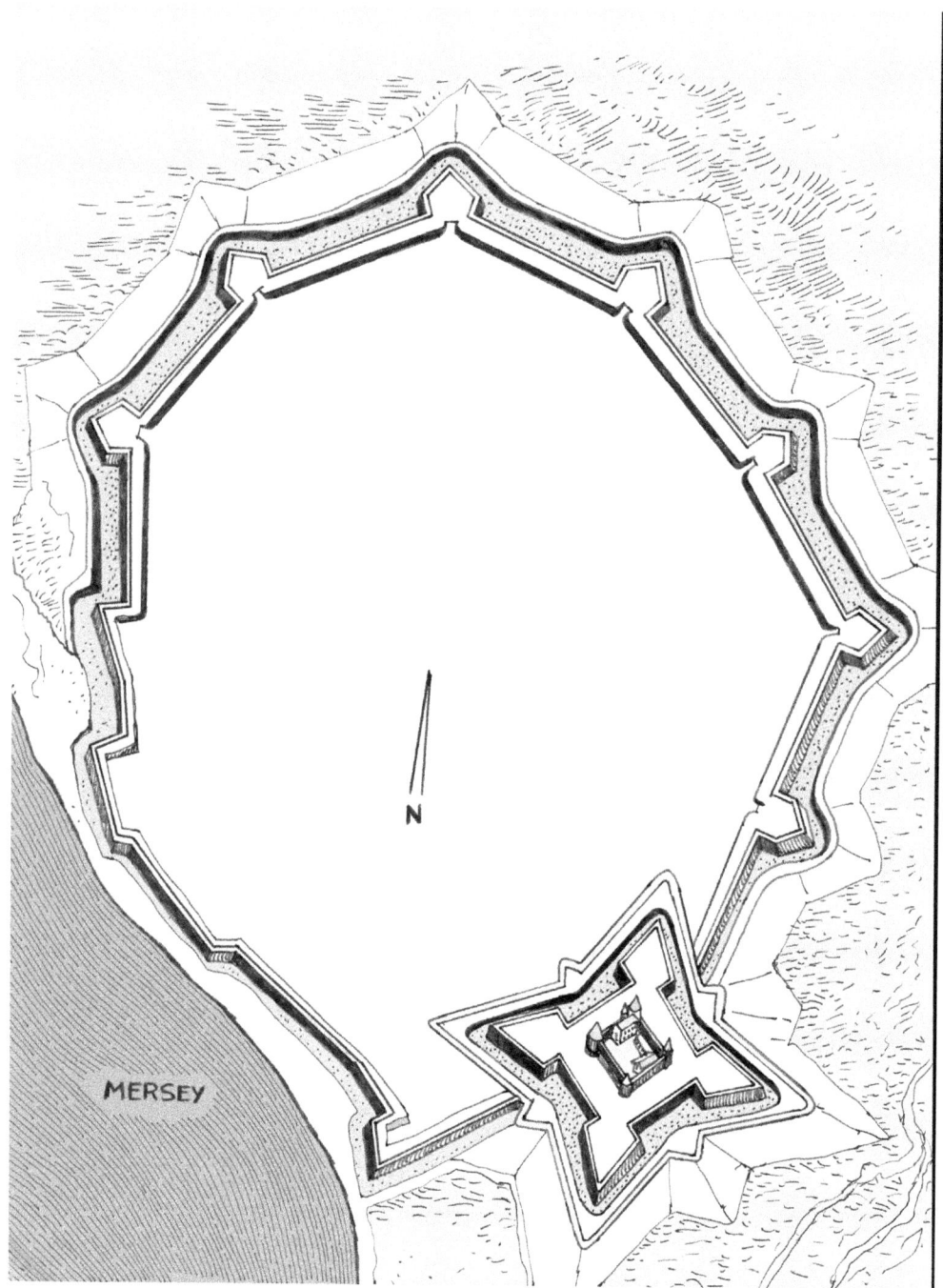

Project for Liverpool.

Part 2. Bastioned Fortifications in the 17th and 18th Centuries

Citadel of Ayr

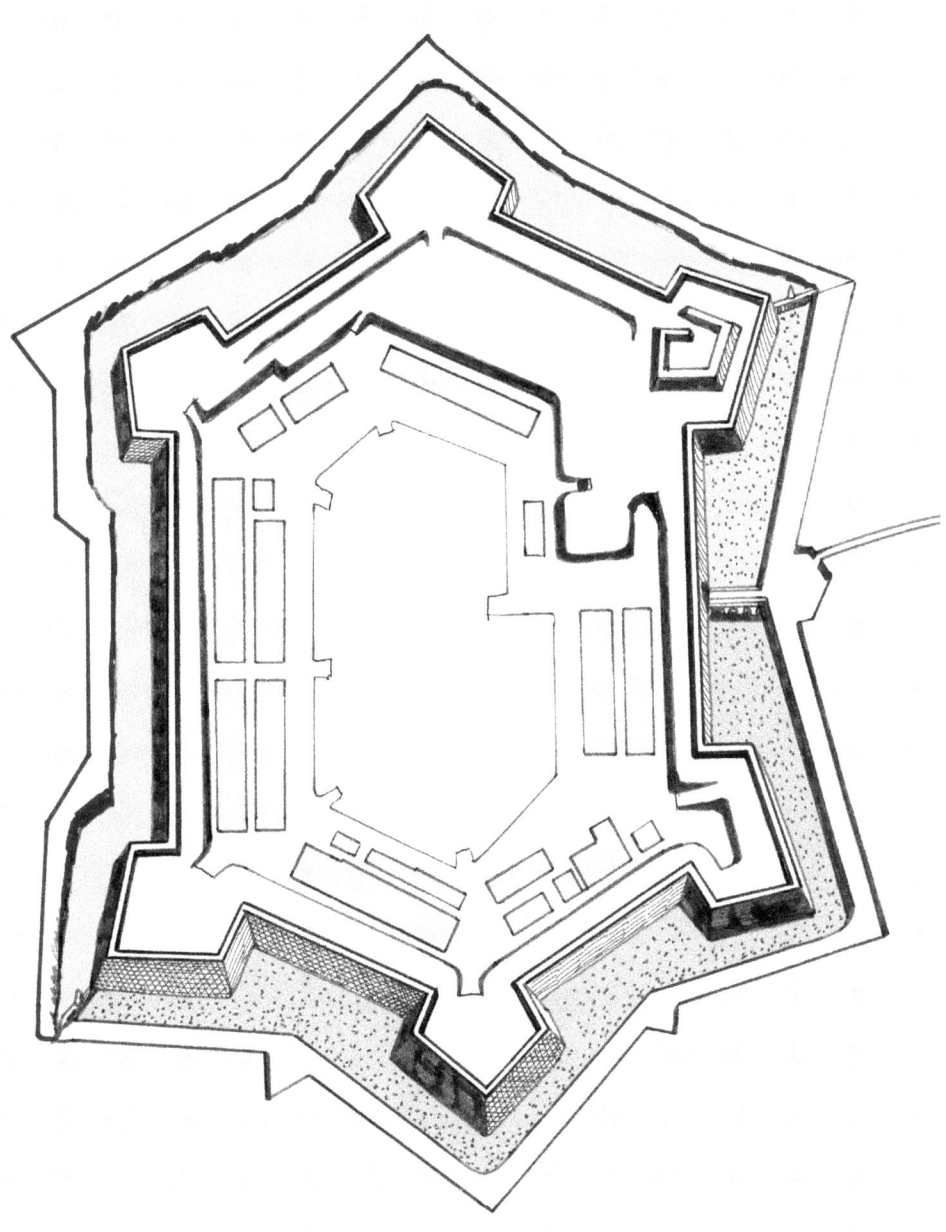

Citadel of Ayr.

In 1652, following military defeat and the declaration of a Commonwealth between England and Scotland, English occupation forces began the construction of fortified posts and citadels intended to control the hostile Scottish population. The English Commonwealth citadels were located in major Lowland towns at the regional and arterial locations of Saint Johnstone (Perth), Inverness, Leith, and Ayr. The best preserved remains of these can be found at Ayr.

The Commonwealth fort at Ayr was designed in 1652 by the chief engineer of the New Model Army, Hans Ewald Tessin, and for the age was a most accomplished work. It was completed in 1654. Sited at the junction of the Firth of Clyde and the outflow of the Ayr River, the fort's plan took the form of a symmetrical elongated hexagon with a bastion at each angle; a floodable outer ditch was provided to the two landward fronts complete with a covered way with re-entrant place of arms and a sloping earthwork glacis. In the middle, there was a large parade square as well as a row of barracks and administrative buildings. The citadel could probably accommodate a garrison of up to 1,000.

Anglo-Dutch Wars

After the English Civil War, a series of armed conflicts broke out between Britain and the Republic of the United Provinces (present-day kingdom of the Netherlands). The first war (1652–54), the second war (1665–67), and the third war (1672–74) were caused by competition in trade and mercantile interests. The fourth war (1780–84) was caused by Dutch interference in supporting the American Revolution. All these wars were fought at sea and in overseas colonies. In the end, these conflicts spelled the decline of the Dutch Republic and announced the emergence of Britain as a maritime and colonial world power.

After the civil wars, the building of fortifications continued in the following periods, but the days of the private baronial fortified residence was over, and henceforth, all fortifications were purely military, decided, funded, and constructed by official governments in response or as anticipation to threats to the nation as a whole.

The late 17th and 18th centuries were an era of formalism and codification. By then, the story of military architecture was dominated by the bastioned system, which became a highly scientific subject designed and built by specialist personnel using professional jargon, sophisticated engineering knowledge, and elaborate geometry.

Coastal Defense

As European powers spread their influence across the globe, they also exported their technologies and their style of fortification. The growth of navies led to the construction of naval fortresses and fortified harbors both at home and in the colonies. These naval strongholds were designed to protect fleets and provide them with secure bases from which to operate and in which to seek refuge. Henceforth, the majority of fortifications built in Britain were coastal defense works closely linked with the Royal Navy and especially intended to repulse attack from the sea.

Coastal fortification, a distinct branch of permanent military architecture, received little recognition before the 19th century. Simply, it was considered quite similar to land defenses, which just happened to have one or two sides facing the sea where ships could bombard the shore, and attacking soldiers, raiders, or marines might land by means of rowing boats. As a result, a work placed in a position to defend a beach, a straight, or the approaches to a port differed very little from a land fort.

Part 2. Bastioned Fortifications in the 17th and 18th Centuries 55

Portsmouth

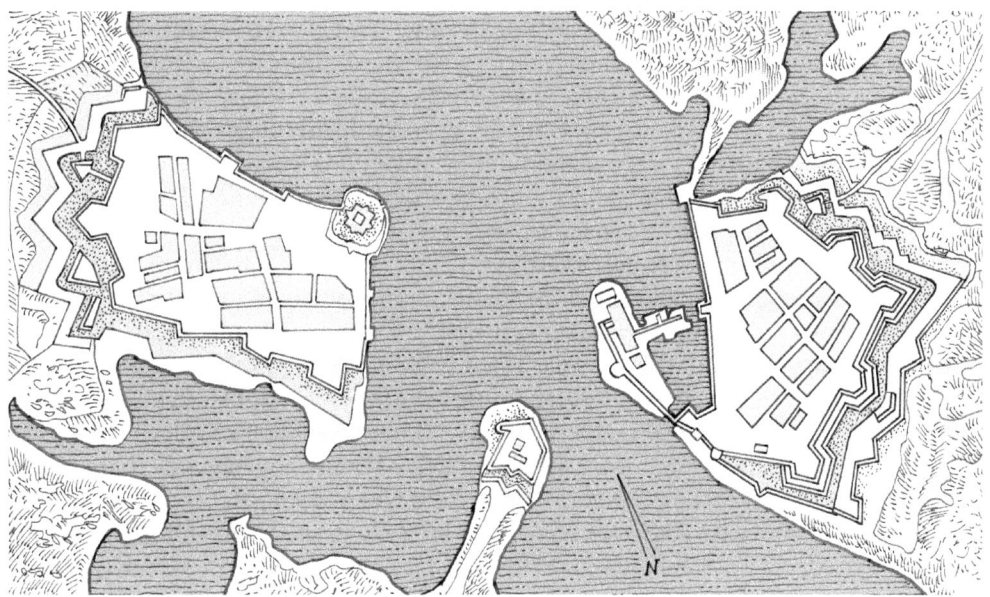

Gosport (left) and Portsmouth (right) in the 1660s.

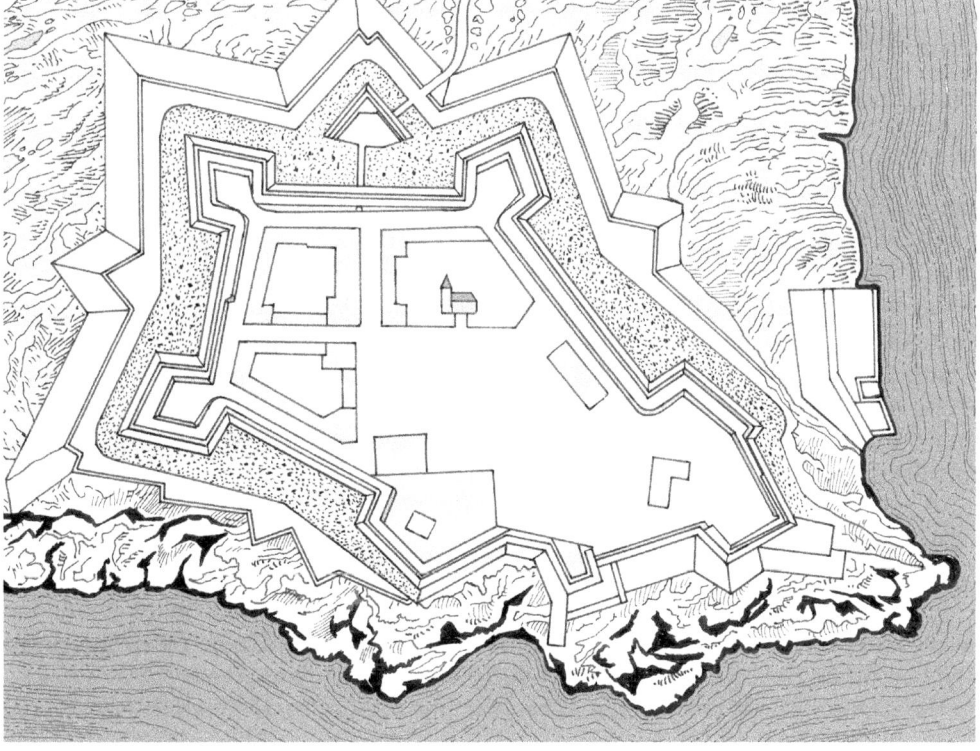

Citadel of Plymouth, c. 1667. Built following the English Civil War, the citadel of Plymouth was one of England's principal fortresses featuring outstanding examples of 17th-century baroque bastioned architecture.

In the 1660s, when the Dutch threat gave rise to new fears of attack, Portsmouth—the most important dockyard on the south coast—was inspected in 1662 by the still-active and omnipresent Sir Bernard de Gomme. De Gomme was responsible for a major rebuilding of the fortifications, starting with new outworks in front of the outdated Elizabethan defenses. This first phase of de Gomme's reinforcement of the fortifications included a covered way, a second ditch, and a demilune at the Landport Gate. His plans for the outworks were carried out between 1665 and 1669. A second phase of fortification continued in the 1670s, when he redesigned the inner Elizabethan bastioned trace with walls revetted in stone. This work was carried out from 1677 to 1685. The Elizabethan bastions were remodeled (especially Pembroke Bastion which was considered too shallow), but the form of the fortifications remained essentially the same. Instead of arrowheads, the bastions had a faussebraye in the flanks to increase their firepower, and a cavalier was added on the Town Bastion. De Gomme built other fortifications to protect the dockyard at Portsmouth; a battery was built stretching from the Point Gate to the Round Tower, and the dockyard itself, which lay to the north of the town, was enclosed by an earthwork bastioned trace. On the opposite side of the channel, de Gomme built a fort on Gosport Point (called Fort Blockhouse).

De Gomme also gave advice on fortifying Harwich, located near Felixstowe in Suffolk at the mouth of the river Orwell. Upon finding Landguard Fort much decayed, de Gomme ordered the fort strengthened as soon as possible with enlarged bastions and restored earthwork ramparts, the construction of a faussebraye at the foot of the rampart all the way round the fort, while a part of the ditches were revetted with brick.

These improvements were put to the test in July 1667.

Dutch Raid on the Medway, 1667

In July 1667, the Dutch fleet launched a daring and devastating raid. They entered the mouth of the Thames, sailed up the Medway, attacking and burning the unfinished fort at Sheerness. The British coastal defense system collapsed like a deck of card as the only fortification left on the Medway was the obsolete Elizabethan Upnor Castle, which proved ineffective in stopping the Dutch ships. As a result, part of the English fleet, moored below Chatham, was destroyed or captured including HM flagship *Royal Charles*. The feelings of national shame and outrage that followed this disaster were if possible heightened by the realization that the enemy could sail up the Thames and attack London. Needless to say, this humiliation caused a complete rethinking of the fortifications of the Thames and Medway Estuaries.

New Coastal Fortifications

Following the calamitous Dutch raid, British engineers and Sir Bernard de Gomme made designs for a reorganization of the coastal defenses, notably a new fort, Sheerness, and two large coastal gun batteries at Gillingham and Cockham Wood.

Cockham Wood Fort was built (along with the demolished Gillingham Fort) on the north bank of the river Medway as a result of the devastating Dutch raid of 1667. It was designed by Sir Bernard de Gomme, chief engineer to Charles II, to help defend Chatham dockyard from an attack by enemy ships sailing up the river Medway. The fort, completed by 1669, consisted of two tiers of gun platforms built into the hillside and a

Part 2. Bastioned Fortifications in the 17th and 18th Centuries

three-story tower serving as defensive quarters for the garrison. The fort was defended by earthworks, a dry ditch, and at the rear by a ravelin. It never saw action, though, because henceforth, the strength of the British Royal Navy was such that no enemy vessel ever again sailed up the river Medway. Cockham Wood Fort fell into disrepair in the course of the 18th century and fell gradually in ruins. Today, one can still see parts of the wall and arches of the lower gun tier on the river side, remnants of the ramparts, and vestiges of the tower.

Cockham Wood Battery, c. 1669.

Landguard

Considering the old Landguard Fort (located at the mouth of the river Orwell outside Felixstowe in Suffolk) not worth repairing, a new fort was constructed just to the south of the old one. The new Landguard Fort was compact in design, having a pentagonal plan, and a triangular seaward battery defended on the landward side by two demibastions. The new fort was reinforced again in 1717, enlarged in 1745, and adjacent batteries were added in the 1750s. In the 1870s, the fort was radically transformed into a modern work with a large circular battery. It remained in military use until World War I. The British army left the fort in 1960. It is now in the guardianship of English Heritage, was restored in the 1990s, now houses a military museum, and is open to the public.

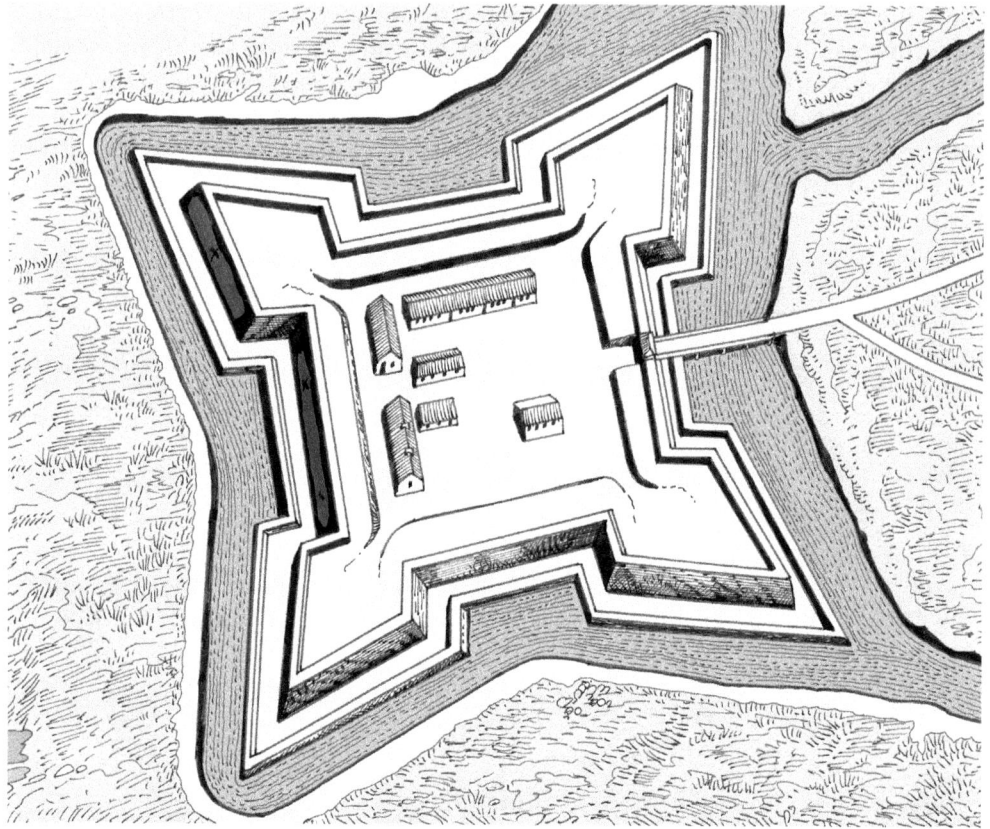

Landguard Fort, Harwich, c. 1667.

Fort Tilbury

Fort Tilbury on the left bank of the Thames opposite Gravesend remains to this day largely unaltered and displays an excellent example of British 17th-century fortification in the Dutch fashion. After the devastating Dutch naval raid on the mouth of the Thames in 1667, King Charles II of England decided to fortify the access to his capital. Fort Tilbury was designed by the Dutch engineer Bernard de Gomme in 1670. It was fully constructed following Dutch bastioned fortification principles. The fort consists of a regular pentagon with five bastions, is fitted with a ravelin and a faussebraye forming a gun battery directed toward the Thames. The work is surrounded by an envelope, a wet ditch, and a flooding area.

Part 2. Bastioned Fortifications in the 17th and 18th Centuries   59

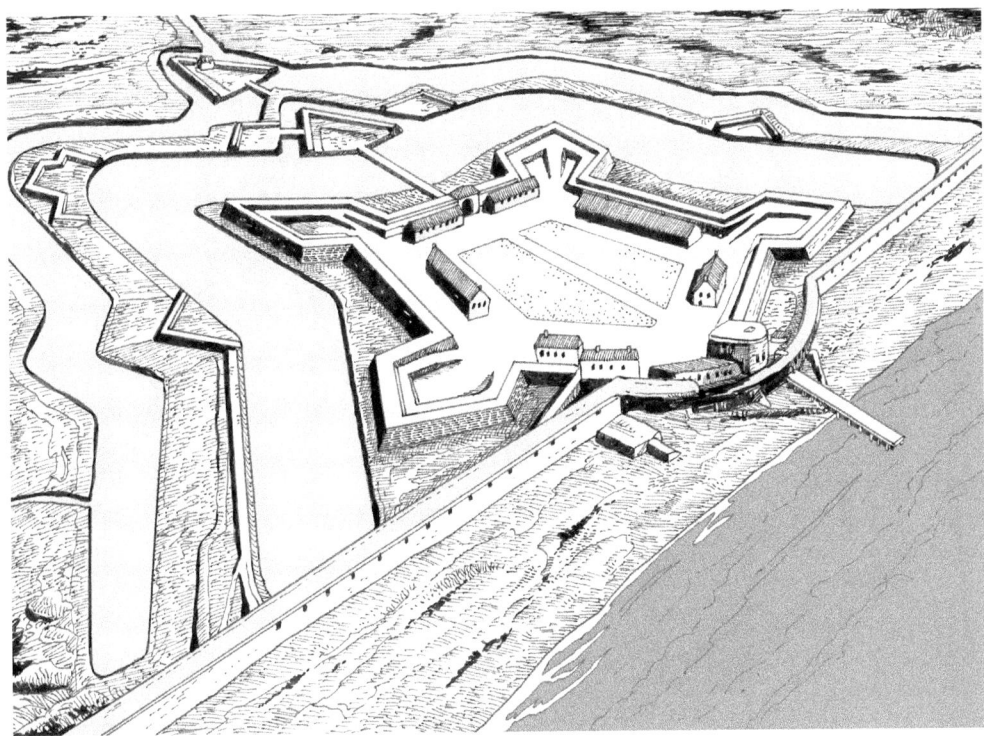

Fort Tilbury.

Fort Charles

Other fortifications built in that period include Fort Charles near Kinale, Ireland. Fort Charles is another of the finest surviving examples of a 17th-century bastioned fort in Britain. The fortress, whose design in 1672 is attributed to the mercenary military engineer Paul Storff de Belleville, has two large bastions overlooking the estuary and three facing inland. Within its walls were the barracks and ancillary facilities to support the fort's garrison. The fort remained in military use until 1921.

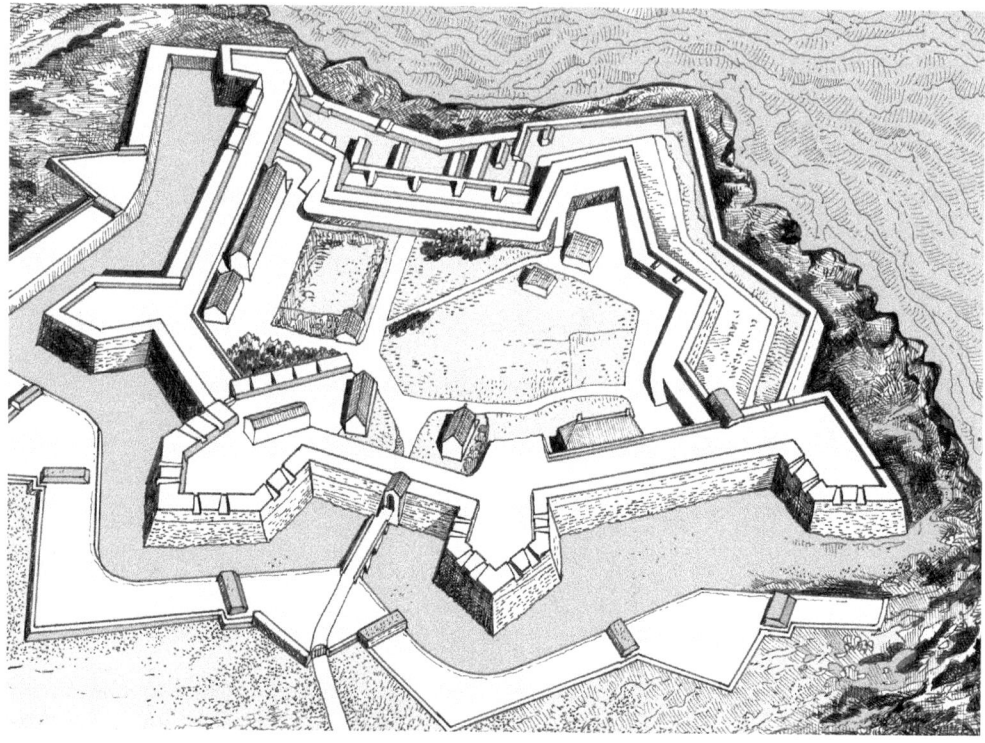

Fort Charles, Kinsale, Ireland.

Hull

The city and port of Hull (official name Kingston-upon-Hull) located in northeastern England at the junction of the Hull and Humber Rivers, was also renovated. By 1681, the engineer Martin Beckman was charged to modernize the outdated fortifications of Hull. Around the old Henrician south blockhouse and castle, he designed a triangular enceinte with revetted ramparts, bastions at each corner, a low faussebraye, and a ditch on the landward side of the citadel. The citadel of Hull was completed in 1685. It served as a detention center for French prisoners of war during the Revolutionary and Napoleonic Wars. It was dismantled in 1864 to make room for the development of the adjacent Victoria Dock.

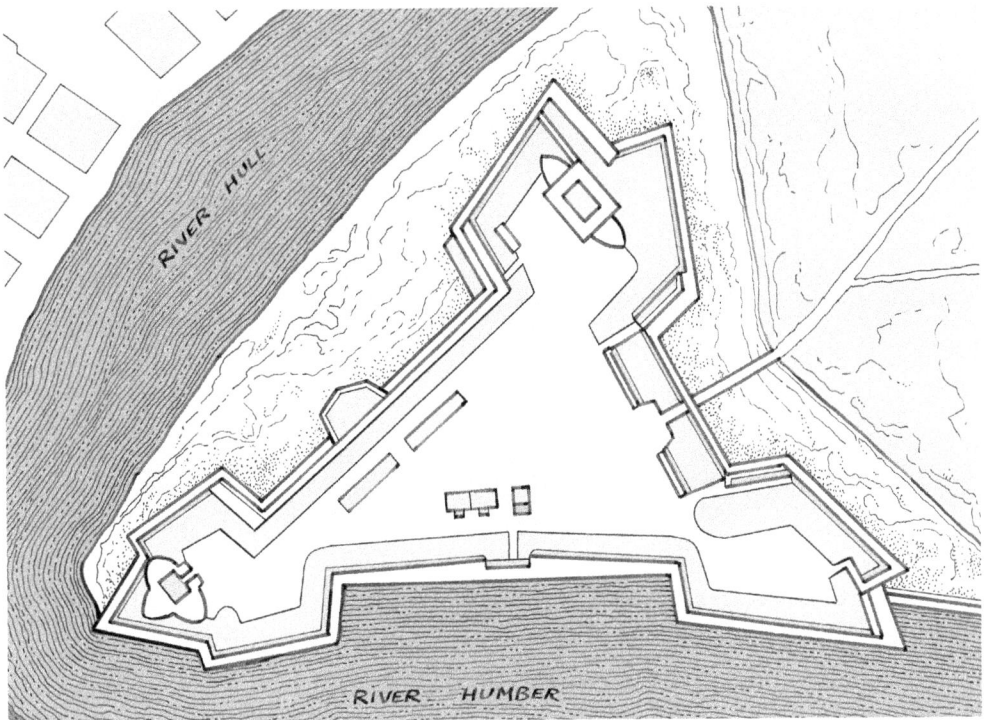

Hull citadel, c. 1682.

Wars Against Louis XIV's France

After the wars against the Dutch when Britain had achieved the dominant trade position it wanted, peace was agreed and Holland and Britain cooperated against France ruled by the aggressive and megalomaniac King Louis XIV. Britain wanted to limit French power, and under the Duke of Marlborough, they won several important victories over the French, notably at Blenheim (1704), Ramillies (1706), Oudenarde (1708), and Malpaquet (1709). By the Treaty of Utrecht in 1713, France was forced to limit its territorial expansion in Europe and had to accept Queen Anne (born 1665, reigned 1702–1714) as the true monarch of Britain instead of James II's son.

Gibraltar

In 1704, during the War of the Spanish Succession against the French, Britain gained the strategic Rock of Gibraltar in southern Spain and henceforth could control the entrance to the Mediterranean Sea. The name comes from the Arabic Jebel-al-Tariq (Tariq's Mountain, named after Tariq ibn-Ziyad who led the first attack on Spain in 711). In antiquity, this place formed one of the limits of the known world and was then called the Pillars of Hercules. The dramatic promontory with its 426 m (1,398 ft) high steep limestone was subsequently ceded to Britain in perpetuity under the clauses of the Treaty of Utrecht in 1713. It became, and remains, an important base for the Royal Navy. It is also an eternal point of contention in Anglo-Spanish relations. In medieval

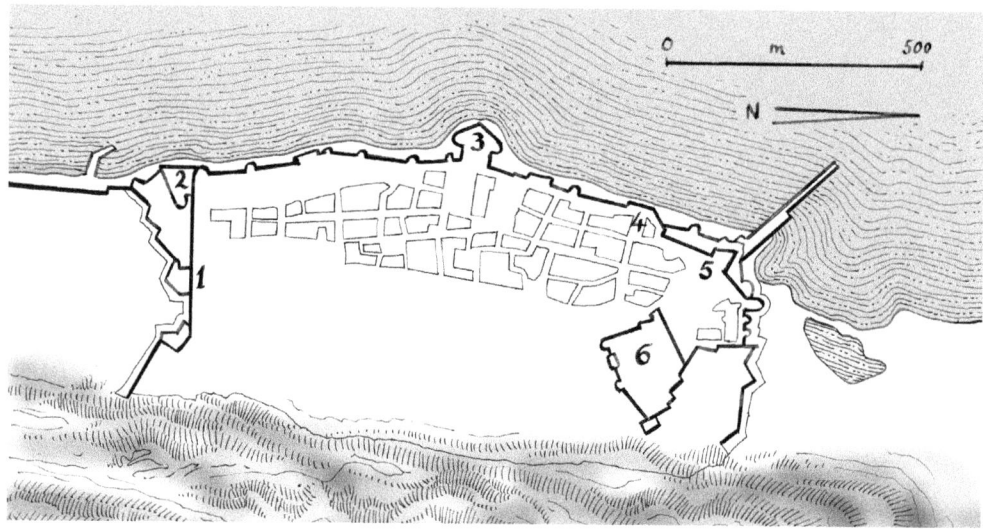

Plan of Gibraltar. The plan shows the northern defenses of Gibraltar c. 1783: (1) flat bastion, (2) south bastion, (3) king's bastion, (4) Orange's bastion, (5) Montague bastion, (6) Old Moorish castle.

times, Gibraltar saw various sieges and changed hands several times before finally becoming part of the Kingdom of Spain at the start of the 16th century. Artillery fortifications were built in the mid–16th century following a raid by Moorish pirates. These were carried out by the Italian engineer Giovanni Battista Calvi and consisted of a wall protecting the south end of the town. This wall, known as the Charles V Wall, extended all the way to the top of the rock. It was stepped with guns mounted in the flanks. The wall blocked any attack by troops landing on the south end of the peninsula, which is what the Moors had done in 1540. In the late 16th century, another wall—called the Philip II Wall—was built behind the Charles V Wall on the upper slopes of the rock as a second line of defense. The next artillery adaptations were designed by another Italian engineer, Giacomo Fratino, who transformed a tower at the northeast corner of the town into a bastion, the Baluarte San Pablo, and built a small bastion at the southwest corner, the Baluarte de Nuestra Señora del Rosario. These bastions were important because they flanked the vulnerable north and south ends of the town as well as the west side, which faced the sea. Finally in the 1620s, the walls on the west side of the town, facing the bay, were improved by the construction of artillery platforms. In 1704, the town surrendered quickly due to the poor state of the fortifications and the low morale of the Spanish defenders. In 1713 at the Treaty of Utrecht, Gibraltar was ceded permanently to Britain. It was a valuable naval base in the Mediterranean, and the British soon set to work improving the fortifications, notably deep casemates excavated in the rocky mountain.

British Military Engineers in the 18th Century

By the end of 17th and during the 18th centuries, bastioned fortifications (created by the Italians, developed and improved by Dutch and French engineers) had come to full maturity. The bastioned system probably reached its most advanced stage of development with Louis XIV's military engineer, the celebrated Sébastien Le Prestre de Vauban (1633–1707). In England, several military architects came to the fore.

John Steed was a British late-17th century military architect. He authored a treatise titled *Fortifications and Military Discipline* published in 1688. The first part of the book deals with the history, principles, and practice of fortification and with descriptions of the various elements. The second part treats observations and instructions about warfare in general.

Steed also designed a bastioned front characterized by long and strong bastion flanks, a curious demilune in the shape of a small hornwork, and large counterguards.

John Lambertus Romer (1680–1754) was a British military engineer, the son of a Dutch engineer who had come to England with William of Orange in the "Glorious Revolution" of 1688. John Romer served as an artilleryman in Flanders and Spain. In 1708, he was appointed assistant engineer to his father at Portsmouth and was employed in works for protecting the shore near the blockhouse. In 1719, he joined the expedition to Vigo, Spain, and took part in the capture of the citadel. On his return home, Romer was appointed engineer in charge of the northern district and Scotland and was involved in defense work at Fort William and Fort George. In the 1730s, Romer was responsible for the design of improved artillery defenses and bastions on the north and west sides of Edinburgh Castle. In 1742, he became director of engineers. During 1745 and 1746, he served under the Duke of Cumberland in the suppression of the Jacobite rebellion and was wounded at the Battle of Culloden in April 1746. Romer retired from service in 1751 and died three years later.

James Gabriel Montresor (1704–1776) was a British military engineer. As a member of the Royal Artillery, Montresor is said to have been present at the 1727 siege of Gibraltar. He was later a bombardier at Gibraltar and was there commissioned an engineer in 1731. He continued to serve in Gibraltar and in the 1740s was named chief engineer of the rock. In 1754, he was appointed chief engineer for General Braddock, went to America, and spent most of the French and Indian War in and around Albany, where his activities included the design and construction of numerous military fortifications, including a new fort at the site previously occupied by Fort William Henry, named Fort George. By 1760, he had risen to the rank of chief engineer and returned to England. He later designed powder magazines at Purfleet and was chief engineer at Chatham. Throughout his career, he also drafted numerous maps and plans of the areas around where he was stationed.

John Muller was a lecturer at the Royal Military Academy. In 1746, he published *A Treatise of Fortification*, which was regarded as an essential and classic work. Muller covered the regular bastioned fortifications constructed by the methods used by Pagan, Vauban, Coehoorn, Belidor, Blondel, and others, explaining how the various elements were constructed and built with detailed plans and drawings. Muller also published in 1757 *The Attack and Defense of Fortified Places* in which he applied mathematical systems to siege warfare, fortification, and artillery.

Jacobite Rebellion

Historical Background

Queen Anne, the last of the Stuarts, died in 1714, and some Tories wanted the deposed James II's son to return to Britain as King James III, provided he gave up

Catholicism. But James Francis Stuart (1688–1766), nicknamed the Old Pretender, was a stubborn man, neither willing to change his religion nor give up his claim to the throne, so he tried to win it by force. In 1715, he started a rebellion against the new king, the Hanoverian George I (born 1660, reigned 1714–1727). George I was a German prince who was distantly related to the English royal family but who was nonetheless the nearest Protestant heir. He was wholly German in habit and culture, did not speak a single word of English, had very little interest in the English people and the kingdom, and his feelings were largely reciprocated by his new subjects. The German king was a man of dogged temper and coarse sensibilities, but he had the good sense to leave the government of Britain entirely in the hands of English politicians and so started the constitutional habit of a chosen ruling prime minister. The 1715 rebellion was a disaster, and the British army had no difficulty in defeating the English and Scottish Jacobites, as pro-Stuart supporters were known. As a result, about 30 citadels and forts were constructed in Scotland to watch over the rebels and continue the repression of the Scots. James's son Charles Edward Stuart attempted to win the throne for his father in 1745–46 in another Jacobite uprising supported by the French, but this failed too. James the Old Pretender died in 1766.

General George Wade

After their victory in 1746, the English government resolved to hold the Highlands like a conquered enemy territory in order to extirpate the whole Highland way of life. To do this, it extended the system of military roads and strengthened the fortress that General Wade had built after 1715. General George Wade (1673–1784) was a British military commander and an important engineer of this period. Born in Ireland, he was commander in chief, and his name is forever attached to the network of military roads he built across the Highlands. Between 1725 and 1737, Wade oversaw the construction of some 250 miles of road, plus 40 bridges, including his most striking legacy, the Tay Bridge at Aberfeldy. Roads linking Perth, Inverness, Stirling, Fort William, and Fort Augustus were primarily intended as military connections to facilitate the disarming and control of the clans, but they also served to open up the country for trade and some measure of development. A traditional doggerel went like this:

> If you had seen these roads before they were made,
> You would have held up your hands and blessed General Wade!

Fort Augustus

A key focal point for Wade's network was Ruthven Barracks, which had been completed near Kingussie in 1721. In 1729, General Wade ordered the construction of a bastioned fort, called Fort Augustus, in a strategic position beside the famous Loch Ness. Named after the son of King George II, William Augustus Duke of Cumberland, Fort Augustus was captured during the second Jacobites rebellion in 1745 but soon was retaken by the government troops after the Battle of Culloden. The fort was still garrisoned when the Caledonian Canal opened in 1822. As peace gradually returned to the region, the need to maintain a garrison diminished, and the fort was sold in 1867 to Lord Lovat who used it as a hunting lodge. A few years later, in 1876, Lord Lovat presented the

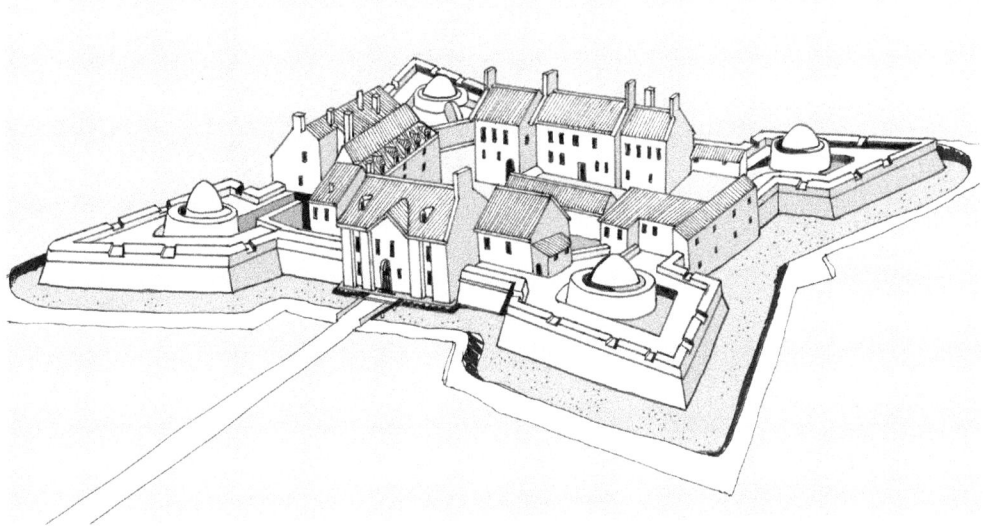

Fort Augustus.

remains of the fort to the Benedictine order for the founding of an abbey and school. The school opened in 1878 and was run by the monks for more than a century until it closed in 1993.

Fort George

General Wade was also responsible for raising a militia called the Highland Watches. The first six companies were formed in 1725, with another four in 1739. These became the Black Watch Regiment and marked the start of the widespread use of Highland troops in the British army that persists to this day. Another fortified key position was Fort George placed on a strategic spit of land projecting into Moray Firth. Fort George was built between 1746 and 1769 on an isolated promontory jutting out west into Moray Firth near Ardesier, about 10 miles northeast of Inverness. Conceived in the immediate aftermath of the 1745 uprising and the nearby Battle of Culloden that concluded it, Fort George was intended to be a once-and-for-all solution to the threat posed by the Highlands and the Jacobites in particular. Designed by William Skinner, the king's military engineer for north Britain, it is a typical bastioned fortress, which—although adapted to the narrowing site—displays a remarkable geometric symmetry designed with scientific precision. On the eastern landside, it comprises a full bastioned front with glacis, covered way with traverses and places of arms, a wide ditch with a triangular ravelin, and two large bastions with the main gate in the middle of the curtain. The massive fort covers 42 acres, and its plan is narrow following the outline of the promontory with two large bastions and curtains in the north and south. The point in the west features a redan and two half bastions arranged as a battery whose guns commanded the waters of the narrow Firth. Fort George was designed to act as a central citadel and depot providing all the facilities of a small town. It comprised a comprehensive range of elegant Georgian-style buildings to accommodate the governor and other officers, an arsenal for the artillery detachment, and large barracks for a garrison of

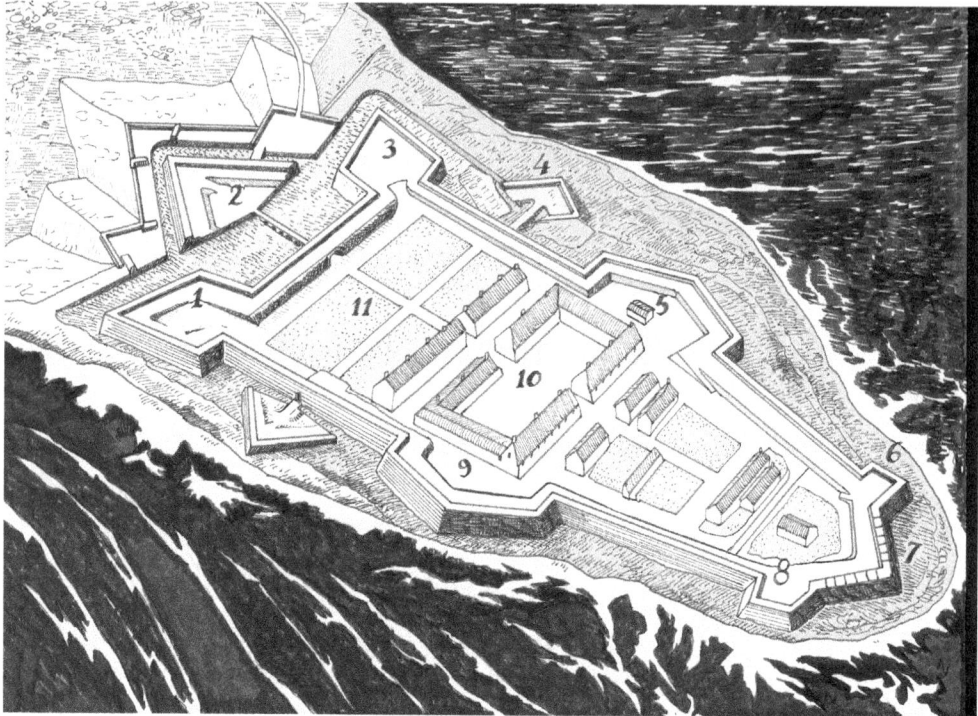

Fort George. The illustration shows the following: (1) Duke of Cumberland's bastion, (2) ravelin with covered way, places-of-arms, and glacis, (3) Prince of Wales's bastion, (4) Sally port protected by a place-of-arms, (5) Prince William-Henry's bastion with powder magazine, (6) Prince Frederick-William's half bastion, (7) point battery, (8) Duke of Marlborough's half bastion, (9) Prince Henry Frederick's bastion, (10) barracks and service buildings, (11) parade ground.

about 1,600 men. There was also a bakehouse, brewhouse, chapel, provision stores, powder magazine, and numerous stores. Like so many British fortifications, Fort George was never tested by enemy attack. It remains today, with very little alteration, just how it was at the time of its completion in 1769. A more advanced descendant of bastioned works like Berwick and Tilbury, Fort George survives unchanged as one of the most complete bastioned artillery fortifications in Britain protruding into Moray Firth like a great stone ship.

The mighty fortress is an excellent example of 18th-century British military engineering at its best. Today, the fortress houses a museum, but parts are still used and operational as army barracks.

Fort Cumberland

Fort Cumberland, located east of the naval port of Portsmouth in south England, was constructed in 1747–48 to guard the entrance to Langstone Harbour and protect the Royal Navy Dockyard by preventing enemy from landing and attacking from the landward side. It was a pentagonal bastioned fort made of masonry and earth. The fort was continuously improved and modernized in 1785–1812, in the 1850s and 1860s, in the 1880s and 1890s, and until 1964 when it lost all military use.

Part 2. Bastioned Fortifications in the 17th and 18th Centuries 67

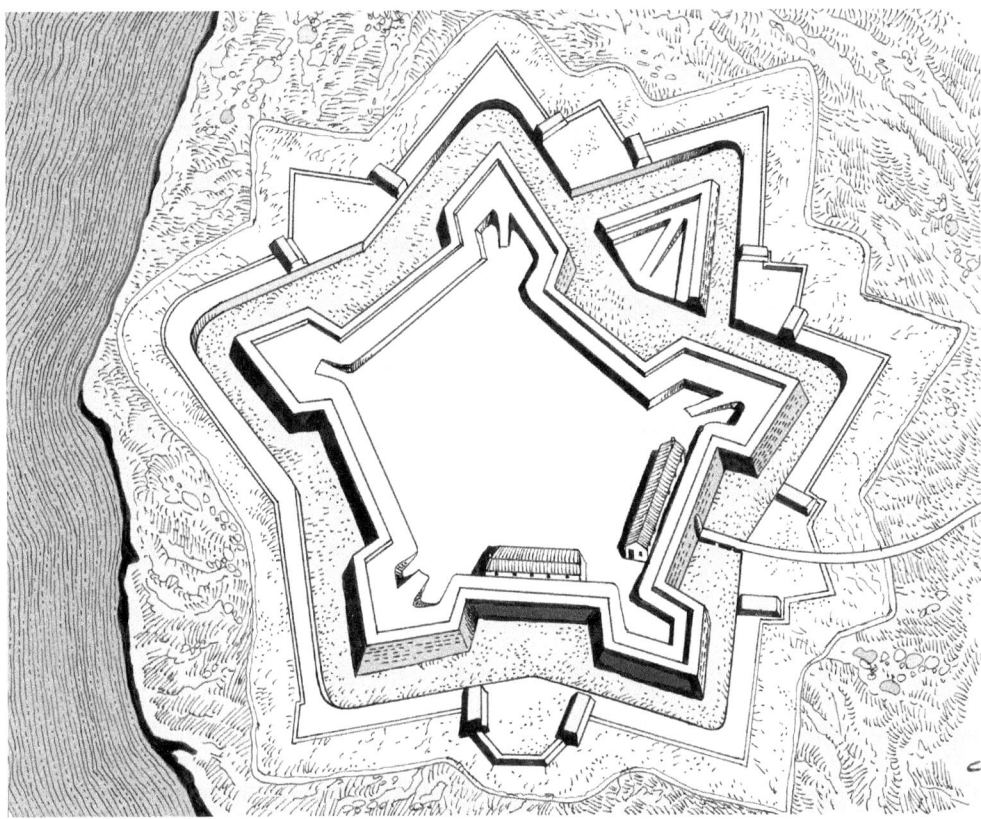

Fort Cumberland, c. 1780.

British Fortifications in North America

Historical Background

The capture or control of foreign lands in the new discovered world, notably North and Central America, was important for Europe's economic development but disastrous for the pre–Columbian natives who were decimated. In the middle of the 18th century, Britain had a small empire abroad, much smaller than Spain and Portugal, but of far greater variety. On the Atlantic coast of North America, Britain controlled 13 colonies designated New England including present-day Massachusetts, New Hampshire, Rhode Island, Connecticut, New York, New Jersey, Pennsylvania, Delaware, Maryland, Virginia, North Carolina, South Carolina, and Georgia. These territories, acquired between 1607 and 1732, were however disputed and threatened by the rival French who were implanted in Nouvelle France (Canada) and Louisiana. In order to hold their territories against both external aggression and internal subversion, fortifications were of paramount importance. Indeed, both the British and the French (totally ignoring the rights and claims of Native Americans or "Indians") said they owned the so-called Louisiana territory. At that time, "Pelican State" Louisiana was not one single state between Texas and Mississippi but included large expanses of lands west of the Appalachian

Timber fort. During the pioneering era of North America, many outposts and trading places on the frontiers were forts featuring palisade walls and flanking blockhouses at the angles.

Mountains and the Mississippi River from Canada to the Gulf of Mexico. In 1720, Louisiana included large parts of Montana, Wyoming, North and South Dakota, Minnesota, Colorado, Nebraska, Iowa, Kansas, Missouri, Oklahoma, Arkansas, Tennessee, Mississippi, and present-day Louisiana. Both France and Britain wanted it for strategic reasons, for the fur trade, and for possible future settlement.

Fortifications were essential assets in the "cold war" between France and England. Tensions grew in those colonies with disputes, skirmishes, and local armed conflicts until open war broke out in 1754.

The French and Indian War

The French and Indian War is the common U.S. name for the war between Great Britain and France in North America from 1754 to 1763. In 1756, the war erupted into a wide conflict known as the Seven Years' War and thus came to be regarded as the North American theater of that war. The war was fought primarily along the frontiers between the British colonies from Virginia to Nova Scotia and began with a dispute over the confluence of the Allegheny and Monongahela Rivers, the site of present-day Pittsburgh, Pennsylvania. The dispute resulted in the Battle of Jumonville Glen in May 1754. British attempts at expeditions in 1755, 1756, and 1757 in the frontier areas of Pennsylvania and New York all failed due to a combination of poor management, internal divisions, and effective French and Indian Iroquois offense. The capture of Fort Beauséjour on the border separating Nova Scotia from Acadia was followed by a British policy of deportation of its French inhabitants. After the disastrous 1757 campaigns, the British government fell and William Pitt came to power, while France was unwilling to risk large convoys to aid the limited forces it had in New France. Pitt significantly increased British military resources in the colonies, and between 1758 and 1760, the British successfully penetrated the heartland of New France with Montreal finally falling in September 1760. The outcome was one of the most significant developments in a century of Anglo-French conflict. To compensate its ally, Spain, for its loss of Florida to the British, France ceded control of French Louisiana west of the Mississippi. France's colonial presence north of the Caribbean was reduced to the tiny islands of Saint Pierre and Miquelon, confirming Britain's position as the dominant colonial power in the eastern half of North America.

New Amsterdam

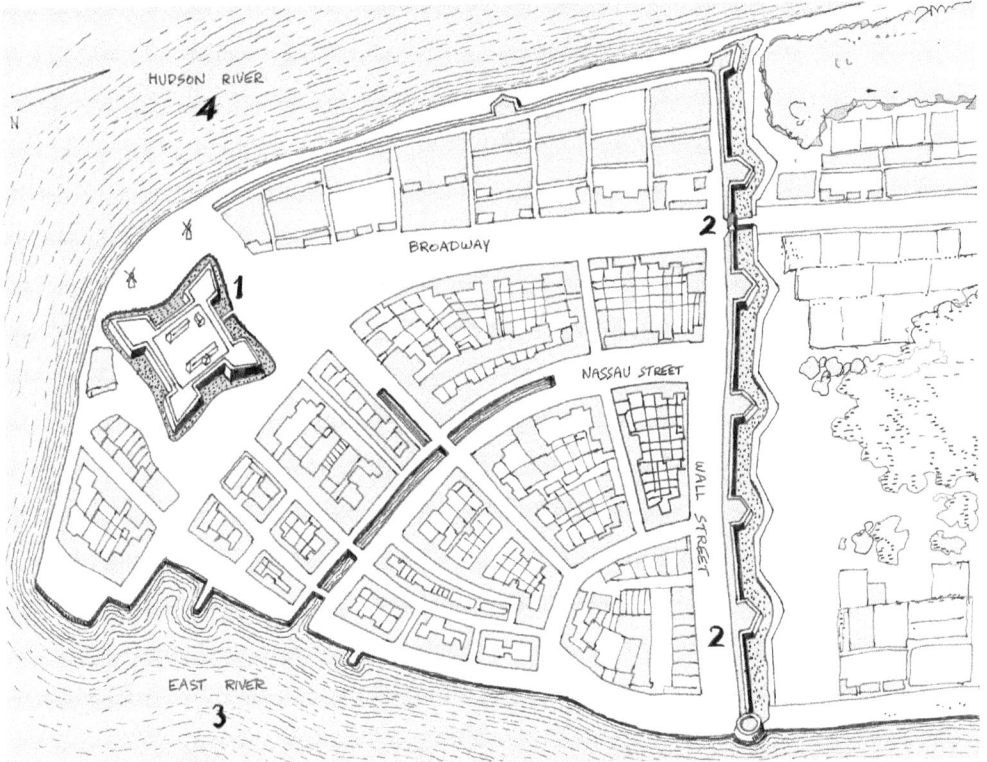

New Amsterdam. The map (simplified after the Castello Plan from 1660) shows Fort Amsterdam (1), the fortifications (2)—today Wall Street, the East River (3), and the Hudson River (4).

The first native New Yorkers were the Lenapes, an Algonquin people who hunted, fished, and farmed in the area between the Delaware and Hudson Rivers. Europeans began to explore the region at the beginning of the 16th century—among the first was Giovanni da Verrazzano, an Italian who sailed up and down the Atlantic coast in search of a route to Asia—but no newcomers settled there until 1624. That year, the Dutch West India Company sent some 30 families to live and work in a tiny settlement on Nutten Island (today's Governors Island) that they called Nieuw Amsterdam. In 1626, the settlement's governor general, Peter Minuit, purchased the much larger Manhattan Island from the natives for 60 guilders. In 1664, the English showed up in battleships, ready for a nasty fight. The Dutch governor Peter Stuyvesant avoided bloodshed by surrendering the port without a shot. New Amsterdam was lost, and the English promptly renamed the colony New York, after Charles II's brother James Stuart, the Duke of York. New York City gained prominence in the 18th century as a major trading port in the British 13 Colonies.

Charleston, South Carolina, United States

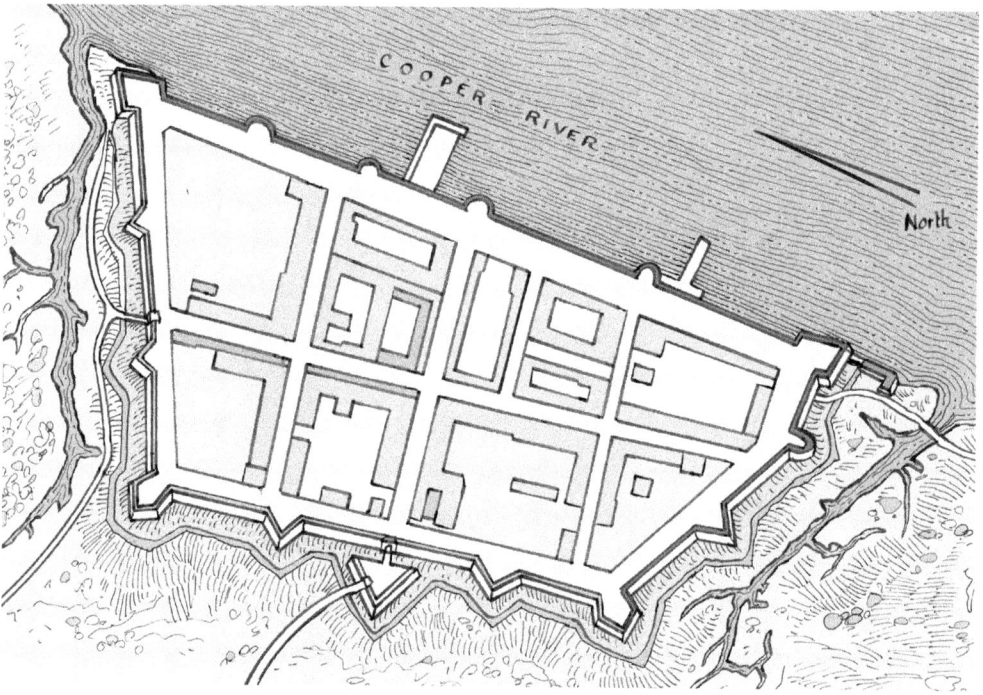

Charleston, South Carolina, c. 1675.

After Charles II of England (1630–1685) was restored to the English throne following Oliver Cromwell's Protectorate, he granted in 1663 the chartered Province of Carolina in North America to eight of his loyal friends, known as the Lords Proprietors, who founded Charles Towne in 1670 a few miles northwest of the present-day city center. By 1680, the settlement was enlarged by additional immigrants from England, Barbados, and the Province of Virginia. As the capital of the Carolina colony, Charles Towne was a center for further expansion. The early settlement was often subject to attack from sea and land, including periodic attacks by pirates and assaults from Spain and France (both of whom contested England's claims to the region). These were combined and aggravated with raids by Native Americans, so fortifications were built. By the mid-18th century, Charles Towne had become a bustling trade center, the hub of the Atlantic trade and slave trade for the Southern colonies. Charles Towne was also the wealthiest and largest city south of Philadelphia. By 1770, it was the fourth largest port in the colonies, after Boston, New York, and Philadelphia.

Louisbourg

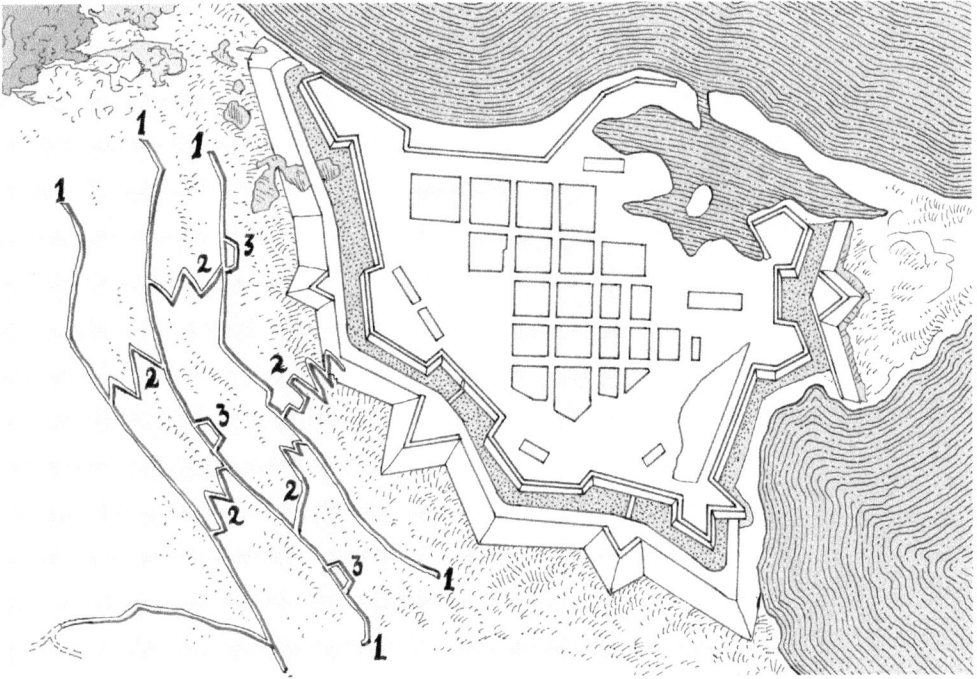

Siege of Louisbourg, June 1745. The illustration shows the main approaching excavated works[JL1] of a 18th-century classical siege: parallels (1), communication sapes (2), and gun batteries (3).

The siege of Louisbourg located on the island of Île-Royale (now called Cape Breton Island) was an episode of the War of the Austrian Succession, known as King George's War in the British colonies. The mutual declarations of war between France and Britain in 1744 were seen as an opportunity by British colonists in New England who were increasingly wary of the threat Louisbourg posed to their fishing fleets working the Grand Banks of Newfoundland. The wariness bordered on an almost fanatical paranoia or a religious fervor, stirred by the general anti–French sentiment shared among most British colonists at the time. British forces attacked the fortified town on April 30, 1745, and finally captured it by the end of June. Louisbourg was handed back to France in 1748 as a result of the Treaty of Aix-la-Chapelle.

Fort Lawrence

Fort Lawrence was located several kilometers west of Amherst, Nova Scotia. In 1750, a British expeditionary force under Major Charles Lawrence arrived at Beaubassin, the region comprising the Tantramar Marshes on the Isthmus of Chignecto. The village was ordered burned by a French priest, Jean-Louis Le Loutre, to ensure that the British could not profit from its seizure. The British forces soon found they were outnumbered by Acadian French colonists and native Mi'kmaq tribesmen. Lawrence's troops retreated but returned in September 1750 in greater numbers and began the

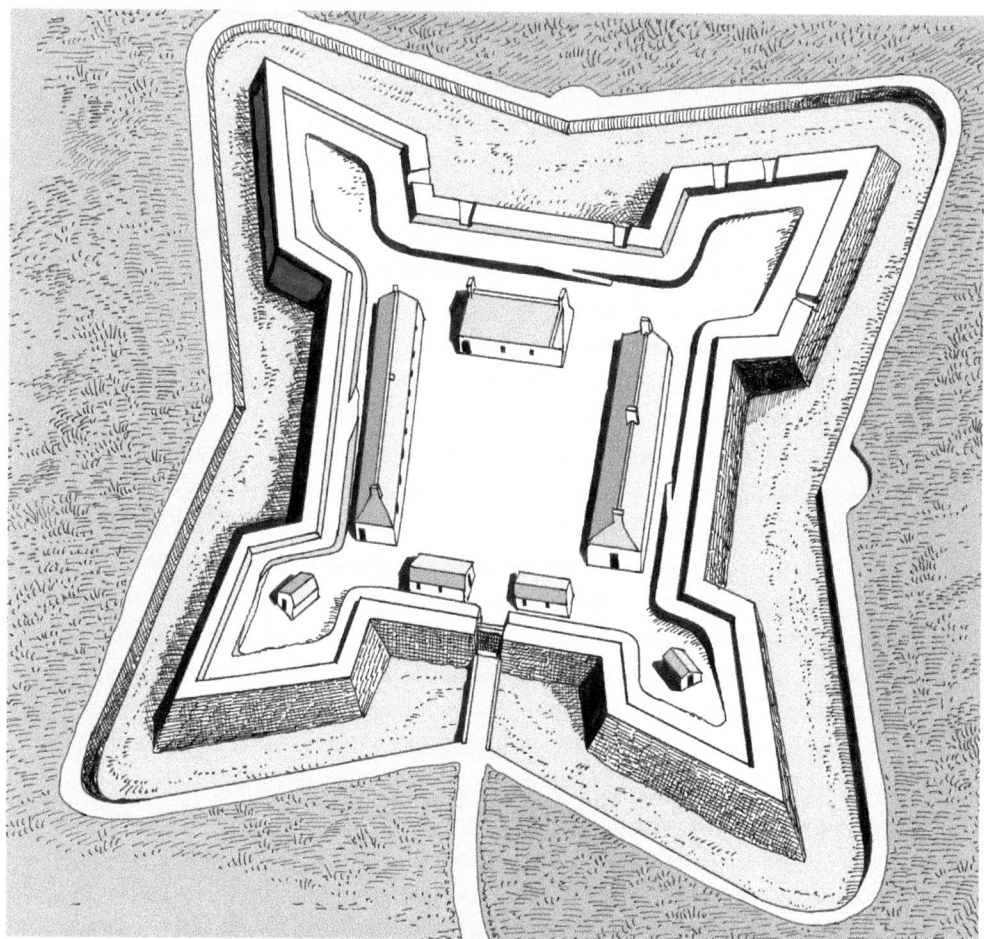

Fort Lawrence, c. 1755.

construction of a palisade fort on a ridge immediately east of the Missaguash River. The work on the fort proceeded rapidly, and the facility was completed within weeks. After the capture of the nearby French Fort Beauséjour, Fort Lawrence was abandoned and destroyed by fire in October 1756.

Part 2. Bastioned Fortifications in the 17th and 18th Centuries

Fort Beauséjour

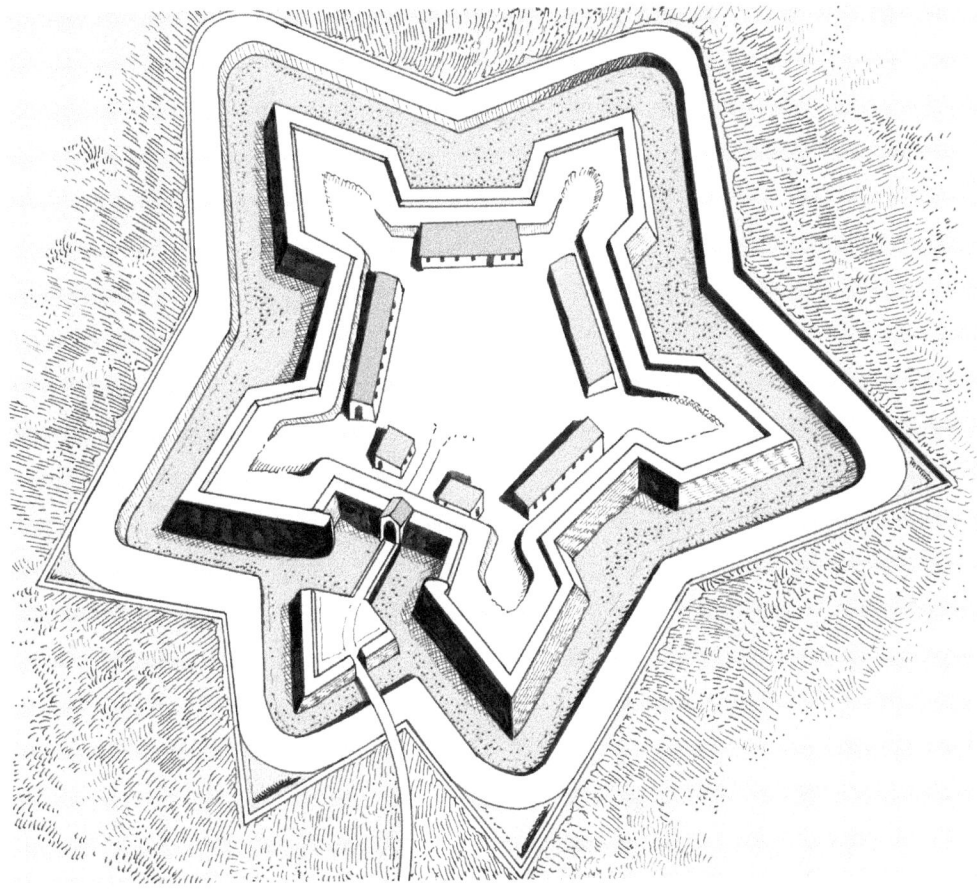

Fort Beauséjour.

Located near Aulac, New Brunswick, Canada, the fort was built by the French in 1750. The Battle of Fort Beauséjour marked the opening of the British offensive in the French and Indian War. Beginning June 3, 1755, a British army under Lieutenant Colonel Robert Monckton staged out of nearby Fort Lawrence and besieged the small French garrison at Fort Beauséjour. Their aim was the Isthmus of Chignecto whose control was crucial to the French because it was the only gateway between Quebec and Louisbourg during the winter months. After two weeks of siege, the fort's commander, Louis Du Pont Duchambon de Vergor, capitulated on June 16. Beauséjour was renamed fort Cumberland, and abandoned and destroyed a year later.

Fort William Henry

Fort William Henry, located near Lake George in the province of New York, was built in 1755 by Major General William Johnson and named in honor of two royal grandsons of King George II. The fort had the shape of a rectangle with four bastions at the

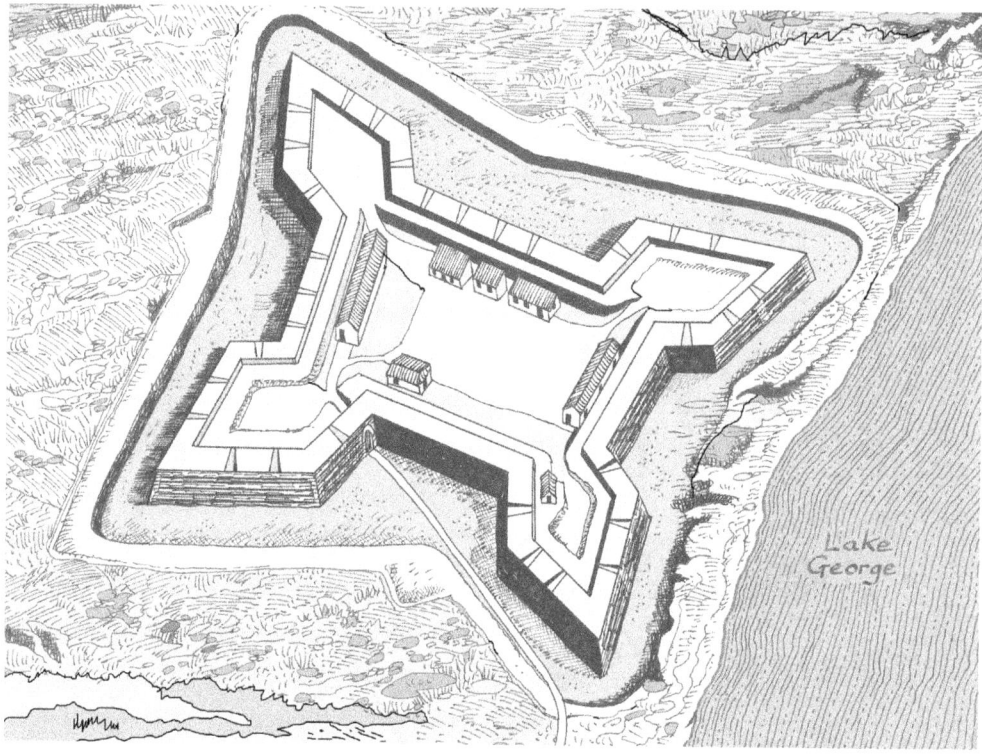

Fort William Henry.

corners but had the particularity of being made of logs. It repulsed several attacks from French troops for two years, until August 1757, when General Montcalm mustered a force of 12,500 French regulars and Indian allies to attack the fort. After six days, the log fortress battered by the French cannonade finally surrendered, and the fort was burned by the French.

Fort Loudon

Located in present-day Monroe County, Tennessee, near the towns of the Overhill Cherokee, this fort was built in 1756 by the British colony of South Carolina, naming it after John Campbell, 4th Earl of Loudoun. With the outbreak of the French and Indian War, the Overhill Cherokee were attacked by French-allied Shawnee and requested the construction of a fort. Fort Loudon was built a few miles downstream from the Cherokee capital Chota. Its purpose was to defend the Cherokee and British settlers on the frontier, to maintain the Cherokee-British alliance, and to guard against French attempts to gain influence among the Cherokee. It also served as a diplomatic and trading outpost. The bastioned fort consisted of a palisade built on top of sloping earth works. Inside the fort, there were a guard house near the gate, a house for the officers, a barrack for the soldiers, a powder magazine, several service buildings, a well, and a parade ground. After Fort Loudon was attacked and burned in 1760, the site was abandoned for nearly two centuries. In the 1930s, the fort was reconstructed and designated a U.S. National Historic Landmark in 1965.

Part 2. Bastioned Fortifications in the 17th and 18th Centuries 75

Fort Loudon.

Fort Ligonier

Named after Field Marshal Sir John Ligonier, this fort was located about 50 miles southeast of Pittsburgh, Pennsylvania. Built in 1758, it protected a post and the passage to the new Fort Pitt, a vital link in the British communication and supply lines. Fort Ligonier was made of wood and had the form of a rectangle with bastions at each corner. The fort was decommissioned from active service in March 1766 after the conclusion of the French and Indian War.

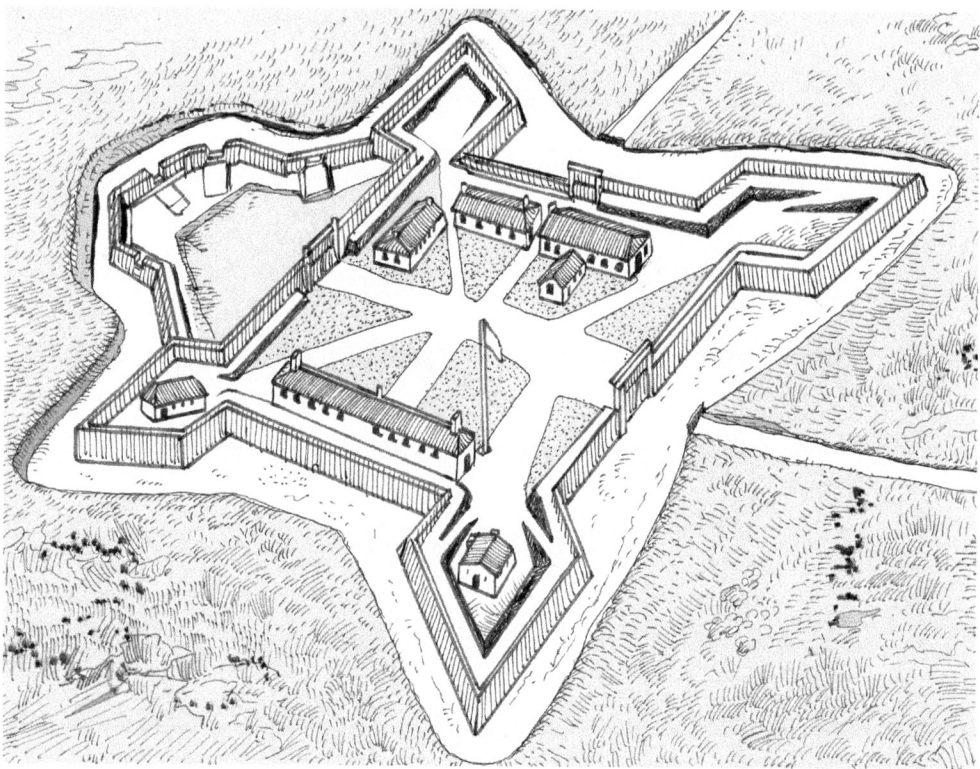

Fort Ligonier. The illustration, based on a drawing by Charles Stotz, shows how the fort might have looked like in 1759.

Fort Pitt

During the late 1740s, William Trent, an Englishman engaged in the fur trade with Ohio Country Indians, built a trading post at the headwaters of the Ohio River (modern-day Pittsburgh). William Trent and the other English traders quickly prospered. They could easily trade with Ohio Country natives and others in northwestern Pennsylvania via the two rivers—the Allegheny and the Monongahela—that came together here to form the Ohio River. In the early 1750s, the French attempted to deny England access to the Ohio Country. In 1754, a French military force captured Trent's outpost and began to construct a stronghold which they named Fort Duquesne. The French also captured several other English settlements in western Pennsylvania. France's seizure of land that the English and their colonists claimed would eventually lead to the French and Indian War. Between 1754 and 1758, the British struggled to recapture their former possessions. Finally, in 1758, they were victorious. After securing Fort Duquesne, the English renamed it Fort Pitt in honor of Whig prime minister William Pitt the Elder (1708–1768). Fort Pitt remained under England's control until the American Revolution, when the colonists took possession of it. The fort gave birth to a settlement called Pittsborough, later named Pittsburgh. Following the American Revolution, the village of Pittsburgh continued to grow and today has become the second largest city in the state of Pennsylvania and the county seat of Allegheny County.

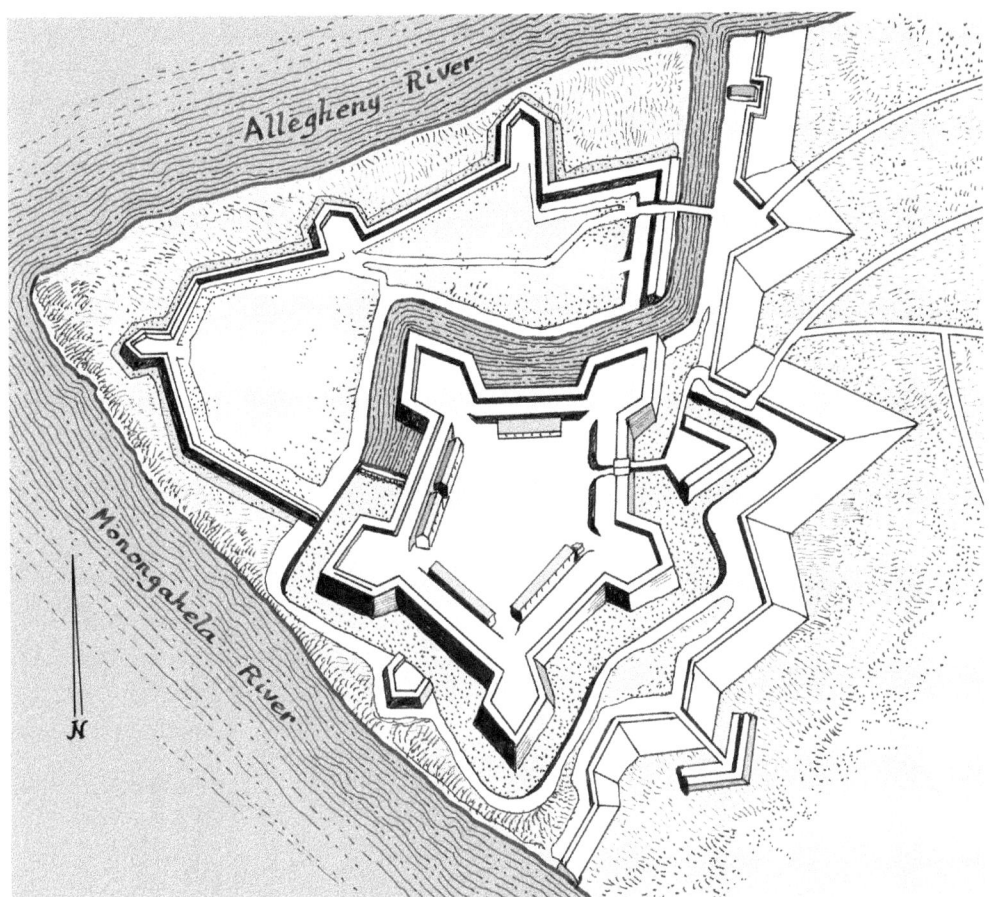

Fort Pitt (Fort Duquesne, Ohio).

American War of Independence

Although the French threat had evaporated after 1763, British North American colonists were unhappy over their share of the winnings in the 1754–1763 Anglo-French War. Besides, they were burdened with taxes imposed by Britain. This eventually led to the American Revolution, the first successful colonial war of independence against a European power. The American Revolutionary War (1775–1783) or American War of Independence began as a conflict between Britain and 13 former British colonies in North America and concluded in a European war between several European great powers. The American War of Independence was a contest of ideologies, animated by an urge to emancipation based (in theory) on equal rights for all men, and government by the general will and not by that of a foreign king. Rejecting British authority, American colonists formally declared their independence as a new sovereign nation in 1776. France's government under King Louis XVI at first secretly provided supplies, ammunition, and weapons to the rebels. In early 1778, France openly entered the war siding with the insurgents against Britain. Spain and the Dutch Republic—French allies— also went to war against Britain. Over the next two years, they threatened to invade

Great Britain and severely tested British military strength with campaigns in Europe, including attacks on Minorca and Gibraltar, and an escalating global naval war. Spain's involvement culminated in the expulsion of British armies from West Florida, securing the American colonies' southern flank.

Throughout the war, the British were able to use their naval superiority to capture and occupy American coastal cities, but control of the countryside (where 90 percent of the population lived) largely eluded them because of the relatively small size of their ground forces. The American War of Independence was essentially a war of attrition, and ultimately, the French involvement proved decisive, with a naval victory in the Chesapeake leading to the surrender of a British army at Yorktown in 1781. In 1783, the Treaty of Paris ended the war and recognized the sovereignty of the United States of America over the original 13 colonies plus the territories bounded by what is now Canada to the north, Florida to the south, and the Mississippi River to the west.

Fort Ticonderoga

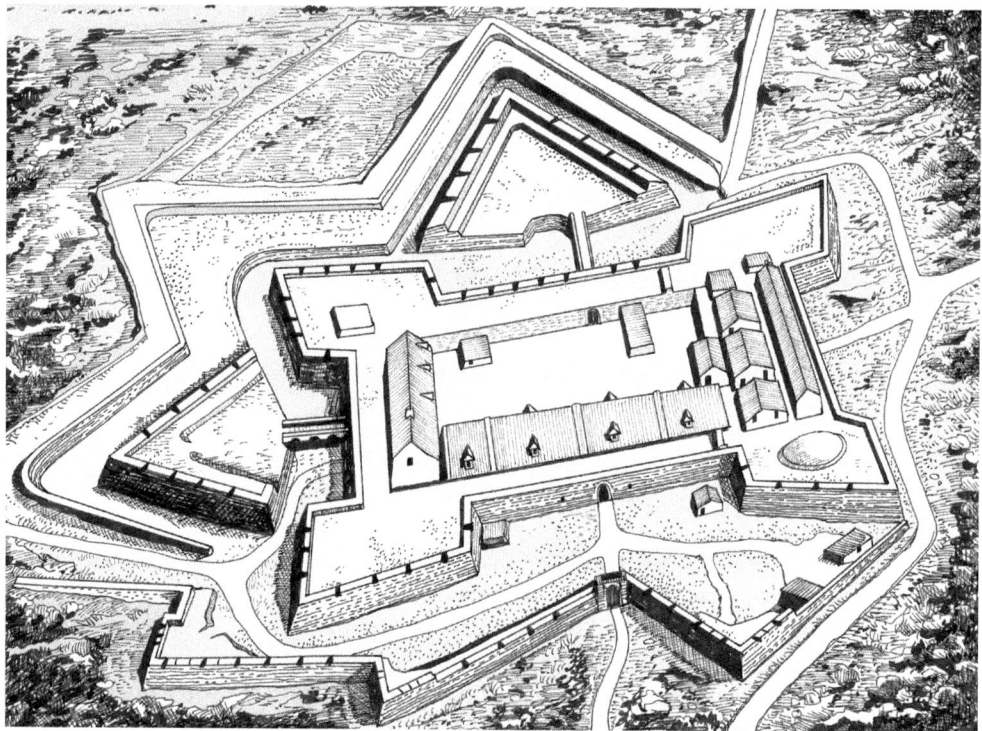

Fort Ticonderoga.

Fort Ticonderoga, originally named Fort Carillon, was built by the French between 1755 and 1759. Located at the narrows near the south end of Lake Champlain in upstate New York, it was one of a series of forts the French had built to control and protect their trade around Lake Champlain. Fort Carillon was captured by the British during the French and Indian War, renamed Ticonderoga, and rebuilt in the 1760s. At the outset of the American Revolution, a small company of British soldiers still manned the fort. In May 1775, Ethan Allen, Benedict Arnold, and the Green Mountain Boys crossed Lake

Champlain from Vermont and at dawn surprised and captured the sleeping garrison—the first American victory of the Revolutionary War. Later, the fort was abandoned and fell into ruins until restoration and reconstruction in 1908. Today, Fort Ticonderoga is open to the public and stands as a good example of 18th-century fortification.

Fort Niagara

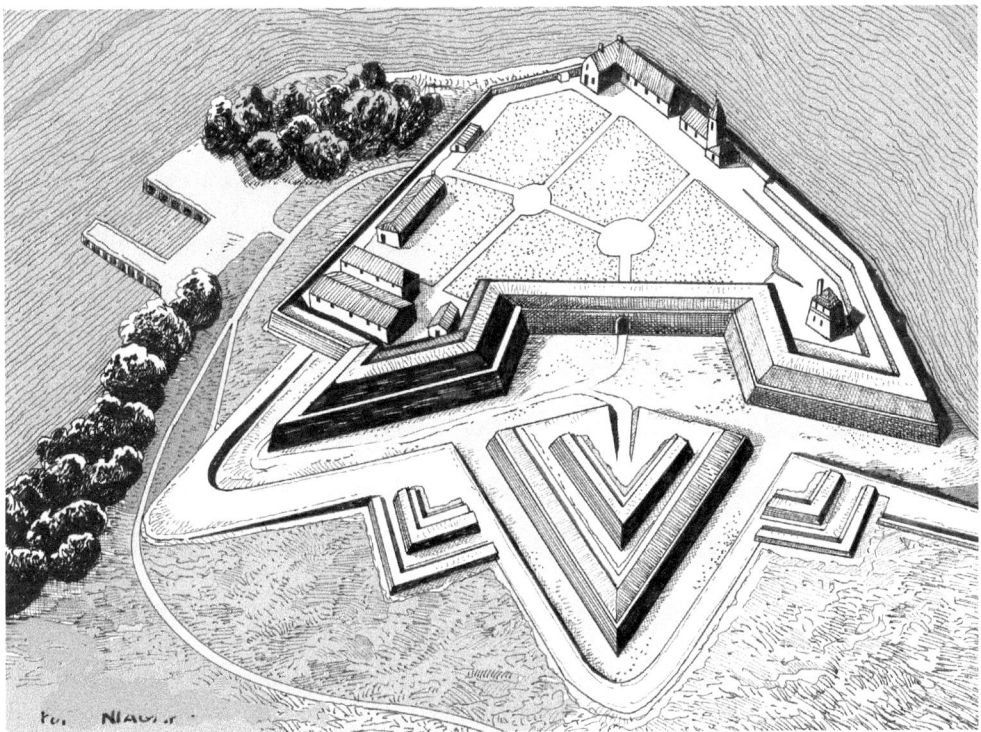

Fort Niagara.

Fort Niagara is located near Youngstown, New York, on the eastern bank of the Niagara River at its mouth on Lake Ontario. The fort originated from a first structure, called Fort Conti, built in 1678 by the French explorer René-Robert Cavelier, Sieur de La Salle (1643–1687), the man who claimed the entire Mississippi River basin for France. The fort was expanded to its present size in 1755 due to increased tension between French and British colonial interests. The fort played a significant part in the French and Indian War and fell to the British after a 19-day siege in July 1759, called the Battle of Fort Niagara. The fort remained in British hands for the next 37 years. Fort Niagara served as the Loyalist base in New York during the American Revolutionary War. Although Fort Niagara was ceded to the United States after the Treaty of Paris in 1783, the region remained effectively under British control for 13 years. American forces occupied the fort in 1796. Today, Fort Niagara has been renovated and is a State Historic Site.

Siege of Yorktown

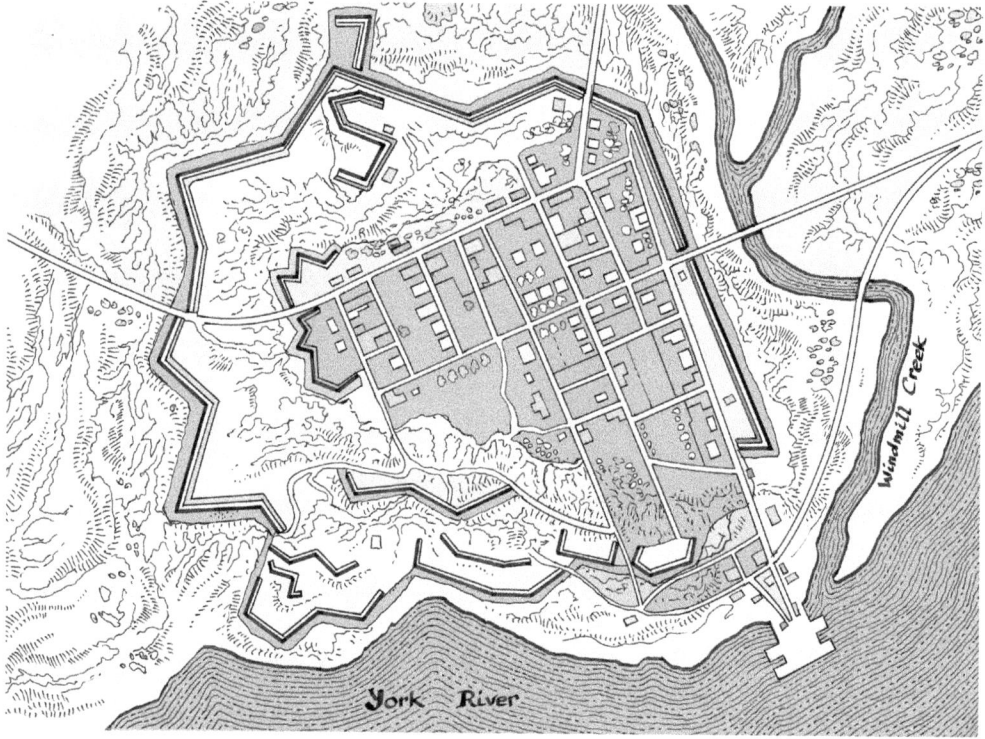

Yorktown, 1781.

The siege of Yorktown or Battle of Yorktown in 1781 was a decisive victory by combined assault of American forces headed by Major General George Washington and French forces led by General Comte de Rochambeau over a British army commanded by Lieutenant General Lord Cornwallis. It proved to be the last major land battle of the American Revolutionary War. Indeed, the surrender of Cornwallis's army prompted the British government eventually to negotiate an end to the conflict. After two days of negotiation, the surrender ceremony took place on October 19, with Cornwallis being absent since he claimed to be ill. With the capture of over 8,000 British soldiers, negotiations between the United States and Great Britain began, resulting in the Treaty of Paris in 1783 and the independence of a federal republic named the United States of America.

Fort Erie

Fort Erie, located on the shores of Lake Erie in Ontario, was first settled in the 1790s by United Empire Loyalists. British forces established a post here at the end of the war with France in 1746. Early forts were wooden, but a stone fort was begun in 1805. Early settlers sought refuge in this area from the Americans. The fort was originally built in 1764, damaged by flood and ice in 1779, and destroyed by a storm in 1802. A second fort was begun a short time later but was unfinished when war broke out in 1812.

Part 2. Bastioned Fortifications in the 17th and 18th Centuries 81

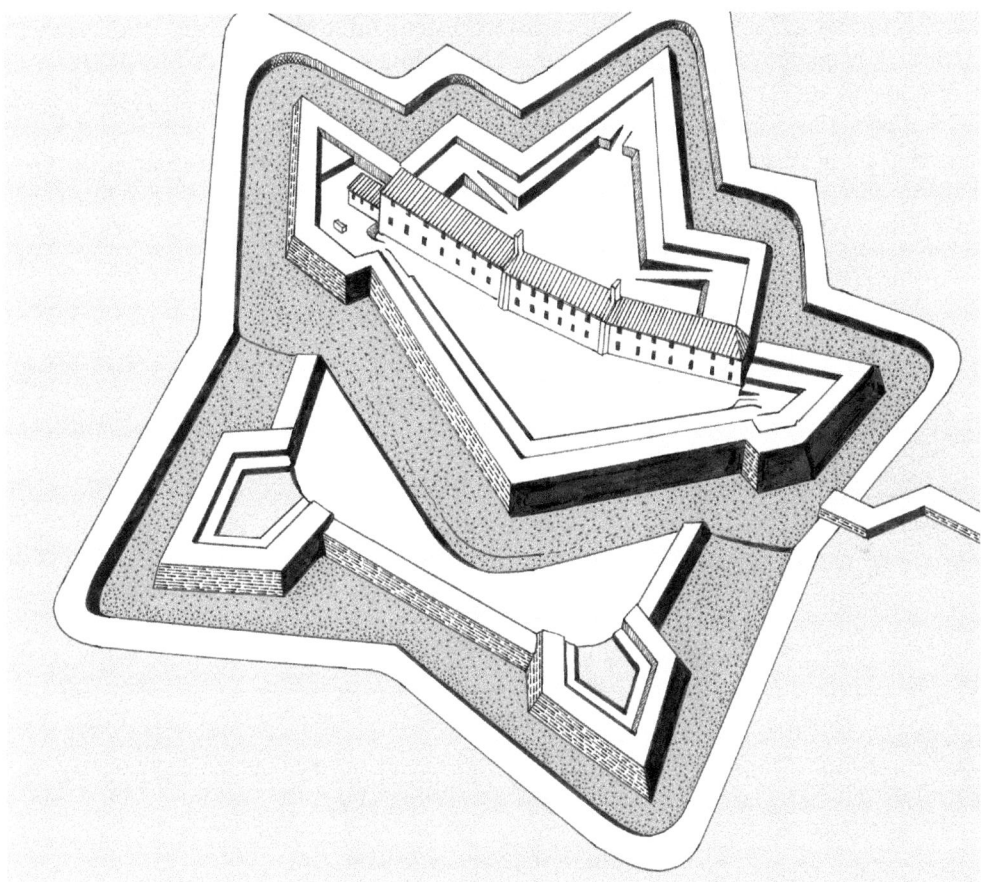

Fort Erie.

The so-called War of 1812 was a conflict between the newly created United States and Britain. Lasting from June 1812 to February 1815, it was caused by restrictions on American commercial activities resulting from the British blockade of French and allied ports during the Napoleonic Wars. It was also prompted by British and Canadian support for Native Americans trying to resist westward expansion. It was ended by a treaty that restored all conquered territories to their owners before the outbreak of war.

In July 1814, Fort Erie was captured by General Jacob Brown's U.S. force and extensively rebuilt by the Americans. Later, the fort was abandoned and fell into ruins. Old Fort Erie was restored in 1939 and today contains relics and artefacts of the war of 1812 and military equipment used by both the Canadian and American soldiers. Old Fort Erie is now open to the public daily during summer.

British Fortifications in the West Indies

British possessions in the West Indies (islands and islets in the Caribbean Sea, i.e., Leeward Islands, Windward Islands, Jamaica, Cayman Islands, and Caicos Islands) were of great interest for the culture of tropical product such as coffee, tobacco, and sugar whose consumption had become a fashion and a flourishing business. The growing sugar economy of the West Indies increased the demand for black slaves deported from Africa. By 1645, for example, there were 40,000 white settlers and 6,000 black slaves in Barbados. By 1685, the balance had changed with only 20,000 white settlers but 46,000 black slaves. The sugar producers and importers used their wealth and influence to make sure that the government would not abolish slavery. Fortifications were also built at the most important ports and production centers against attack from other rival European colonial powers and pirates.

Fort Charlotte (Bahamas)

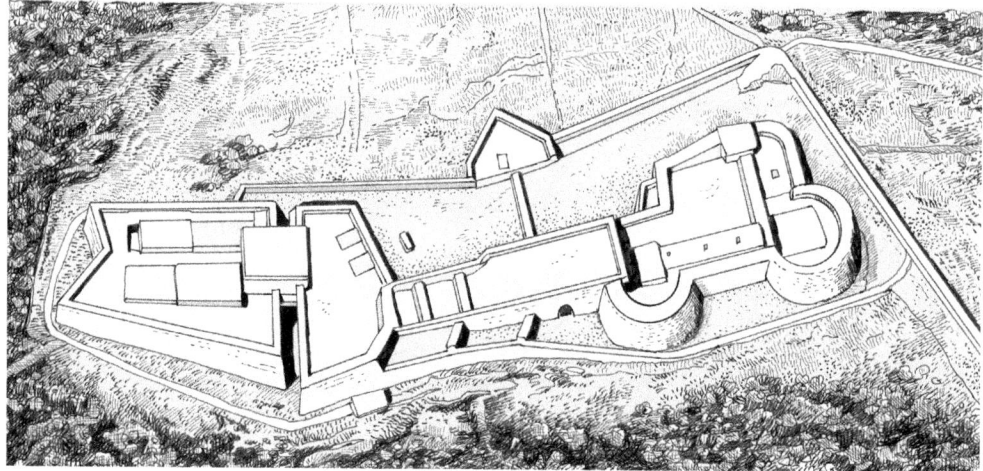

Fort Charlotte, Nassau, Bahamas.

The Bahamas were populated by English colonists since 1648. Nassau, first established as Charlestown in 1666, was christened Nassau in 1695. Wrecked ships became a livelihood for the city's settlers. If bad weather and poor maps did not bring enough salvage ashore, the wreckers would put lights on the reefs to lure ships to their doom. None of this rogue activity was approved of by the faraway English government, but the English did put a seal of approval on the beginnings of piracy. In the 17th century, England was constantly at war and the Royal Navy had its hands full, so letters of marque (government commission authorizing attack on enemy ships) were given to local pirate captains. Then called privateers, pirates operated from the Bahamas. Piracy quickly became rampant, and in Nassau a sort of privateer's republic was established. Edward Teach, better known as Blackbeard, declared himself Nassau's magistrate. Calico Jack Rackham, Anne Bonney, and Mary Read were among many infamous pirates of the Caribbean based here. When England signed peace treaties with its enemies, the privateers (who had far exceeded the limits of their letter of marque) officially became outlaws. In 1718, their republic came to an end when England sent Governor Woodes Rogers to Nassau, who expelled the pirates and restored commerce, law, and order. Built in 1789 by Lord Dunmore and named in honor of the wife of King George III, Fort Charlotte was intended to protect the western entrance of Nassau harbor. It includes a ditch, drawbridge, ramparts, and two low massive gun towers.

Part 2. Bastioned Fortifications in the 17th and 18th Centuries

Fort Fincastle, Nassau, Bahamas

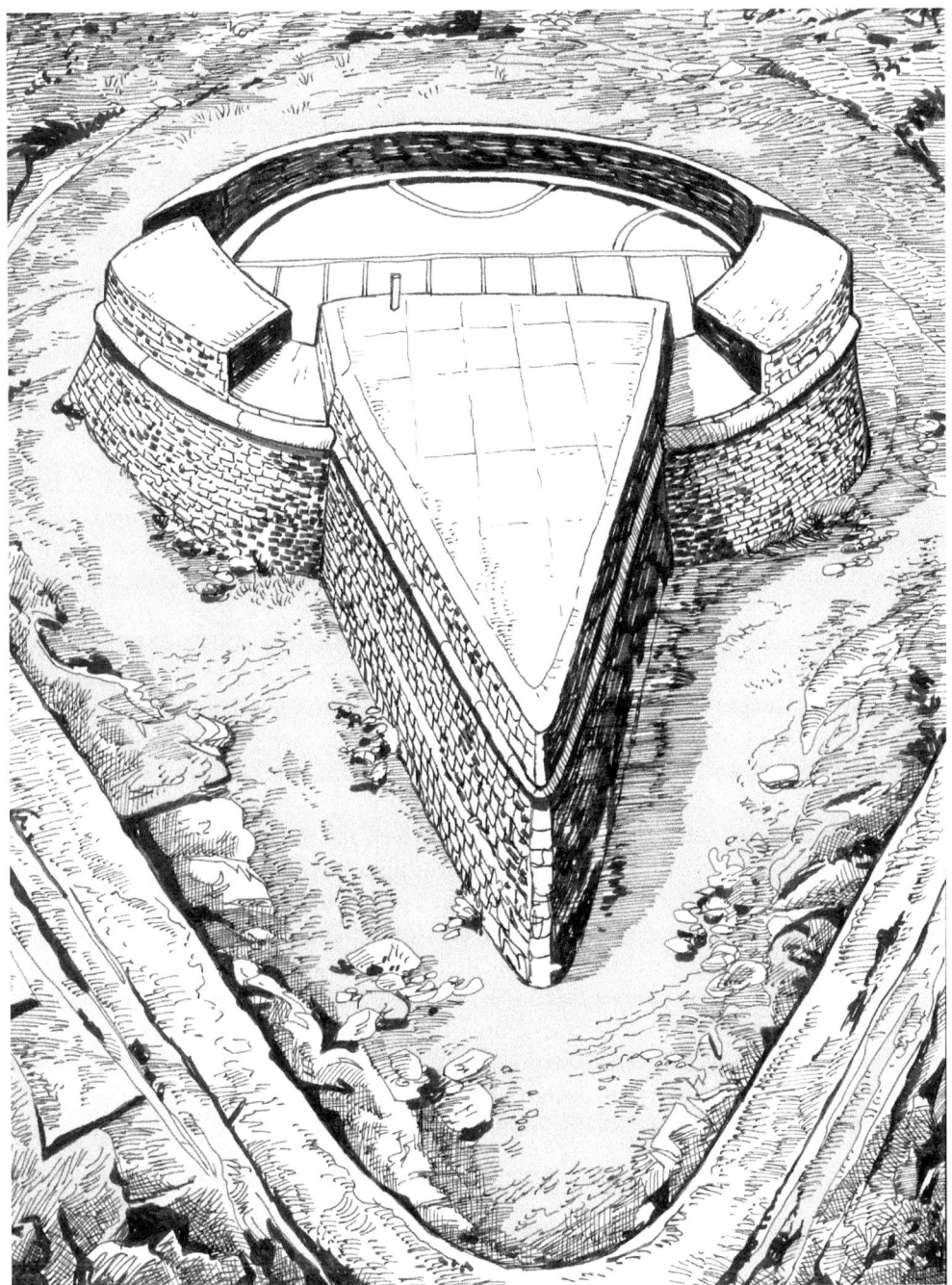

Fort Fincastle, Nassau, Bahamas.

English Puritans known as "Eleutheran Adventurers" arrived in the Bahamas in 1649 in search of religious freedom. During the late 1600s to early 1700s, many privateers and pirates used the Bahamas as base, whose close proximity to Spanish shipping lanes made for the perfect spot to attack and plunder merchant ships. Established around 1670 as a commercial port, Nassau was overrun by lawless pirates and seafaring men. By 1718, the king of England appointed Woodes Rogers to serve as the royal governor. His job was to restore order and put an end to piracy. Rogers offered amnesty to those who surrendered, and those who resisted would be hanged. After the destruction of piracy, the Bahamas became a prosperous British colony, and several forts were built to protect the islands. Located in the city of Nassau on the island of New Providence, Fort Fincastle was built in the early 1790 by Lord Dunmore, Viscount Fincastle on Society Hill overlooking Nassau and defending Paradise Island and the eastern approaches to New Providence. It served as a lighthouse until September 1817 when it was replaced by the lighthouse on Paradise Island. It was subsequently used as a signal station.

British Fortifications in India and Asia

In this period, Britain established its first trading settlements in India on both the west and east coasts. At first, the British were only interested in trade and did not interfere in Indian politics. However, competition with France later resulted in direct efforts to control Indian politics either by alliance or by the conquest of Indian princely states.

With the advent of the East India Company, the British established trading posts along the coast. The East Indian Company, originally chartered in 1600 by Queen Elisabeth I as the Governor and Company of Merchants of London Trading into the East Indies, eventually came to rule large areas of India with its own private armies, exercising military power, and assuming administrative functions. The company was dissolved in 1874.

The need for security against local rulers as well as other European rival nations led to the construction of forts at each post. Mumbai fort, Fort William in Calcutta, and Fort St. George in Chennai were the main urban bastioned fortifications constructed. These cities developed from the small townships outside the forts. Parsimony of the East India Company, nonavailability of trained engineers, and use of local materials and artisans resulted in the simple design and construction initially. The vulnerability of these earlier forts, hostilities with the French, and the growing might of the company resulted in stronger and more complex designs for the second round of construction, with the design of Fort St. George reflecting the influences of the French classical bastioned fortifications advocated by Louis XIV's greatest engineer, Vauban.

Fort Marlborough

Fort Marlborough, located at Bengkulu City in the province of Bengkulu (Bencoolen in English) on the southwest coast of Sumatra, Indonesia, was built between 1713 and 1719 by the British East Indian Company under Governor General Joseph Callet. The square fort had a length of 44 meters and featured bastions at each corner. The main entrance facing south was protected by a demilune. Populated by English colonists since 1648, Sumatra was ceded to the Dutch in 1824.

Part 2. Bastioned Fortifications in the 17th and 18th Centuries 85

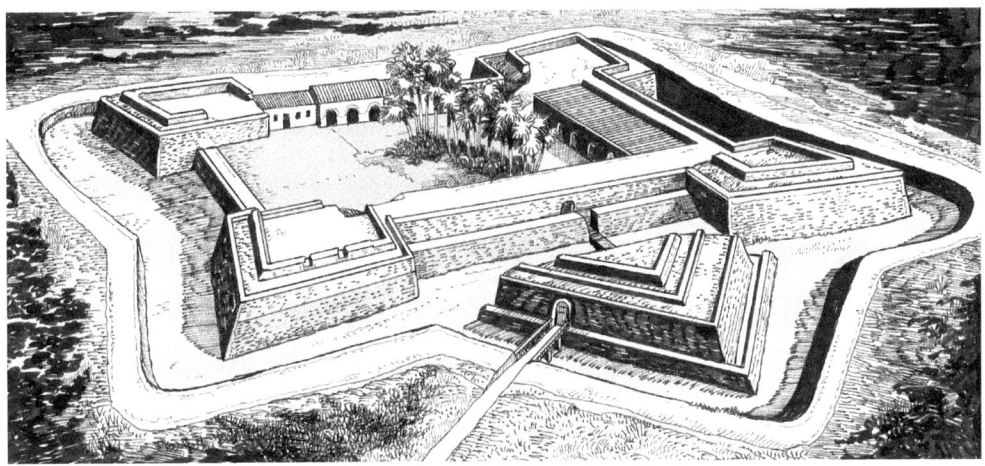

Fort Marlborough.

Madras

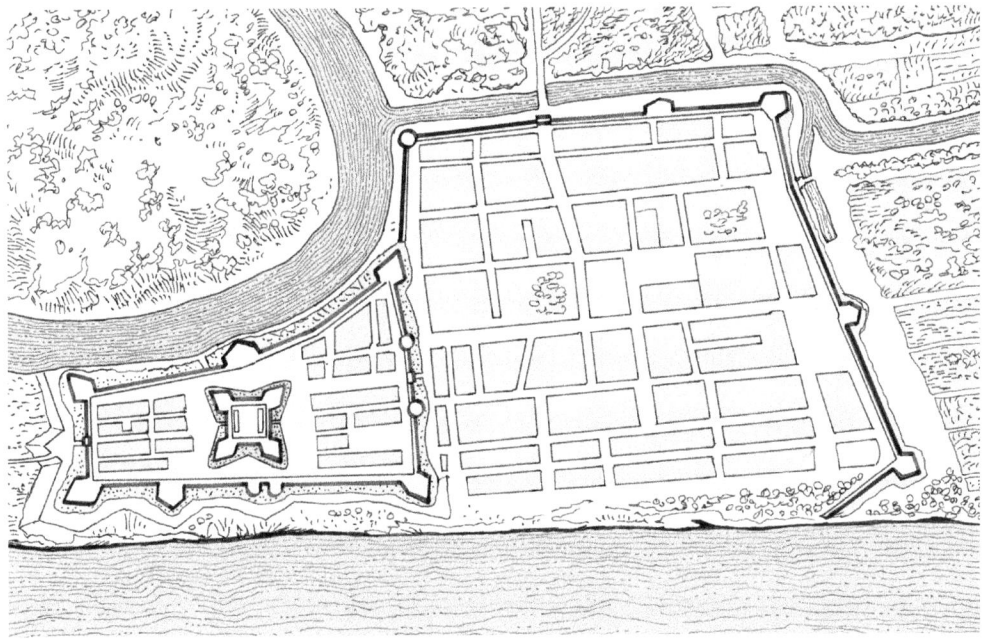

Madras in 1726.

Madras (aka Chennai) is located in southeastern India facing the Gulf of Bengal. The origin of Madras as we know it dates back to a few centuries—about 350 years. Prior to that, small villages existed for several centuries long before the Europeans arrived here. As usual in this part of the world, the Portuguese were the first Europeans to reach the shores of Madras. In 1639, Francis Day and Andrew Cogan of the British East India Company established a fort, factory, and trading post. The settlement was completed in 1640 and became known as Fort St. George. Outside of the fort was George Town.

Between 1668 and 1749, the East India Company expanded its power and helped transform Madras into a major commercial center. King James II granted George Town a municipal charter in 1688, making Madras the oldest municipality in India. In the middle of the 18th century, the French challenged the supremacy of the British in India resulting in the Carnatic Wars between 1758 and 1761. In the first Carnatic War, the French captured Madras but had to return it to the British. The settlement around the fort and the new town rose in stature and power with the rise of British Empire. Toward the end of the 19th century, Madras was clearly established as an important hub in south India. Today, Madras (renamed Chennai since 1996) is the fourth largest city in India.

Fort St. George (Madras)

Fort St. George (Madras).

Fort St. George (or historically, White Town) is the name of the first British fortress in India, founded in 1639, in the coastal city of Madras. The construction of the fort provided the impetus for further settlements and trading activity in what was originally a no man's land. The British East India Company, which had entered India around 1600 for trading activities, had begun licensed trading at Surat, which was its initial bastion. However, to secure its trade lines and commercial interests in the spice trade, the company felt the necessity of a port closer to the Straits of Malacca. It succeeded in purchasing a piece of coastal land, where it began construction of a harbor and a fort. The fort was completed on April 23, coinciding with St. George's Day, celebrated in honor of St. George, the patron saint of England. The fort, hence christened Fort St. George, faced the sea and a few fishing villages and soon became the hub of merchant activity. It

Part 2. Bastioned Fortifications in the 17th and 18th Centuries 87

gave birth to a new settlement area called George Town (historically referred to as Black Town), which grew to envelop the villages and led to the formation of the city of Madras. It also helped establish British influence over the Carnatic region and keep at bay the kings of Arcot and Srirangapatna as well as the French forces based at Pondicherry.

The fort is a stronghold with 6-meter-high walls that withstood a number of assaults in the 18th century. It briefly came into the possession of the French from 1746 to 1749 but was restored to the British under the Treaty of Aix-la-Chapelle, which ended the War of Austrian Succession.

The Fort St. George complex housed the administrative buildings of the government of Tamil Nadu until March 2010. The legislature of Tamil Nadu and the secretariat (with headquarters of various government departments) was situated in the fort. The fort itself was open to the public but only certain areas. The main building or the secretariat was open only to government officials and the police. The cannons and the moat that guarded this old building have been left untouched. In 2010, the legislature and the secretariat moved to a new location and plans have been announced to convert the old assembly complex into a library for the Central Institute of Classical Tamil.

Fort William

Fort William (Calcutta).

Fort William, located in Calcutta on the eastern banks of the river Hooghly, was built during the early years of the Bengal presidency of British India. It was named after King William III of England. The construction of the fortress was started by Robert

Clive in 1758, after the Battle of Plassey (1757), and completed in 1781. Fort William is of stupendous dimensions and spread over an area of 532 hectares. It is built of brick and mortar, and its bastioned design displays a fine alluring symmetry with bastions facing landward and tenaille front towards the Hooghly River. It is surrounded by a dry moat 9 m deep and 15 m wide, which can be flooded. Today, this remarkable fort located in the periphery of the lush green Maidan is the property of the Indian army.

PART 3

British Fortifications During the Napoleonic Era

Napoleonic Wars

The Napoleonic Wars (1803–1815) were a series of conflicts between Napoleon's French Empire and several European coalitions. Before the megalomaniac "Boney" was defeated and ousted for good in 1815, the French threatened to invade Britain. Since 1801, it was obvious that preparations were made, and the French emperor concentrated a large army at Boulogne and in other major ports in the Channel and the North Sea.

A tangible expression of Britain's concern for an invasion and its determination to fight to the finish was pursuing the policy that the best defense is offensive. The Royal

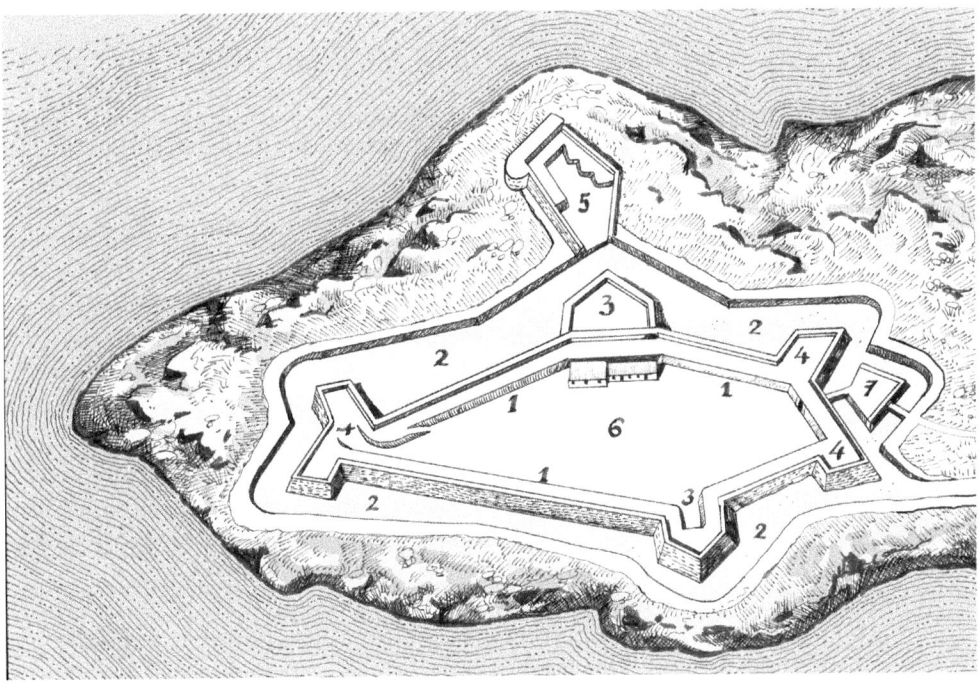

Fort Regent, St. Helier, Jersey Island. Fort Regent was constructed between 1806 and 1814. The fort included curtain walls (1), ditches (2), bastions (3), half bastions (4), and a redoubt/battery (5). There was a parade ground (6) in the middle, and the entrance was defended by a ravelin or demilune (7).

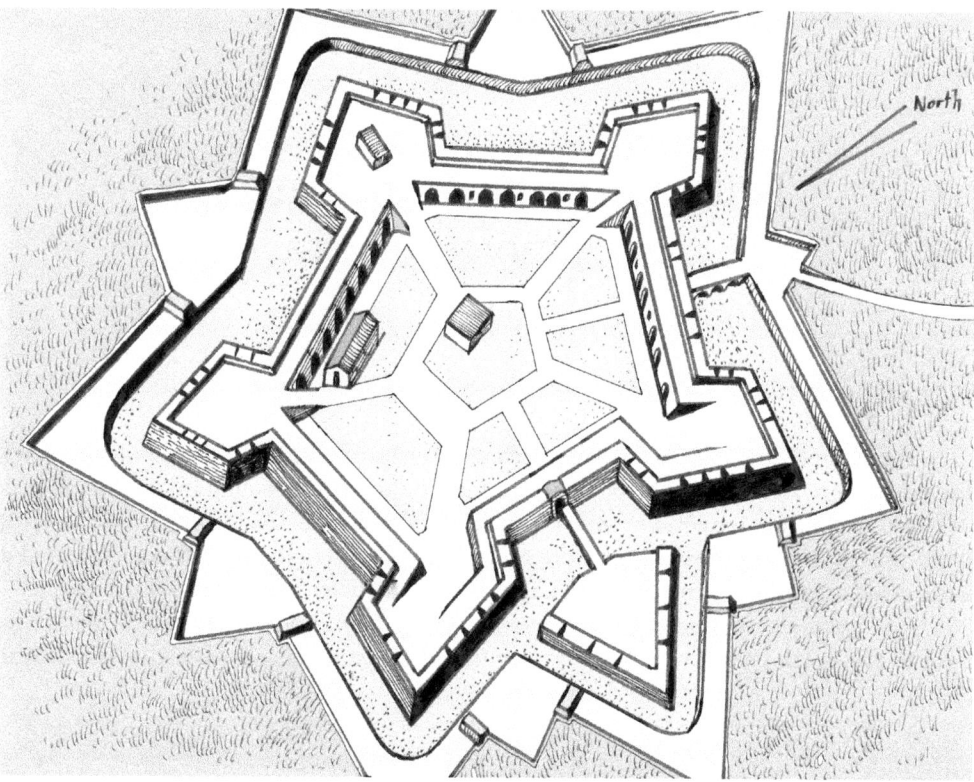

Fort Cumberland, Portsmouth, Hampshire. Fort Cumberland was built between 1785 and 1810 in the far east of Portsmouth at Eastney Point. It was intended to protect the access to Langstone Harbor.

Navy blockaded French harbors for disrupting trade and hunted the French ships in coastal waters. But even the powerful and large British fleet could not be everywhere at once. The navy alone was not enough, and a new generation of permanent defenses was needed. Debates took place in 1803 and early 1804 and certain points emerged and were agreed on. This led, again, to the reinforcement of coastal fortifications intended to turn southern England—as always the most accessible part to a French invasion by the most direct route from Calais, Boulogne, and Dunkirk—into a large armed entrenched camp. The French fleet was utterly defeated and destroyed at the naval battle of Trafalgar in October 1805. Napoleon never did invade Britain, so the English Napoleonic coastal fortresses were never attacked, and the great expenditure were seen by many as a folly. However, in retrospect, their success as a deterrent might also be argued.

Martello Towers

The designs chosen were circular "sea towers" called Martello towers and circular redoubts. On experiment, it was indeed found that shots that glanced off circular walls caused less damage than those hitting the flat surface of a square or rectangular fort. But for a few exceptions like Fort Regent on Jersey island, the bastioned system was not used probably because there was no time and too high cost would have been involved.

Besides, by the Napoleonic era, the increased range of the French artillery (greatly modernized and standardized by Lieutenant General J.B. Vaquette de Gribeauval in the 1770s) made close-range flanking fire less critical. Also, the towers were intended only to delay the invasion, to win time, allowing the army and the navy to gather and deal with the invaders.

On the whole, the towers' masonry were strong, and rooms inside the towers and redoubts were vaulted so as to be bombproof (withstand bombardment). However, there was no standardization. Fortifications of this period leads one to the conclusion that British engineers were reluctant to accept uniform solutions, preferring to adopt a somewhat makeshift scheme adapted to each particular case. This would become one of the hallmarks of British fortifications: irregular dispositions intended to suit peculiar situations and adapted to particular locations.

An interesting development in British fortification was the introduction of the so-called Martello towers adapted to the use of artillery during the French Revolution and the Napoleonic era. In a sense, they were a return to a basic type present in virtually all periods: the cylindrical tower or small keep. The term *Martello* is derived from Mortella Point at the port of San Fiurenzu (Saint-Florent in French) west of Bastia, Corsica, where a Genoese watchtower had been built in the 16th century against pirates from North Africa. In 1794, this ancient work (manned by only a few men operating two 18-pounder and one 6-pounder cannons) was attacked from the sea by two British ships, HMS frigate *Juno* (32 guns), and HMS battleship *Fortitude* (74 guns). Much to the British sailors' surprise, the ships were repulsed with serious losses, and it had to be left to the marines to take the tower from the landside after a two-hour heavy and continuous bombardment. The tower had thus proved so resistant to artillery that the British army and naval officers involved in the attack were convinced that this was an outstanding type of defense. Soon, the British started the construction of such coastal towers, notably at Simonstown harbor in Cape Town, South Africa, in 1796. In 1803, when war with Napoleonic France started again, one of the proposals for the defense of England's south coast consisted of a line of sea towers in the Mortella Point style—or Martello towers as they were misspelled. In 1805, although Napoleon had abandoned all desire to launch an attack on southern England, the idea was revived, and Martello towers were built by independent local builders commissioned and placed under the supervision of Lieutenant General R. Morse of the Royal Engineers. In 1808, there were 73 towers along the 50-mile stretch of the Sussex and Kent coast from Eastborne to Folkestone. In the period 1808–1812, another 29 towers were constructed from Clacton in Essex to Aldeburgh in Suffolk. By 1812, when the immediate threat of a French invasion was over, the whole coastal defense system was completed with 103 Martello towers together with redoubts, forts, and coastal batteries protecting nearly 200 miles of the most vulnerable parts of the south and east English coast.

The tower design varied from place to place but basically comprised two floors and a roof platform. The external appearance of a Martello tower was that of an inverted flowerpot with a pronounced inward batter or slope to the wall. The towers were based on the observation that a strong tower of circular plan armed with only one, two, or three guns mounted on the terrassed roof could withstand attack by greatly superior forces and armament. On average, a tower was about 33 feet high to resist escalade and between 48 and 55 feet in diameter at the base. In fact, some towers were slightly oval or elliptical with a bigger thickness facing the sea where the enemy was supposed to come from.

A unique and curious exception is the extant tower at Aldeburgh, Suffolk, which has a quatrefoil plan (resembling a four-leaf clover). Martello towers were built of brick rendered with stucco and bonded with a mixture of lime, sand, hot tallow, and ash. Many towers were covered with a plaster cover and many featured a ditch. Most had a single entrance placed on the landward side. For defense against landing parties, this doorway was very often at upper floor level (about six meters above ground) and approached by a ladder or a small drawbridge across the ditch. Some entrances were defended by a medieval brattice. The tower had no or few windows or openings at ground floor level, only a few loopholes and winding air holes for ventilation. The few openings faced landward and were deeply recessed to prevent shots entering them. Internally, there were usually two or three stories. The arrangement was simple including a strong, round pillar at the center from basement to a vaulted roof for extra stability and strength. Radial walls from this pillar divided the space into several rooms. The windowless and secure lower floor contained one or more storerooms for water, supply, and ammunitions, as each tower was planned to be able to withstand a siege. In the ammunition store, all the fittings were of wood or nonferrous metal to prevent sparks and thus explosions. On the upper floor, there were accommodation for the garrison (between 15 and 24 men) and a room for the officer in charge. The vaulted ceiling of the upper story supported a ten-feet thick roof, which included a combat terrasse. This was often armed with a single 24- or 32-pounder gun, which could pivot on a rail through 360°. Some towers were fitted with a furnace for hot shot so dreaded by wooden-hulled sailships. Many towers stood alone as isolated strongholds to protect any part of the coast vulnerable to enemy landings, and many were so sited about 550 meters (600 yards) from each other that they formed a chain with the guns of each tower overlapping the arc of fire of its neighbors. So in theory at least, each tower could offer its neighbors supporting fire and no enemy ship could approach the British shores without coming under devastating fire. The system of coastal Martello watchtowers turned the south of Britain into a large armed entrenched camp.

Some towers were garrisoned by regular troops, others by part-time militias paid and equipped by the county authorities and so releasing army units for service elsewhere. Even in peacetime, it was ensured that somebody was always on watch, while every militiaman reported for weekly training and of course in emergencies.

The British Martello towers, just like Henry VIII's coastal forts, were never put to the test, and their guns never fired a shot against any invader. In several ways, the period 1803–1805 was similar to 1539–1540, and the invasion for which defenses had been built never came. One of the reasons for the continuing building program was caused by bureaucratic process which took long to set in train and could not be quickly halted. After all, contracts had been signed, people's livelihoods and industrial profits were at stake, and work had been put in motion—work that could not be abandoned after everybody had said how important it was. The fact that not one Martello tower fired a shot in anger at a foreign invader might illustrate the theory that the best defense is that which deters the enemy from attempting any attack at all.

Of the original 103 Martello towers, about 43 survive, some of them in ruinous condition. However, moves have been made to restore a number of them and to open them to the public. The rest has gone, collapsed into the sea or demolished.

Martello towers were also built in the Channel Islands (Fort Grey, Fort Hommet, and Fort Saumarez on Guernsey, as well as Icho tower, Portelet, Lewis, Kempt, and

Collette on Jersey), Scotland (e.g., at Leith Harbour and Longhope), Wales (e.g., Pembroke Dock), and the sensitive area of Ireland (e.g., Dublin Bay and Cork Harbour). Similar designs were also constructed in British colonial possessions all around the world.

Martello towers were not recommended for inland fortresses, but their construction as coastal watchtowers or strongholds continued as late as the 1850s when artillery development made clear that the design had become obsolete.

It should be noted that in the early 19th century, the U.S. government built several Martello towers copied from or influenced by the British designs constructed in Canada. American towers are to be found at the following locations: two are at Key West, Florida; others were built at the harbors of Portsmouth, New Hampshire, Charleston, South Carolina, and New York City; two more Martello towers stood on Tybee Island, Georgia, and Bayou Dupre, Louisiana.

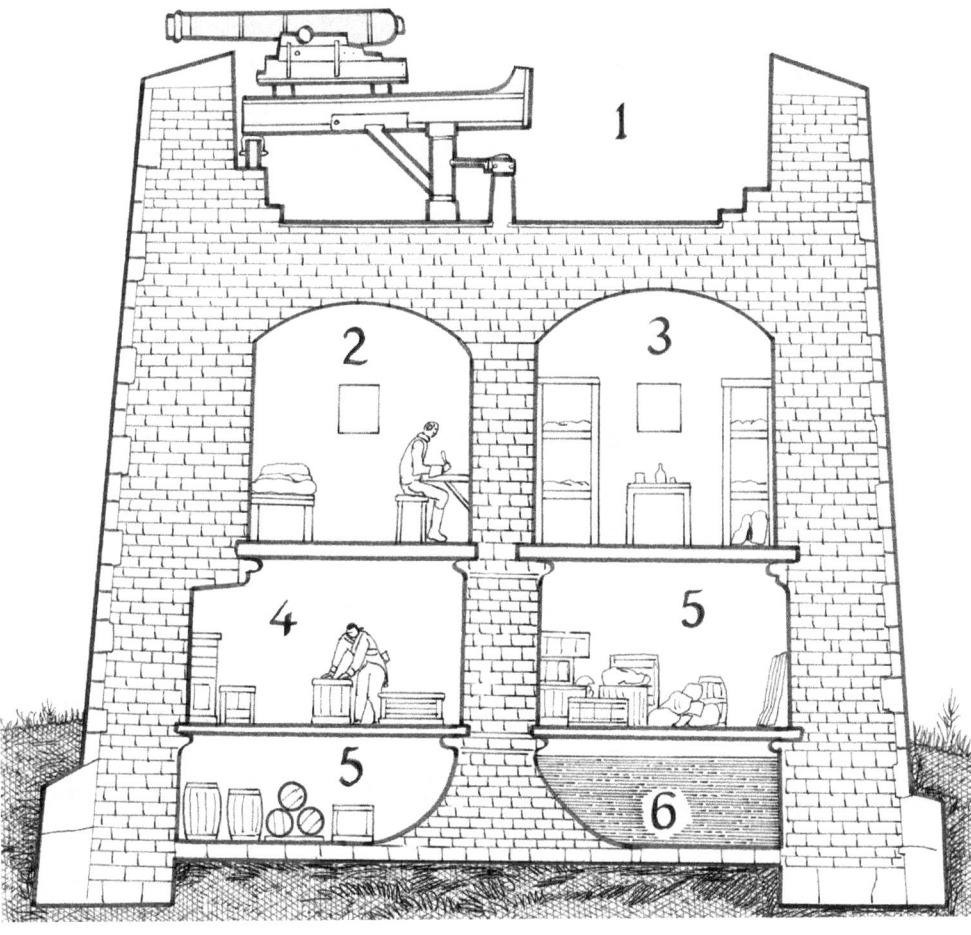

Cross-section of Martello tower. The 24- or 32-pounder gun was placed on the top terrasse (1). The tower also included a room for the commanding officer (2), one or two chambers for the garrison (3), an ammunition store (4), a food store (5), and a cistern (6) collecting rainwater.

Martello tower at Clacton-on-Sea, Essex.

Part 3. British Fortifications During the Napoleonic Era

Martello tower on Magilligan Point at Limavady near Londonderry, Ireland. The tower was intended to defend the narrow entrance to the bay of Lough Foyle.

Aldeburgh tower. The quite unique quatrefoil Aldeburgh tower is located in the Orford Ness peninsula on the coast of Suffolk.

Martello tower at Bantry Bay, Ireland.

Coastal Forts

The Napoleonic era saw the return of the tower, not only in small form as the previously described Martello towers and the French *tours modèles* but also for the design of larger forts. Influenced by the theorical works of the French marquis Marc René de Montalembert (1714–1800), one saw the revival of concepts like the massive roundish artillery tower, the general use of stone and masonry for vaulted casemates, and an increased use of caponiers, subterranean passages, and riflemen' galleries concealing the defenders from enemy sight and fire.

Martello towers were the first line of defenses, but it was clear that defense of harbors had to be strengthened. Included in the anti–French invasion scheme were three much larger circular forts (aka redoubts) that were constructed at Eastbourne, Dymchurch, and Harwich. These ports, if captured, would have offered the opportunity for a speedy disembarkation and buildup of weapons and supplies. The forts acted as command posts and supply depots for the smaller Martello towers in their vicinity as well as being powerful artillery firebases in their own right. The brick-built circular redoubts were intended to delay and obstruct a French amphibious assault long enough for the British forces to intervene. They included a ditch, artillery casemates, storerooms, magazines, and barracks, all arched and vaulted, arranged in separate compartments placed within the circular curtain. The massive roof was strong enough to resist exploding shells and supported an open terrasse for additional cannons and mortars.

The garrisons of the Martello towers and the redoubts were generally mixed with a few regular troops and part-time local militiamen paid by the county.

Eastbourne Redoubt

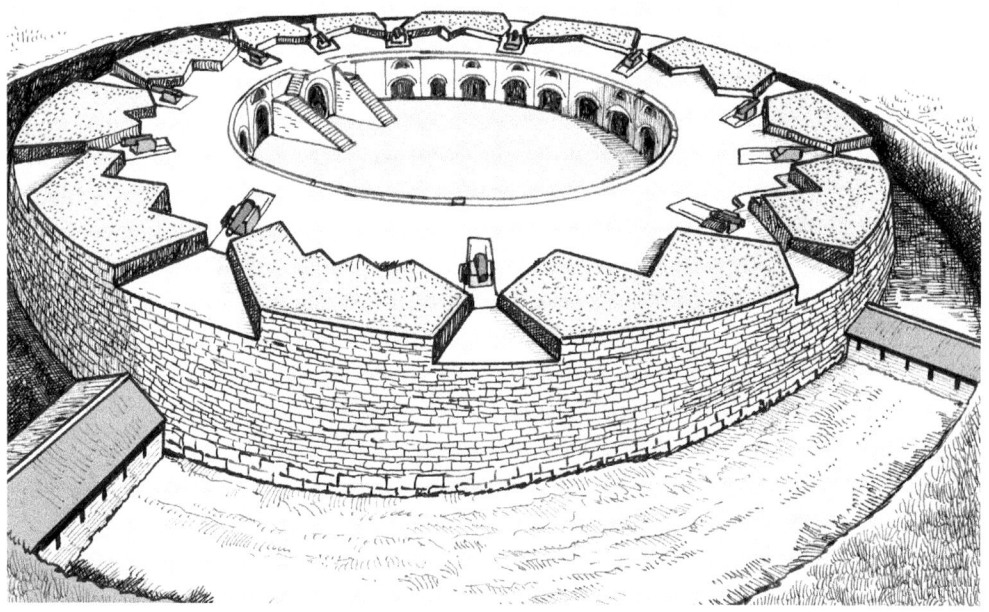

Eastbourne Redoubt.

Caponier in the ditch of Eastbourne Redoubt. A caponier is a small and low-profile flanking work built across a dry moat; it is projecting at the foot of a wall or of a tower. The illustration shows (1) counterscarp (the vertical or nearly vertical outer wall of the ditch), (2) dry ditch, (3) caponier with firing loops, (4) scarp (the inner wall of the redoubt).

Eastbourne Redoubt, located on Royal Parade in Eastbourne on the coast of East Sussex, was constructed about the same time as the Martello towers (between 1804 and 1810) and for the same purpose: to delay and obstruct a French amphibious assault long enough for the British field army and Royal Navy to engage and repulse the invader. Eastbourne Redoubt is a brick-built circular fort 220 feet in diameter surrounded by a ditch 25 feet wide and 24 feet deep. It includes 24-gun casemates, storerooms, magazines, and barracks whose arched vaults supported a broad artillery platform protected with breastwork and pierced with embrasures for 11 guns. The casemated barracks under the platform were separate compartments whose arched vaults supported a massive roof that could resist enemy projectiles and could support the weight and stress of the fort's guns. They could accommodate about 350 soldiers. Each room for the troop included folding beds, overhead shelves, and special stands for rifles or muskets. Each dormitory was heated by a stove and had its own food store and water supply, underfloor cisterns being filled by rainwater piped from the roof. The gunners reached the top of the platform via internal staircases from the barrack casemates below so that they would not be exposed to enemy projectiles falling into the open area in the middle of the fort. The single main entrance to the fort was across a bridge that linked the redoubt with the inland face. The exterior wall was a single massive circular construction devoid of all openings except for a single gateway and drawbridge over the ditch. The ditch itself was defended by five caponiers—fortified corridors projecting from the fort allowing flanking fire in the ditch. The caponiers had musket embrasures on each side. Eastbourne Redoubt was not a stronghold for offensive operations but rather an artillery firebase, and its size inevitably resulted in its employment as a local military headquarters. Today, the fort is a museum housing collections from the Royal Sussex Regiment, the Queen's Royal Irish Hussars, and the Sussex Combined Services Collection.

Dymchurch Redoubt

A similar fort is Dymchurch Redoubt, located between Dymchurch and Hythe in Kent. It was built between 1804 and 1812 to support a chain of 21 Martello towers that stretched between Hythe in Kent and Rye in Sussex, and to act as a supply depot for

Dymchurch Redoubt. Eastbourne, Dymchurch, and Harwich redoubts are thick masonry and ramparted forts with a very low profile in order to present only a small target for enemy attackers.

them. It specifically protected the sluices that were the key to the drainage of Romney Marsh in Kent (see below, Royal Military Canal). By the time it was finished, the invasion threat was over. Dymchurch Redoubt is circular in form and built of brick with granite and sandstone dressings, measuring up to 68 meters in diameter and stands 12 meters above the bottom of its 9-meter-wide dry moat. It lacks the caponiers or musketry galleries of the otherwise similar Eastbourne Redoubt. Beyond the moat, an earth bank or glacis helped to protect the masonry from artillery fire. Built on two stories, the upper floor had open emplacements for ten 24-pounder guns mounted on wooden traversing platforms. The lower floor featured 24 vaulted barrack and storage casemates, which opened onto a circular parade ground. They were designed to accommodate 350 officers and men. Entry was originally via a wooden footbridge supported by stilts, which could be collapsed in an emergency.

Harwich Redoubt

Another circular fort similar in design and purpose to Eastbourne and Dymchurch is Harwich Redoubt situated on the high ground halfway between the town of Harwich and Beacon Hill in Essex. Built in 1808–1810, it was a circular fort approximately 200 feet in diameter, with a central parade ground of 85 feet diameter. The redoubt is sunk into the hill so that, although it is placed in an elevated position, its silhouette is concealed by the deep dry ditch that runs around it. The redoubt's ten 24-pounder guns had an effective range of over a mile and covered the harbor entrance, the landward approach to Harwich, and the harbor itself. Ammunition was lifted to the guns from the magazines below using hoists. Under the single circular wall, there were the usual storerooms, service places, and vaulted barrack facing the central circular courtyard capable of accommodating a garrison of 300 men. On top of the buildings lies the main gun platform, which has view of the open sea, Landguard Point, Harwich town, the harbor, and Beacon Hill. The purpose of the redoubt was to provide a strong point in the Harwich defenses and a command center for local forces in the event of a French landing. Its strategic location meant that it could be used to coordinate all operations with the other fortifications around the harbor, including the Martello towers at Shotley and Felixstowe, and Landguard Fort.

Dover Western Heights

Facing the famous Norman medieval castle of Dover, Western Heights is the area of high ground to the west of Dover overlooking the town. The impressive position on the ridge includes a series of forts, strongholds, coastal batteries, and ditches intended to protect the key port of Dover and prevent an enemy from using it as a bridgehead for invading England. It was indeed feared that Napoleon would land at Dover or in the region of Dover as it was the nearest English port and town to France. At the same time, it was recognized that Dover's fortifications presented a fatal weakness. They were indeed organized to repulse an attack from the sea but offered little serious defenses against an attack from inland, notably from Folkestone. The weakest point was Western Heights that overlook the medieval castle, the town, and the port. The obvious solution was to turn Dover's weakest point into a strong fortress. The first earthwork

Part 3. British Fortifications During the Napoleonic Era

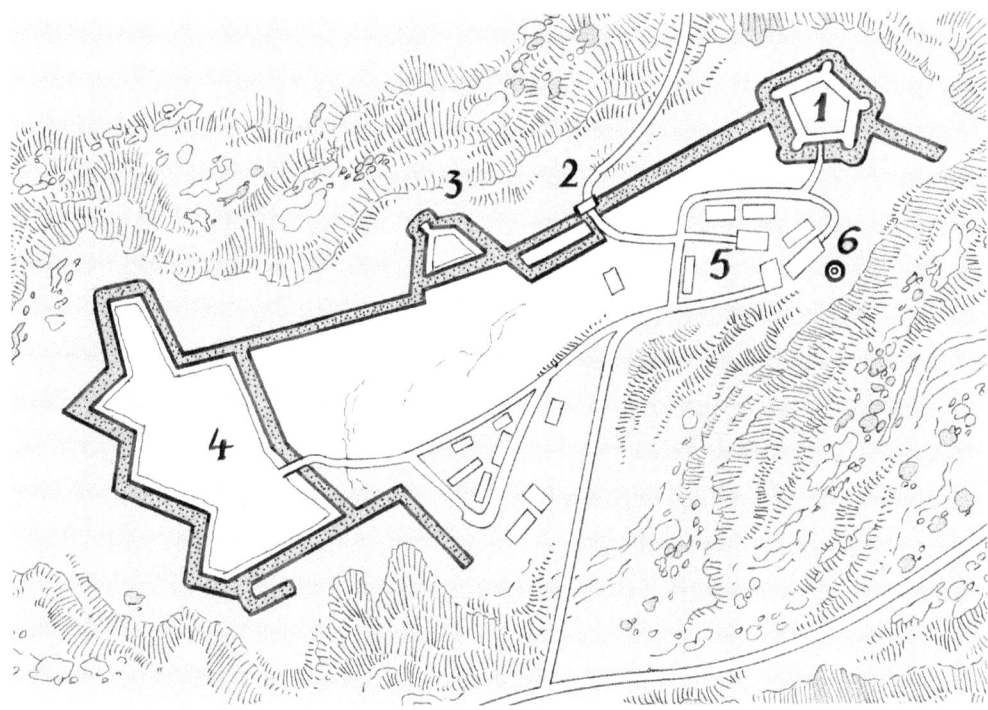

Map of the Dover Western Heights: (1) drop redoubt, (2) north entrance, (3) north center bastion, (4) citadel, (5) barracks, (6) grand shaft.

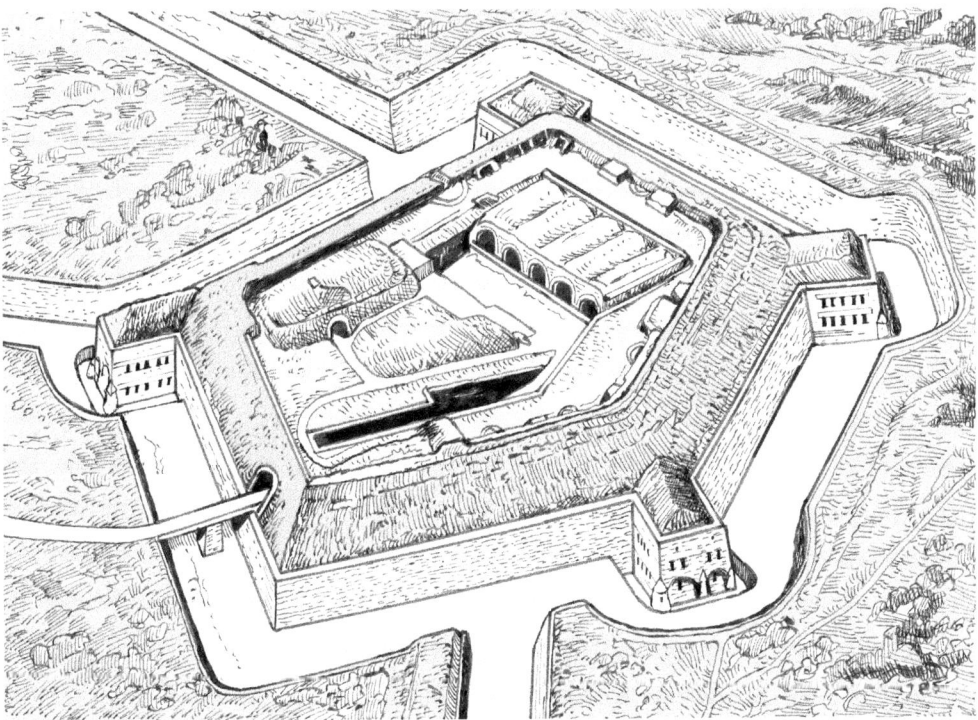

Drop redoubt. The redoubt is a pentagonal fort with caponier at four of its corners.

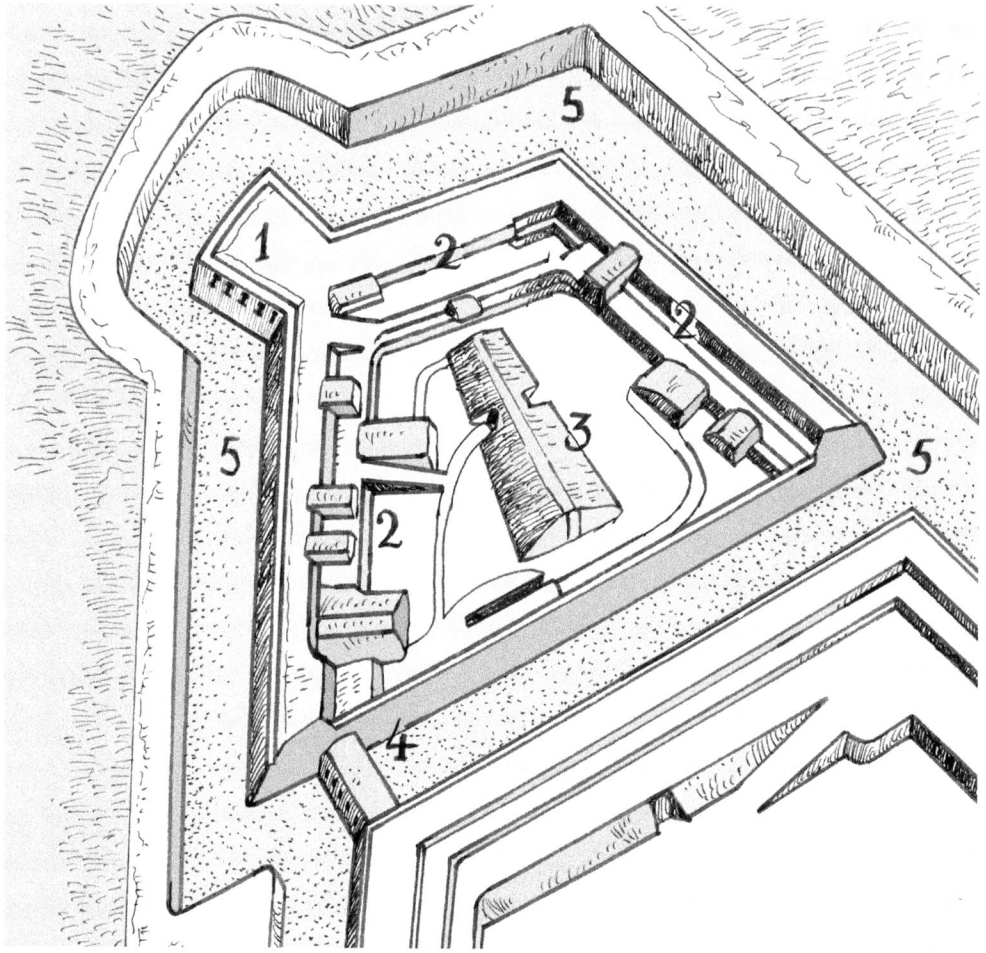

North center bastion. The illustration shows the following elements of the detached fort when it was completed in 1860: a caponier (1), artillery emplacements (2), a thick traverse (3), a caponier (4) across the ditch at the back, and a dry ditch (5) that connected to the drop redoubt and the citadel.

fortifications were established in 1779, and important defense works were added in from 1800 to 1810 to repulse a possible French landing on the south coast and to deny the capture of the port. These works were the Citadel, the North Centre Bastion, and Drop Redoubt. They still exist today and are now a local nature reserve, but parts are accessible having been dug out, vegetation removed, and partially restored. They form the most spectacular example of Napoleonic period fortifications in Britain and display just how architectural beauty can be achieved through a concern for utility and through the ruthless expression of functionality.

These structures above Dover are connected by a series of deep dry ditches with a total length of four miles called the Lines, which vary from 30 to 50 feet in depth and about 30 feet wide. They provided protection from direct infantry assault. They are revetted with strong brick walls and defended by strategically sited galleries built into the walls and by projecting caponiers for flanking fire. As with most artillery fortifications since the early 16th century, the walls of the Western Heights works were

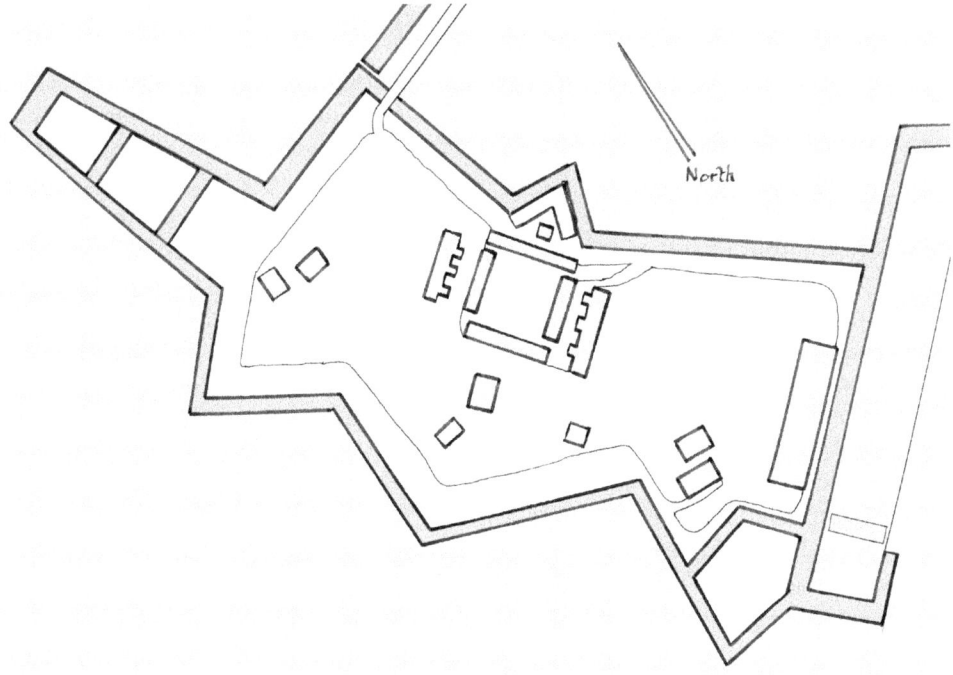

Plan of the Dover citadel.

made thick and ramparted—this means filled with huge masses of earth held in place by masonry so that they could best withstand impact of enemy shots. They also had a very low profile to present a difficult target for enemy gunners. The ambitious fortifications of Dover Western Heights were not completed until the middle of the 19th century.

Drop Redoubt to the northeast is one of the three forts on Western Heights and is linked to the others, the Citadel and the North Centre Bastion. The name *Drop* seems to refer to the remains of a Roman lighthouse, locally known as the Devil's Drop of Mortar. The redoubt is actually a powerful self-contained fort and is arguably the most impressive and immediately noticeable feature on Dover's Western Heights. It was a place of immense strength designed to act independently of the other works. The artillery at the redoubt faced mostly inland; it was intended to counter an invading force attempting to capture Dover from the rear. The construction of the redoubt was carried out in two periods: the first from 1804 to 1808 during the Napoleonic Wars and the second from 1859 to 1864 when a renewed threat of war in Europe in the mid–1850s encouraged the British government to complete and modernize the fortifications of Western Heights.

The core of the defensive system was the impressive Citadel on the highest point of the ridge in the west, which was intended to act as a last defense should the remainder of the Western Heights works fall into enemy hands. Work started in 1780, and continued until 1890. The large citadel included a defended entrance and several large buildings defended by three large redans on its western side and one single large redan on its eastern side. The ramparts consisted of bombproof artillery casemates and barracks stores and magazines and enclosed a parade ground in the middle. The Citadel was surrounded by a broad and deep ditch defended by several caponiers. In the 1950s, the Citadel barracks were used as a prison for young offenders; now it is a detention center for illegal immigrants waiting to be deported. The old officers' mess can still be seen on the top of the hills above Aycliffe.

The North Centre Bastion was a detached redoubt, actually a small interval fort placed between the Citadel and Drop Redoubt. It was intended to add firepower to the northern approaches to Dover, particularly the Folkestone road. Incomplete in 1814, the work was modified in 1859.

With Dover becoming a garrison town, a need for barracks, storerooms, and a hospital for numerous troops and their equipment became obvious. There were two main sites of barracks in Western Heights. The first dates to 1804 and was known as the Grand Shaft Barracks, being located at the top of the Grand Shaft Staircase (see below). They provided accommodation for 59 officers, 1,300 noncommissioned officers, and privates. They were renowned for their light and airy situation, and close to them near Archcliffe Gate was a military hospital with beds for 180. The barracks were demolished to make room for new buildings.

The second large set of barracks was the South Front Barracks, constructed in the 1860s. These were constructed in a huge trench, facing the sea with different floor levels connected to the hill behind by cast iron bridges and galleries. These were not such pleasant barracks to live in being cold and dark.

The need to allow the displacement of many soldiers from the barracks to the harbor level at the foot of the cliff led to the design and construction of the spectacular triple helix Grand Shaft Staircase. Probably designed by Royal engineer Brigadier-General William Twiss, this was constructed between 1804 and 1807. It is 42 meters (140 feet) deep, and 8 meters (26 feet) in diameter. It is a triple spiral staircase, which leads from Western Heights Barracks down to Snargate Street. It is a shortcut that allowed troops to descend and ascend safely, at speed, and in bulk the 200 steps in case of a landing on the shore. Its design is bold, simple but ingenious, and extremely functional. It is a unique piece of military engineering with a sculptural and almost abstract beauty that transcends its purely utilitarian purpose. It consists of two vertical concentric hollow brick cylinders. Between them were built three intertwined staircases of Purbeck limestone. The inner cylinder was provided with window apertures for light and ventilation. At the bottom, where the staircases meet, a sloping corridor leads to Snargate Street. The construction of the Grand Shaft was a difficult affair as drilling and digging in the cliff's chalk and clay was dangerous.

The Saint Martin's Battery was situated on high ground overlooking the Grand Shaft. This battery was used not only in Napoleonic times but also in both world wars of the 20th century. It commands unrivaled views of the harbor and town and was reached from the Grand Shaft stairway by means of steps known as Saint Martin's steps. Saint Martin is the patron saint of Dover, and he is depicted on the town crest.

At the same time, the outer defenses of the medieval Dover Castle on the eastern ridge were remodeled. The huge Horseshoe, East Arrow, and East Demi-Bastion were added to provide extra gun positions on the east side, while the newly built Constable's Bastion improved protection on the west. At the northern end of the castle, the Spur was strengthened with the addition of a gun battery in the form of a redan. The roof of the medieval keep was made stronger in order to house artillery, and the new Canon's Gateway added to the defenses.

Archcliffe Fort stands on a headland overlooking Dover harbor, known as Archcliffe Point. In 1370, a watchtower surrounded by a chalk bank and ditch had been built on the site of the present Archcliffe Fort. This fortification remained substantially unchanged until 1539 when Henry VIII ordered that a strong bulwark be constructed. Later when the Spanish Armada threatened the south coast, this fort was strengthened. During the 17th and 18th centuries, repairs and improvements were made, including the building of two new guard houses, raising the parapet, and the construction of new barracks. During the Napoleonic Wars, additional attention was given to reinforce the defenses. The entrance was remodeled in 1807–9 and again in 1814–15 when a brick barbican was added. Archcliff Fort, however, became much less important after the Western Heights defenses were completed. In the 1920s, the southern half of the fort was demolished to make way for a railway line. It was decommissioned in 1956, and part of it, notably the barbican, was removed to widen the A20 motorway. Today, what remains of the stronghold is used by the Emmaus Community, a charitable group working to help homeless people by providing accommodation and work for them.

Chatham Dockyard Defenses

The devastating Dutch raid of 1667, and wars with the French in the 18th century led to concerns about the vulnerability of Chatham dockyard. From 1700, there were

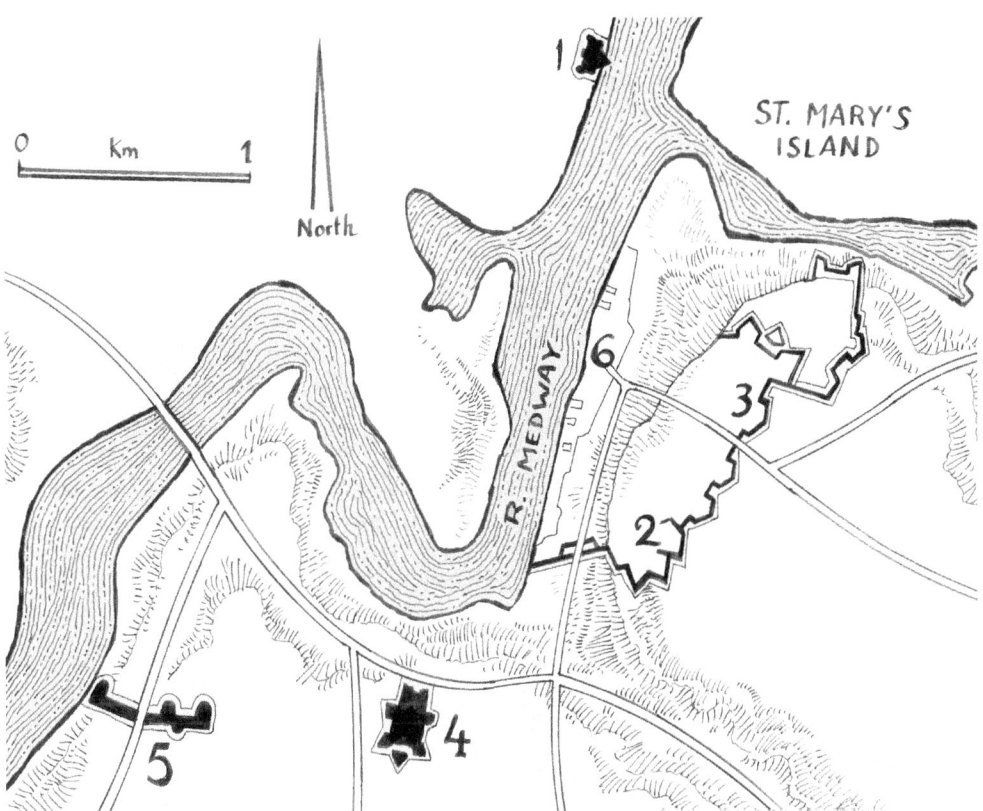

Map of Chatham Napoleonic defenses. (1) Upnor Castle, (2) Fort Amherst, (3) Cumberland Lines, (4) Fort Pitt, (5) Fort Clarence, (6) naval dockyard.

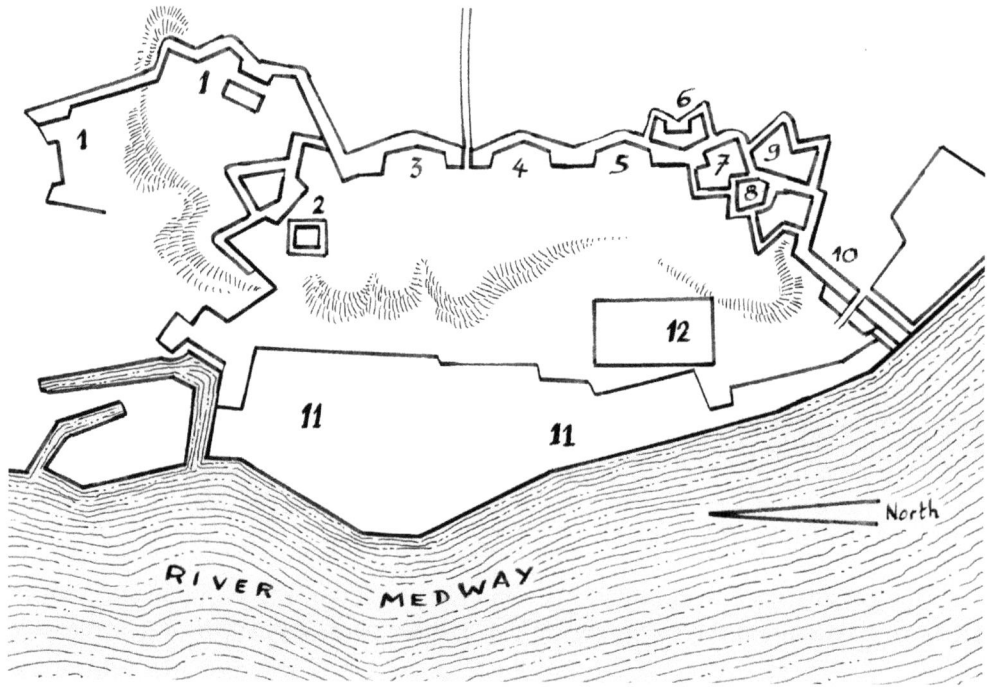

Cumberland Lines in 1812. The plan shows the following: (1) uncompleted new sections of fortifications in the north and casemate barracks; (2) Townsend Redoubt and Prince Frederick bastion; (3) Prince Henry bastion; (4) Prince Edward bastion; (5) king's bastion; (6) hornwork; (7) Prince of Orange bastion; (8) Amherst Redoubt; (9) Prince William bastion; (10) Belvedere battery; (11) Chatham naval dockyard; (12) Brompton barracks.

various plans to fortify Chatham, but all were refused. Eventually in 1756, work started according to new plans drawn up by the Dutch engineer Hugh de Beigg. There was to be a line of bastioned fortification around the dockyard of Chatham and the town of Brompton, nearly two kilometers in length. Two square redoubts, Townsend Redoubt in the north and Amherst Redoubt in the south, formed strong points in the defenses. The lines designed by De Breigg took advantage of the high ground on the landward side of Chatham but only had a shallow ditch, and there was just one demilune in the entire lines. These fortifications were rather long requiring a large garrison and over 1,000 cannons to defend them. During the Napoleonic Wars, Chatham became a vital base for the defense of Britain in case of a French invasion. If a French army did land in Kent, the route to London would be barred by the key crossing point of the river Medway at Chatham. For this reason, the fortifications of Chatham were strengthened. From 1805 to 1812, the highest point in the defenses, Amherst Redoubt, was greatly strengthened and became known as Fort Amherst. This fort was made up of several independent yet complementary works and formed one powerful fortress. The various works of Fort Amherst and their casemates and galleries were all linked together by a system of tunnels that allowed quick and safe access even in case of bombardment.

The bastioned line was also extended to the north to include the enlarged dockyard and the village of St. Mary's. More barracks were constructed at Chatham, which became a major army base from which counterattacks could be launched against a French invasion force. Brompton Barracks was built in 1804–1806 and was originally

Fort Clarence. Fort Clarence, located in Rochester, Kent, was built between 1808 and 1812 as part of the defenses against an expected Napoleonic invasion. The elongated fort was positioned to repel any attack from the River Medway. The depicted Clarence tower resembles something like a red brick medieval castle keep with corner turrets. It was obsolete before it was completed and turned in 1819 into a naval hospital and converted in 1845 into a military prison.

called the Artillery Barracks as it was required to house the gunners who were to man the defenses of Chatham. Provision was made at the barracks for about 1,000 men and nearly 500 horses. In 1812, Major Pasley arrived to create the Engineers Establishment. The artillery eventually moved out and the officer's mess became the corps headquarters mess in 1856 when the Royal Engineers Woolwich Depot closed and the staff moved to Chatham.

The Medway barrier was completed by the construction of two other forts: Fort Pitt (built 1805–1819), a pentagonal bastioned fort, and Fort Clarence (1805–1811), a large brick gun tower with elongated defensive ditches. By 1811, the threat of invasion from France had gone and the Royal Navy was the undisputed champion of the seas, so the fortifications at Chatham were no longer necessary, and works were stopped. The fortifications of Chatham were used in siege warfare training throughout the first half of the 19th century.

Today, there are extensive remains of the fortifications of Chatham. A part of the fortifications has been built over, but sections of the Cumberland lines still exist, although access is limited. They are fenced off and overgrown in many places. Fort Amherst is occupied by a historical society that runs events and carries out restoration work there. The dockyard at Chatham was closed down as a military base in 1984. It is now a major tourist attraction, a maritime museum with historic ships on display and a number of exhibitions.

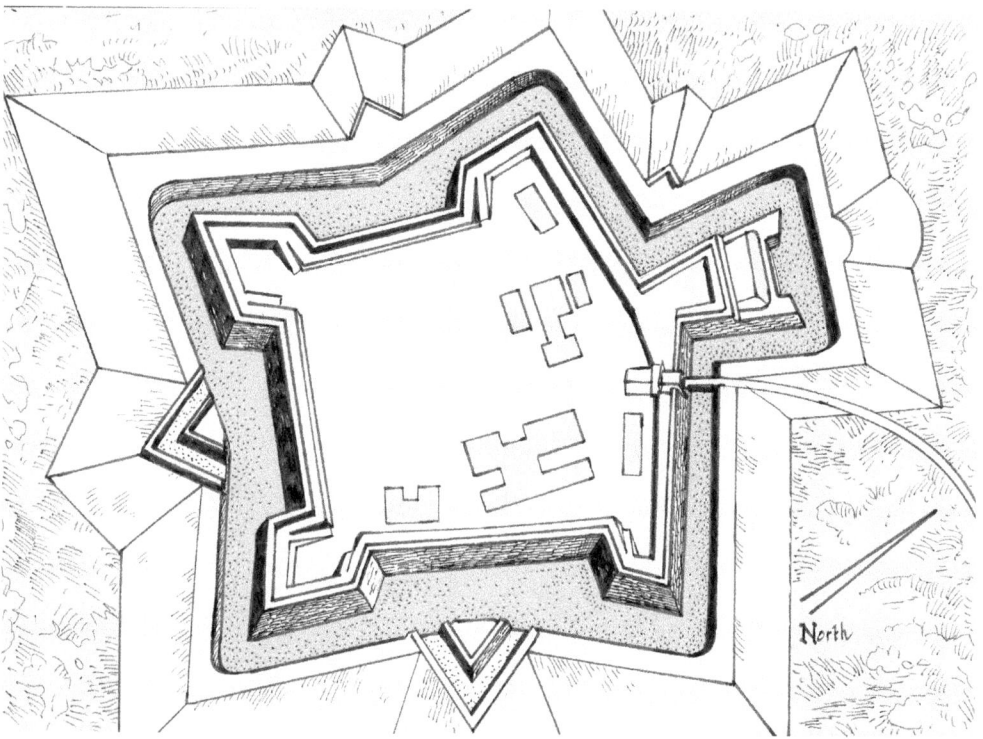

Fort Pitt, Chatham. Fort Pitt was constructed between 1805 and 1819 on a hill overlooking the Medway. Like Fort Clarence, Fort Pitt's life as a defensive work was short. In 1828, it became a depot and a home for invalid soldiers.

The Royal Military Canal

Another major defensive works of the Napoleonic era in Britain was the Royal Military Canal. The Romney Marsh is a sparsely populated, flat, and low-lying stretch of wet coast covering about 100 square miles (260 square kilometers) between Folkestone and Rye in Kent. With parts below sea level and crisscrossed by numerous waterways, this wetland was expected to be a possible landing point for a French invasion. After much deliberation, Lieutenant Colonel John Brown, commandant of the Royal Staff Corps, suggested that a defensive canal be built from Seabrook, near Folkestone around the back of the Romney Marsh, to the river Rother near Rye, a distance of 19 miles. Acting as a large entrenchment, it would be 19 meters wide at the surface, 13.5 meters wide at the bottom, and 3 meters deep. The excavated soil would be piled onto the northern bank to make a breastwork behind which troops could be positioned out of enemy sight and fire. In September 1804, the project was adopted and it was also proposed that the canal be extended from the river Rother to Cliff End, East Sussex, incorporating a part of the river Brede in the process. The total length of the canal would be about 30 miles (48 kilometers), of which 22.5 miles had to be dug. Between October 1804 and May 1805, work proceeded slowly during the harsh winter.

The canal was dug entirely by hand using picks and shovels, and the soil was carried away in wheelbarrows. Once the canal was dug, it was lined with clay. By August 1806, the canal was open from Seabrook to the river Rother in Rye. However, concessions had been made. The original dimensions of the canal were greatly reduced due to increasing cost and problems encountered by the builders and pressures of time so that for most of its length, the canal was half its projected width. In April 1809, the canal was actually completed, but by that time, it was clear that the threat of a French invasion had gone.

Then the military canal—considered a huge waste of public money—became an embarrassment to the government. Despite efforts to utilize the canal commercially, traffic was never heavy, and the opening of the Ashford to Hastings railway line (running along Rye, Appledore, and Hamstreet) in 1851 further decreased its use. Thus, during the 1860s the government took steps to unburden itself from the canal, and stretches were sold to individual owners and companies. The canal was requisitioned by the War Department in 1935 as war in Europe became increasingly likely. In 1940, it was recognized as an effective antitank ditch, and its banks were lined with concrete pillboxes as the nation awaited invasion, this time from Nazi Germany. Fortunately, once again, invasion never came. Today, the canal and its sluices are a pleasant and peaceful place for quiet enjoyment whether walking, strolling, jogging, cycling, fishing, or pleasure sailing. It also provides a home for many forms of wildlife.

The Lines of Torres Vedras

War in the Iberian Peninsula

In his struggle against the British, Napoleon decided in his decree of Berlin in November 1806 to ruin and starve them by establishing a large-scale naval embargo (called the Continental Blockade or Continental System), which forbade all trade and other commercial activities between Britain and all European countries occupied, allied with, or dependent on France. Portugal refused to participate in the embargo, remaining loyal to a long-standing alliance with Britain. Accordingly in 1807, Napoleon decided to invade Portugal. Under the pretext of reinforcing the Franco-Spanish army occupying Portugal, Napoleon invaded Spain as well. He replaced the Spanish king Charles IV with his own brother Joseph and installed his brother-in-law Joachim Murat in Joseph's stead in Naples. However, the majority of the Spanish people remained loyal to their former monarch and rejected the French rulers. This led to a spontaneous and fierce resistance against the French occupiers. The British took advantage of this new front to land troops in Portugal under the leadership of Arthur Wellesley, the future Lord Wellington (1769–1852). The costly and extremely brutal Peninsular War in Spain continued, taking the form of a terribly costly guerrilla. The whole affair was a grave mistake, a crippling move that eroded morale and prestige, and, in Napoleon's own words, an ulcer that drained the lifeblood from the French army. Napoleon was obliged to leave 300,000 of his finest troops to contain Spanish guerrillas as well as British and Portuguese forces commanded by Wellington. French control over the Iberian Peninsula deteriorated and collapsed in 1813. The war went on through allied victories and concluded after Napoleon's abdication in 1814.

Worthy of mention are the lines of Torres Vedras built by Wellington's army in 1809–1811. Designed by Colonel Richard Fletcher and constructed under the leadership of Captain John T. Jones, the lines were named after the municipality of Torres Vedras located north of the capital, Lisbon, in the district Oeste of the Centro region. The lines

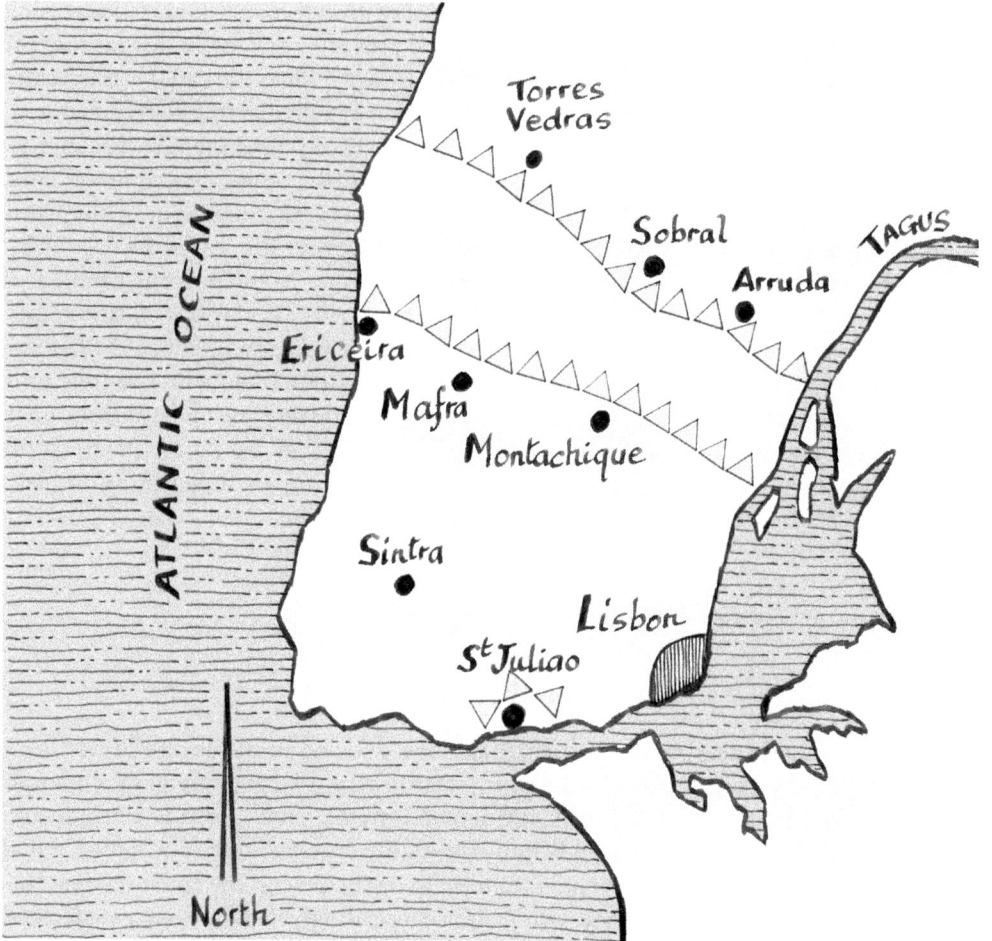

Map of the Lines of Torres Vedras (c. 1810). The lines were a series of detached strongholds across the Estramadura, the region north of Lisbon between the Atlantic Ocean and the River Tagus.

were composed of 114 redoubts, fortlets, and field works entrenchments made of earth, sited with care to taking full advantage of the hilly landscape in order to guard the passes leading to Lisbon that lay some 50 kilometers (30 miles) to the rear. There were actually two major lines stretching across the province of Estramadura from the Atlantic coast to the Tagus River. First, the northern position followed the hills dominating the cities of Cortada, Torres Vedras, Sobral de Monte Agraço, Arruda, and Alhandra. Second, the southern line ran along São Lourenço, Mafra, Cabeço de Montachique, Bucelas, and Alverca do Ribatejo. Both had a length of about 25 kilometers (9.5 miles), were armed with 232 field guns, and manned by 17,500 soldiers, both British and Portuguese. A third small, fortified line covered the port of São Julião da Barra on the Atlantic Ocean west of Lisbon, which was the intended embarkation point in the event of a general retreat. In addition, British gunboats were deployed on the Tagus to prevent any outflanking move up the river. From these entrenched positions, Wellington turned withdrawal into victory. Indeed, Marshal André Masséna with over 100,000 men made a few half-hearted attempts to break through the lines but finally forbore to attack them

Part 3. British Fortifications During the Napoleonic Era

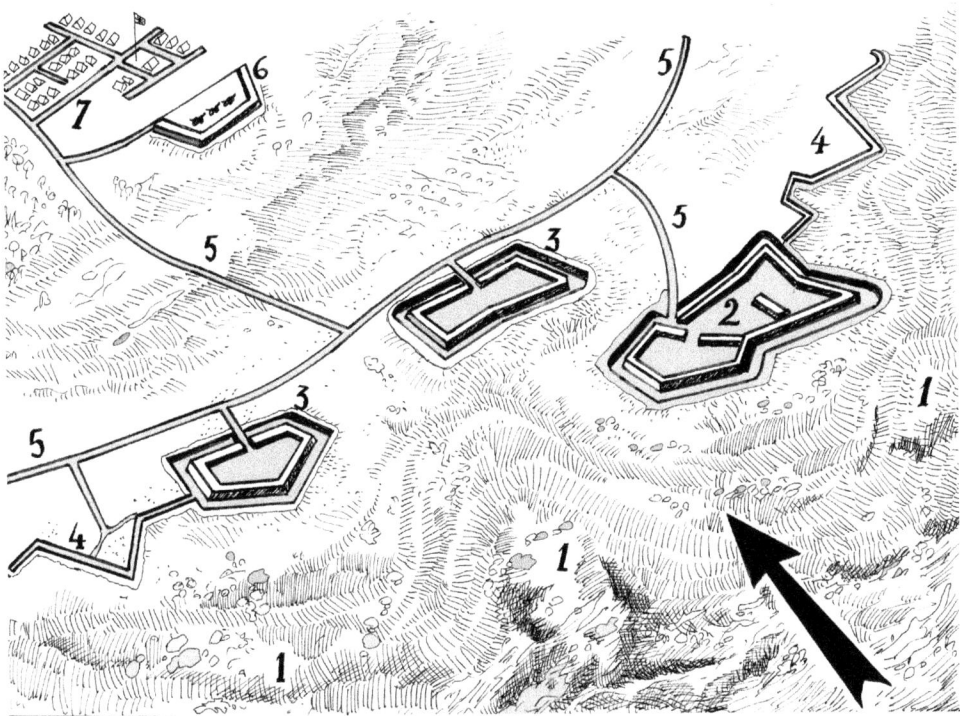

Schematic view of the Torres Vedras field fortifications: (1) open slopes; (2) forts; (3) redoubts; (4) infantry trenches and V-shaped redans; (5) communication tracks, military roads and paths; (6) guns and howitzers support batteries; (7) camps for detachments of Wellington's army ready to counterattack.

knowing full well that it was too formidable an obstacle even for him to overcome. So like other British Napoleonic fortifications, the Lines of Torres Vedras was never used, but here the deterrent aspect cannot be denied. In March–May 1811, Masséna's army (short of supply, depleted by sickness, and harassed by Portuguese guerrillas) started a long retreat north, leaving in its wake a track of burning farms, looted villages, murdered peasants, and other atrocities—one of the most terrible episodes of this ugly war.

Construction of the Line

The works of Torres Vedras differed from ordinary fieldworks in having an unusual degree of strength, plenty of time, and civilian labor available by requisition for their construction. In this respect, they approximated more to semipermanent works. The redoubts, posts, fortlets, batteries, artificial embankments, and infantry emplacements in the Lines of Torres Vedras varied considerably in size, shape, and strength according to their position of importance and the ground on which they were sited. There was little attempt at impressive sophistication like elaborate outworks. However, the simple fortifications of Torres Vedras displayed a remarkable adaptability on the part of Fletcher, Jones, and their engineers who used care and skill on the natural strength and features of the hilly region to make the Lines of Torres Vedras an impenetrable fortress.

The unsophisticated British lines of Torres Vedras announced the later concept of creating a belt of forts backed up by a field army to oppose to an enemy offensive.

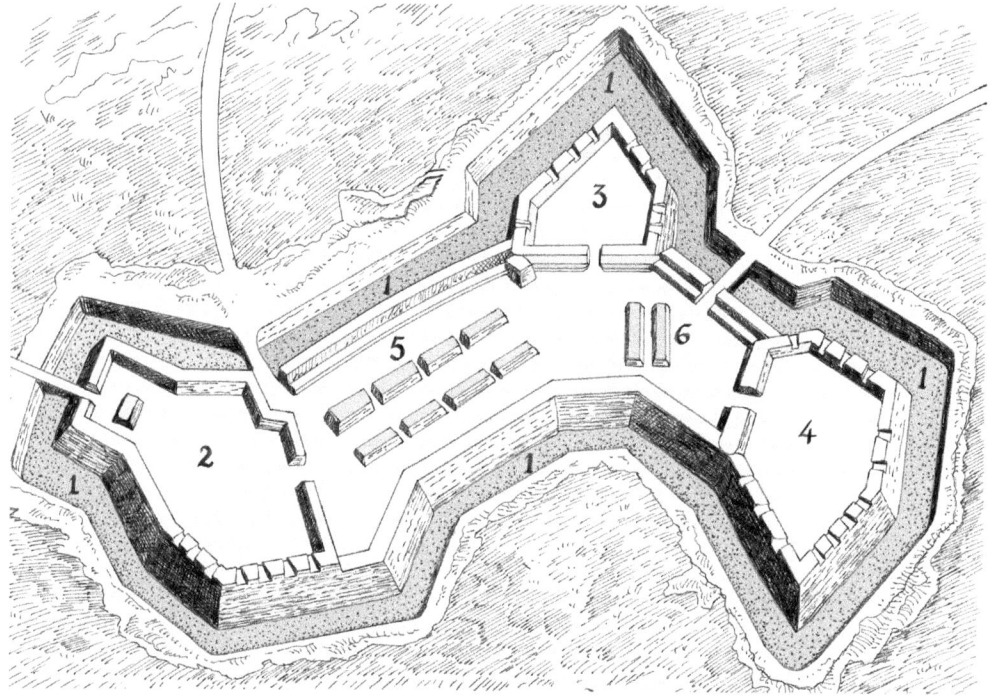

Fort San Vicente. One of the largest works in the Lines of Torres Vedras, Fort San Vicente had a total garrison of 1,720 men. It included the following features: (1) ditch; (2) Fort 20 (manned by 470 infantrymen and gunners operating eight guns); (3) Fort 21 (270 men and nine guns); (4) Fort 22 (380 soldiers and nine guns); (5) bombproof traverse made of earth protecting against the effects of blast from enemy shelling; (6) magazines.

They were an embryonic system of barrier fortresses, a theory that was to find its fullest expression after the Napoleonic Wars in the 19th-century forts.

The construction of the temporary and simple Torres Vedras fortifications was rather cheap compared to permanent and elaborate masonry works, but the cost in human terms and ruined livelihoods was immeasurable. In a rather densely populated region close to Lisbon, the establishment of military works meant the requisition of the population to dig the works. When that was completed, people were forced to evacuate their homes, or if they were allowed to stay, their activities on their own land were considerably reduced because vineyards, orchards, olive groves, grazing meadows, and fields had suffered serious damages. Roads leading to the capital were blocked, and in several cases, even farms, mills, barns, and other agricultural buildings were destroyed on military order. Local civilians had no choice but to comply with the military's ruthless demands. Compensations were paid out to some farmers, but a number of individuals had to bear the cost of the damage themselves.

Today, there remains a sufficient number of redoubts, strongholds, and other fortified points (notably Fort São Vicente) to be visited and through which one can have a satisfying idea of how the lines actually were.

Part 2. Bastioned Fortifications in the 17th and 18th Centuries

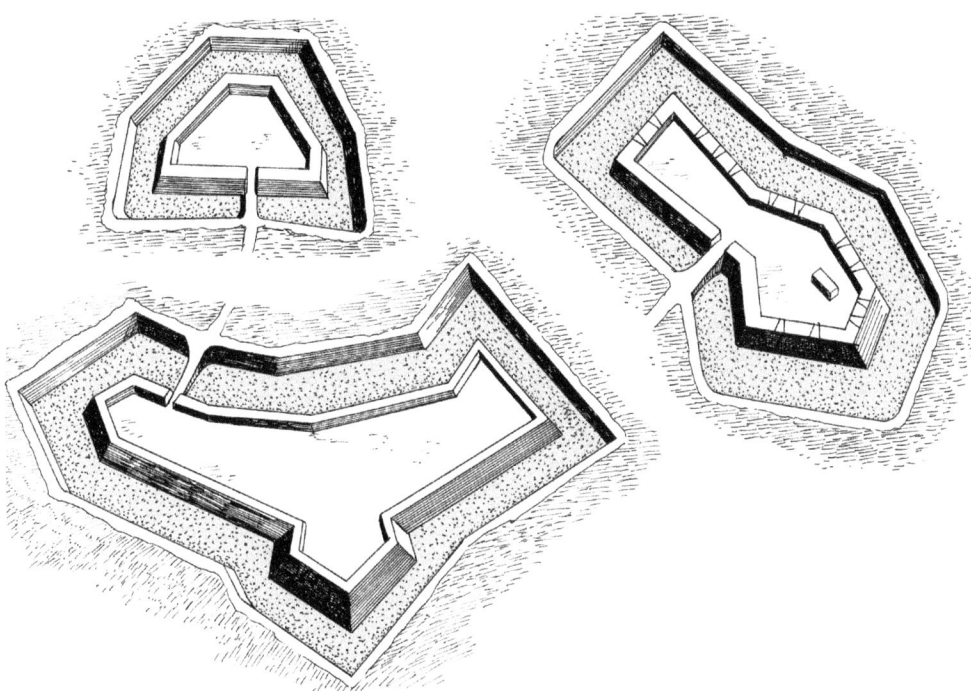

Redoubts at Torres Vedras. Shapes and sizes of forts and redoubts were various, relatively simple, and adapted to the surrounding.

PART 4

Fortifications in Britain in the 19th Century

Britain in the 19th Century

New Fortifications

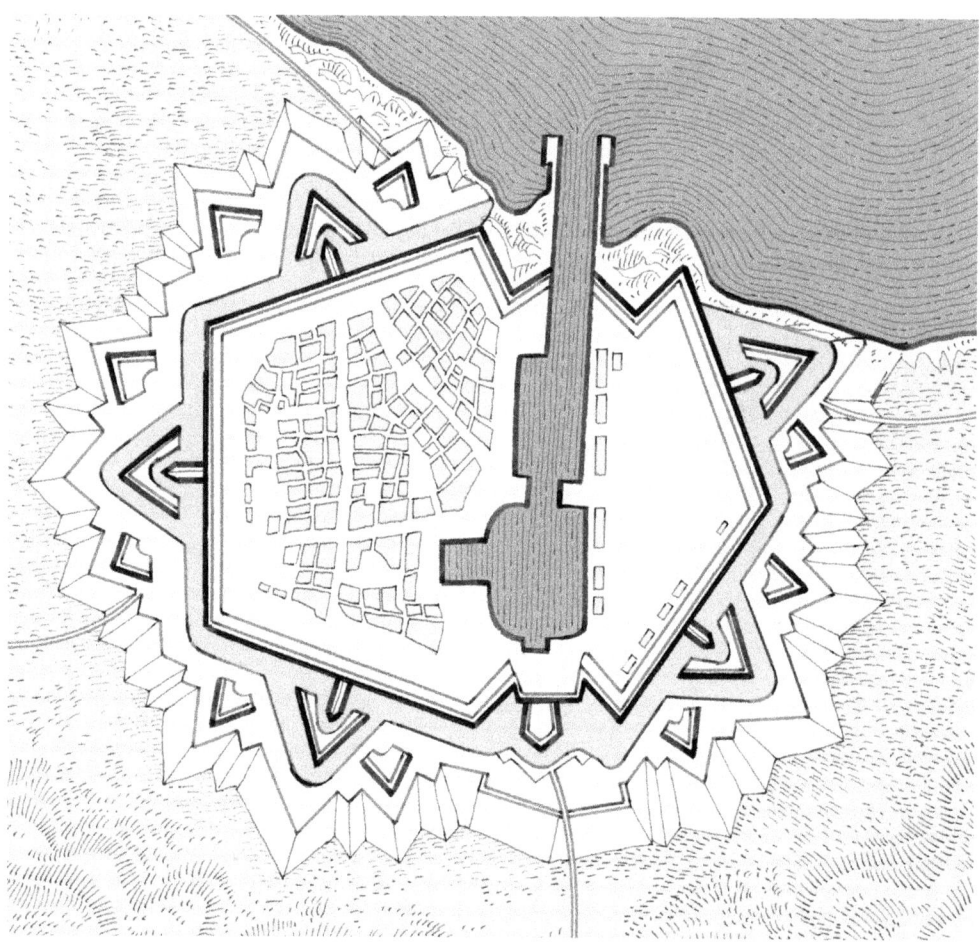

Project of fortification for the military port of Cherbourg (Normandy, France) envisaged by Montalembert designed in 1778.

Part 4. Fortifications in Britain in the 19th Century

After the French dictator Napoleon I had been defeated in 1815 and ousted for good on the small island of Saint Helena where he died in May 1821, Europe breathed freely again and turned to other matters. Western Europe entered into a new industrial, economic, and social organization. This new upheaval was the Industrial Revolution—an economic transformation that gradually but inexorably took the place of the ancient and obsolete social organization.

Nineteenth-century Britain was dominated by the long reign of Queen Victoria (b. 1819, reigned 1837–1901). In that period known as the Victorian era, the massive advance of technology and industrialization was rapidly reshaping both the landscape and the social structure of the whole country, and the establishment of a large colonial empire in Asia, India, and Africa brought Britain to global preeminence.

In the domain of military architecture, the 19th century was a time of feverish weapon escalation punctuated by several sudden crises that made everything existing obsolete and worthless.

New principles of fortifications appeared, many of which were based on theories developed by the French officer René de Montalembert (1714–1800). Published in several books in 1776–1784, Montalembert's new system called *fortification perpendiculaire* was ignored in France but particularly well received and improved by a number of foreign military engineers, notably in Austria, Bavaria, and Prussia. The leading ideas

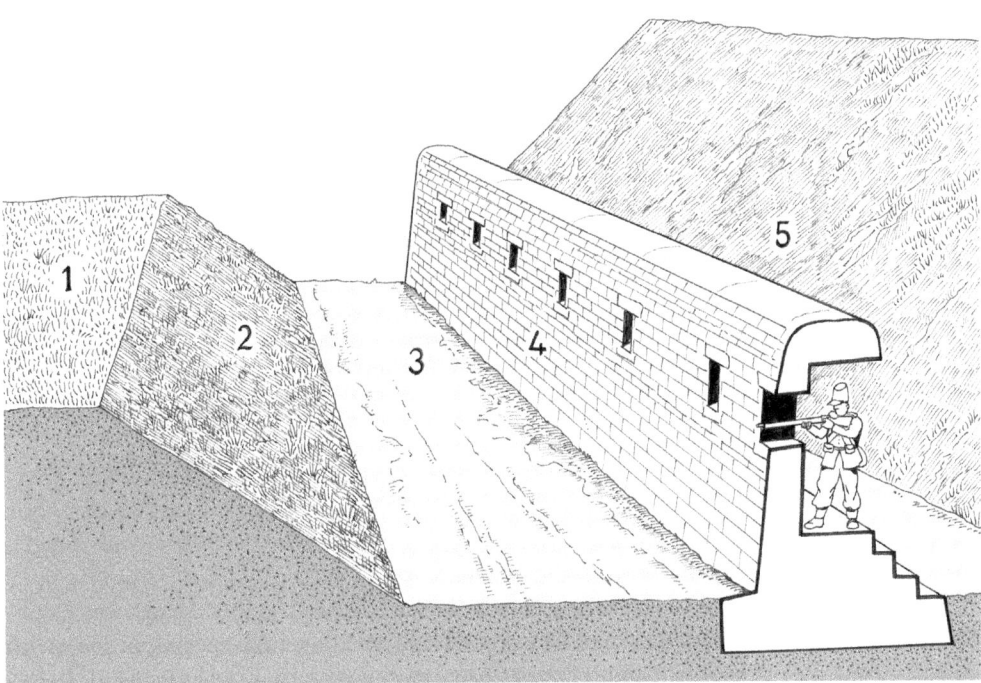

Cross-section of Carnot wall. Attributed to the French engineer and Napoleon's secretary of war Lazare Carnot (1753–1823), this defensive structure was a free-standing stonewall, about 5–7 m high and 1.5–2 m thick. It was intended to protect the ditch at the base of the scarp. (1) glacis; (2) non-revetted sloping counterscarp; (3) ditch; (4) Carnot's wall; (5) non-revetted detached scarp.

advanced by those innovators were actually quite simple: a siege is an artillery duel; the winner of this duel is the party with the best artillery; so defensive artillery must be superior in both quantity (firepower) and quality (protection) than the artillery of the attacker. New artillery fortifications were needed, and this soon led to the gradual appearance of various new methods improved and adopted under the prevalent designation of the polygonal system. According to Montalembert and the new generation of German military engineers, a modern fortress had to be a simple-shaped artillery stronghold with straight or multi-angled (tenaille-shaped) curtains with gun and rifle emplacements on top. In addition, a modern fortress should feature massive cylindrical artillery towers bristling with an overwhelming presence of weapons placed in casemates in several tiers. Those powerful artillery units were thus intended to keep enemy besiegers at a safe distance. Should enemy parties still infiltrate the surrounding terrain, then they would come under close-range fire from ship-shaped caponiers placed across the ditch and replacing the former corner bastions. The basic idea was sound and simple. It could be applied to any sort of ground, level or broken, and to long or short fronts. It greatly influenced 19th-century military architecture. From the 1830s to the 1850s, new designs (including both continuous urban enceintes and rings of detached forts and batteries) were applied especially in Germany. The Germans were indeed prompt to design and built new fortifications based on Montalembert's theory. Modern works appeared as early as 1834 at Germersheim (south of Mannheim in Rhineland-Palatinate), and the new system was gradually developed and employed notably at Coblenz, Mainz, Ulm, Ingolstadt, Linz, Cologne, and other key border positions in the main German kingdoms—namely Bavaria, Prussia, and Austria.

British Engineers

As already discussed, during and after the Napoleonic era, there was a general revival of the large circular artillery tower as advocated by Montalembert. This fashion was illustrated by the French Napoleonic standardized *tours-modèles*, the enormous Maximilian towers in Austria and Italy, and the Dutch *torenforten*. In Britain, this trend was exemplified by the previously described Martello towers and the circular redoubts. Indeed, the improvements and innovations carried out in Germany and continental Europe did not remain unnoticed. There were many skilled British engineers in the 19th century, and dealing with all of them would fill several volumes. Suffice it here to briefly discuss a few of them as examples.

An important figure in British military engineering of the Napoleonic and post–Napoleonic era was Charles William Pasley (1780–1861). As a British soldier who took part to the Napoleonic Wars, he wrote the defining text on the role of the post–American Revolution British Empire in *An Essay on the Military Policy and Institutions of the British Empire*, published in 1810. Pasley created and directed the Royal Engineer Establishment founded in 1812—renamed Royal School of Military Engineering in 1868 with headquarters first at Woolwich and later at Brompton Barracks, Chatham, Kent. Pasley devoted his whole life to the foundation of a complete military engineering science and to the thorough organization and training of the corps of Royal Engineers. He published several instruction treatises and manuals, including *Course of Elementary Fortification* in 1822 and *Simple Practical Treatise on Field Fortification* in 1823, both books becoming the standard manuals of field and permanent fortifications for the British army. In

1841, Pasley was promoted to the rank of major general and was made an inspector general of railways. He was promoted lieutenant general in 1851, made colonel commandant of the Royal Engineers in 1853, and general in 1860. Pasley died in London in April 1861.

The British civilian architect James Fergusson (1808–1886) published in 1849 his *Essay on a Proposed New System of Fortification*, containing a summary of concepts and recommendations of many of his contemporary European engineers. The book covered the design and construction for revetements, casemates, covered way, powder magazines, flanking defense, ammunition stores, garrison quarters, towers, entrenched camps, outposts, and other fortifications. Although he may have been a dilettante on the subject, Fergusson's opinions were, however, sufficiently competent and respected by the military experts for him to be invited to be member of a royal commission of defense.

Lieutenant Henry Yule (1820–1889) of the Bengal Engineers authored in 1851 *Fortification for Officers of the Army and Students of Military History*, a thorough and excellent work including a historical study of the development of military architecture.

Captain August Frederick Lendy (1826–1889) was a much-decorated army officer who had fought in many overseas campaigns. He published *Principles of War* in 1853; *Maxims, Advice, and Instructions in the Art of War* translated from the French in 1857; *Elements of Fortification, Field and Permanent* in 1857; *Treatise on Fortification or Lectures Delivered to Officers Reading for the Staff* in 1862—the most thorough, competent, and all-embracing book, which made a remarkable contribution to the British art of fortification.

Lieutenant General Sir William Jervois (1821–1897) was a British military engineer and diplomat. Jervois joined the British army in 1839 and was sent to South Africa. In 1858, when ranked a major, he was appointed secretary of a Royal Commission created to survey the state and efficiency of British land-based fortifications against naval attack. Jervois inspected Portsmouth, Spithead, the Isle of Wight, Plymouth, Portland, Pembroke Dock, Dover, Chatham, and the Medway. In his report, published in February 1860, Jervois considered several plans for the defenses around London. He also inspected and oversaw the design of the so-called Palmerston Forts and surveyed fortifications in Canada, South Australia, Gibraltar, and the Adaman Islands. In 1877 and 1878, he published the so-called Jervois-Scratchley reports about British colonial defenses. In the 1880s, many of his recommendations were implemented by the various colonial governments. From 1875 to 1888, he was, consecutively, governor of the Straits Settlements, governor of South Australia, and governor of New Zealand.

New Fortifications

One of the most obvious effects of the Industrial Revolution after 1815 was the simultaneous development of improved attack weapons, notably artillery range, which had been significantly increased. Several coastal forts and batteries freshly constructed or improved in 1825–1860 were typically beefy artillery towers loosely based on Montalembert's concepts. They often consisted of large crescent-shaped, oval, or cylindrical structures. They were made of thick brick and stone masonry in order to be shellproof. They were generally one or two stories high and enclosed by a ditch (or a moat) and a large flat and bare glacis. There was a central parade ground in the middle, and the ring of rampart housed quarters for the soldiers as well as supplies, ammunition stores, and everything needed by a garrison. The numerous artillery pieces were emplaced inside vaulted casemates (firing chambers) within the ramparts. The drawbacks to the casemate were well known: confined space in which to operate guns, issues with blast, noise, smoke, and toxic fumes, as well as restricted field of fire and poor view for observation. So artillery was also deployed on top of the building on a terassed open platform protected by a thick breastwork.

Fort Perch Rock

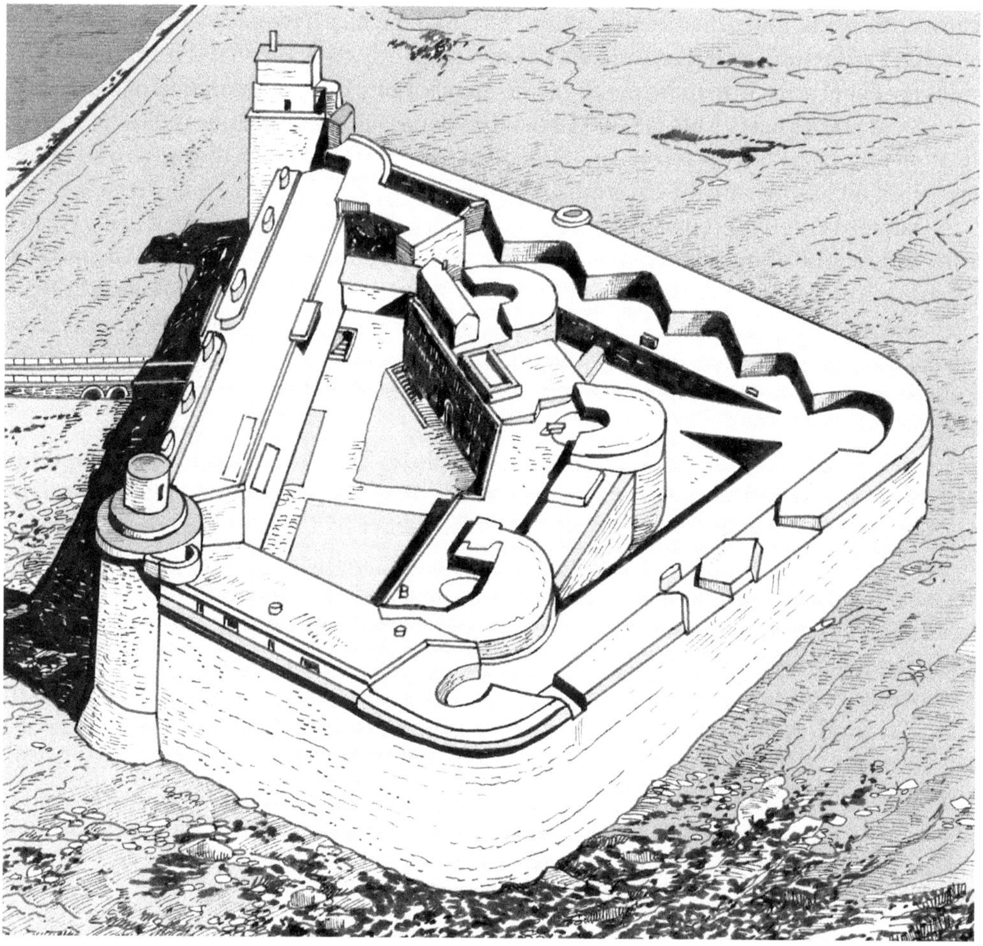

Fort Perch Rock (New Brighton).

In 1803, the British authorities were not only concerned about a possible invasion by the French in southern England but also about an attack on Liverpool. Therefore, it was decided to build a fort on the beach at New Brighton at the northeastern tip of the Wirral Peninsula, which occupies a strategic position at the entrance to the Mersey estuary. Because of difficulties in finding funds, the construction did not start before 1826. Three years later, the lozenge-shaped fort was completed. Designed by Captain John Sikes Kitson of the Royal Engineers, it could accommodate a garrison of about 100 men. Armament included 18 guns, 16 of which were 32-pounders facing the Rock Channel that was the main entrance to the Mersey at that time. Ships sailing to Liverpool had to pass 900 yards from the guns, and it soon earned the nickname of "Little Gibraltar of the Mersey." The fort was decommissioned by the War Office in 1956 and passed through various hands until it became a museum and home of cultural events.

Pembroke Dock

Pembroke Dock gun tower.

Located at Pembroke in Pembrokeshire in Wales, the southwest gun tower was built in 1851 in order to protect the approach to Milton Haven Waterway. Actually composed of three intertwined towers, the building has a height of 13 meters, and its roundish walls made of brick and gray limestones are up to 2.7 meters (9 feet) thick. The interior is divided into three levels including gun casemates, musket positions, storerooms, a kitchen, a water pump, a washroom, and accommodation for the garrison consisting of 1 officer and 33 men. In the basement, there was a water tank with a capacity of 12,500 gallons. The top of the roof is arranged as an open platform armed with 12-pound brass howitzers and guns on traversing mounts. At high tide, access to the fort is via a long walkway. Although it was disarmed in 1882, the tower continued to serve as a military fortress in World Wars I and II. The South Pembrokeshire District Council acquired the edifice in 1975, and it was refurbished in the mid–'90s, becoming the Gun Tower Museum operated by the Pembroke Dock Museum Trust until 2017.

Littlehampton Redoubt

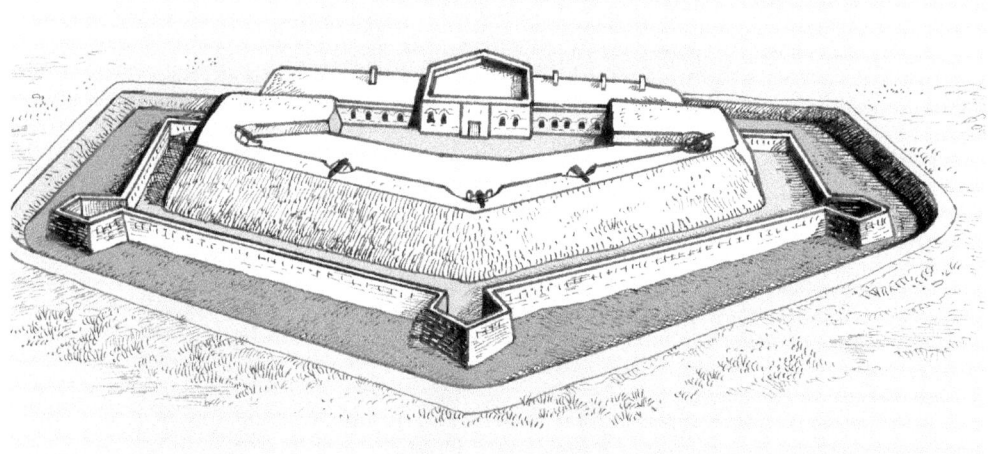

Littlehampton Redoubt.

Littlehampton Redoubt, usually known as Littlehampton Fort, was built in 1854 to protect the entrance to the river Arun at Littlehampton on the south coast of England (West Sussex) against a possible attack by the French under Emperor Napoleon III. The new fort built on the west bank consisted of a platform from which cannon could sweep the harbor mouth, with a barrack behind and a surrounding defensive ditch and wall. The fort was an innovative military structure, incorporating the new feature of a Carnot wall for close-range defense and a fortified barrack block at the rear. In 1861, the fort could accommodate 70 men and officers operating three 68-pounder and two 32-pounder cannons all emplaced on the rampart. Its active use as a fort was short for only about 20 years, owing to technical advances in armaments, but it was a precursor of the Palmerston Forts. Having had various uses since decommissioning in 1891, it is now abandoned, fenced, and in a ruinous and overgrown state.

Fort Landguard

Located at Felixstowe in Suffolk, Fort Landguard was intended to defend the mouth of the river Orwell and the approach to Harwich harbor. The current fort was built in the 18th century in bastioned style and modified in the 19th century with the addition of several gun batteries. It was manned through both world wars and played an important antiaircraft role during World War II. In 1951, two of the old gun casemates were converted into a Cold War control room. The fort was disarmed and closed in 1956. In 1997,

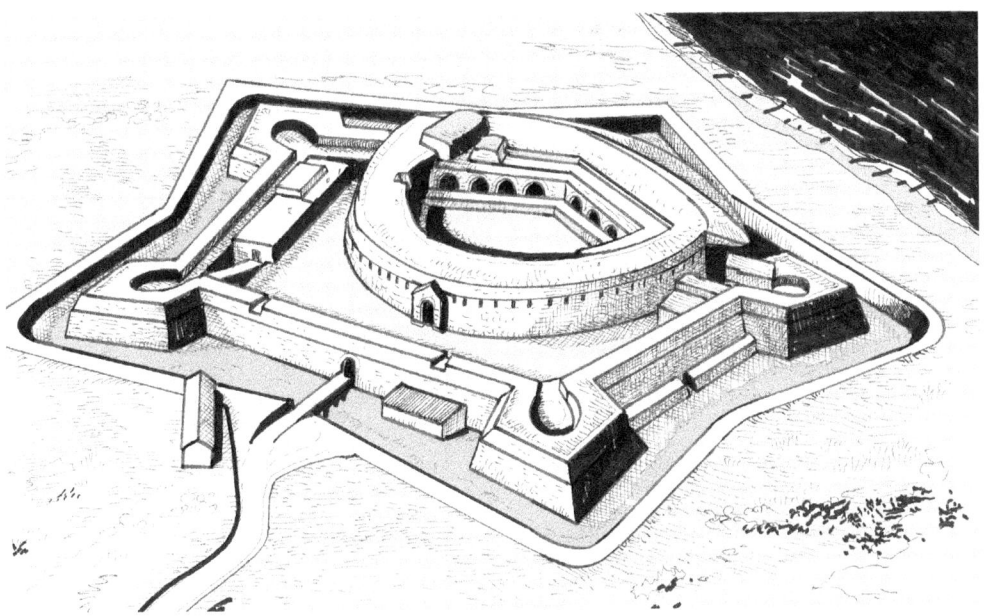

Fort Landguard.

the fort was structurally consolidated by English Heritage into whose care it had been placed and is maintained and opened to the public.

Rifled Artillery

While military engineers sought, designed, and developed new works of fortifications, headways in ordnance and ammunitions were making them obsolete. In the 1860s, unexpected advances were made in the domain of metallurgy, notably the introduction of the Bessemer converter patented in 1856. The result was an abundance of reliable and strong metals like steel that could be used to make large and powerful cannons. For over three centuries, if for minor improvements, the development of artillery had been constant but not revolutionary. Suddenly in the 1860s, tremendous advances in metallurgy and ballistics allowed the appearance of new far-reaching weapons known as rifled guns. Firing pointed shells spin-stabilized by a spirally grooved barrel, rifled cannons could achieve much longer range, much better precision and accuracy, and much more hard-hitting power than any previous smooth-bored gun. Besides, the prevailing solid cannon ball was gradually replaced with a single round (combining propellent charge and projectile filled with explosive). After 1842, effective breech loading was introduced. With the ammunition inserted at the breech rather than through the muzzle, the rate of fire was considerably increased.

Used in combination with armored steamship and rapidly adapted to siege and infantry support roles, the highly destructive rifled artillery rendered most existing coastal defenses inadequate and vulnerable. Besides, the American Civil War (1861–1865) saw the introduction of innovative weapons, e.g., exploding shells starting to replace solid shot, accurate breech-loading rifled artillery, repeating carbines, armored trains, and heavy artillery mounted on railroad. It amply demonstrated that entrenched infantry, backed by artillery, could inflict debilitating losses on attacking forces. At sea, the American Civil War witnessed the first use of torpedo boats, primitive forms of submarine, torpedoes and sea mines, as well as iron-plated steam-powered ships. In the domain of permanent fortification, the US Civil War clearly established that exposed masonry was incapable of withstanding the repeated fire of modern rifled artillery but highlighted that armor plating and thick earthworks were remarkably resilient. The rapid, unexpected, and revolutionary developments in rifled artillery indeed turned the eternal contest between attack and defense into a desperate and expensive race.

Armored Steamship

The Industrial Revolution allowed radical changes in technology, which revolutionized both land and sea warfare. After the Crimean War (1854–1856) started the age of the ironclad warship propelled by steam. France was first in the field, launching its first sea-going armored battleship in 1859. In 1862, the British launched a similar ironclad steam frigate named *Royal Oak*. A year later, they developed the armored *Valiant* with 34 guns and the *Minotaur* with 50 guns. In 1871 appeared HMS *Devastation*, one of the first ships completely deprived of sail with guns placed in turrets.

The adoption of steam-powered warship changed the whole geography of maritime strategy. Whereas timber-hulled sailships were at the mercy of the wind, the new steamship could navigate to some extent without caring about the elements, and this was a decisive advantage in speed and maneuverability especially in relatively narrow and shallow waters like the English Channel, the North Sea, and the Mediterranean Sea. The use of metal in naval construction made it possible to break free from the limitations imposed by wind and sail. It also greatly decreased the danger of fire caused by incendiary projectiles. Within a few decades, developments were prodigious. Weights and displacements were enormously increased. Thick armor and long-range heavy rifled guns emplaced in rotating turrets made the steamship a much stronger vessel than wooden sailing ship. However, steamships were not without disadvantages. Like sailships, they still rolled in the waves, making accurate gunfire difficult. But the main drawback was that steam-powered warship had to be regularly resupplied with coal. This acted as a tether keeping a fleet more and more dependent on supply bases. As a result, any industrially developed nation with global naval ambition needed to possess coaling bases and arsenals all over the world. The steam warship stimulated colonial expansion, imperialism, militarism, and navalism. A great state had to have a powerful battle fleet to protect its homeland and colonial empire. In spite of this disadvantage, the ironclad battleship became a symbol of national pride expressing technological achievement, reputation, prestige, influence, worldwide reach, and immense destructive power. Frantic competition in speed, size of guns, thickness of armor, and the number of warships developed between imperialist rivals Britain, France, and Russia, soon followed by Germany, Japan, and the United States. The new ironclad steamship with its maneuverability, defensive capability, and powerful offensive armament represented, of course, a major challenge to all nations with a seaboard. Britain's defenses relying on the Royal Navy, and fortified naval bases were particularly exposed. With the rise of the iron ship, British coastal artillery proper started with new fortifications for the protection of both home and every corner of the empire.

Napoleon III's Threat

In France, the seizure of power by the revengeful warmonger and Anglophobic Louis-Napoleon Bonaparte (1808–1873) contained a real danger for Britain. Louis Napoleon Bonaparte (Boney's nephew) had been elected president of the Second Republic in 1848. In 1852, he staged a coup, established a military dictature known as the Second Empire, and took the title of emperor of the French under the name Napoleon III. This new aggressive leadership was thus regarded with suspicion, and again, something like a fear developed in Britain, the more so after the Crimean War (1853–1856) when Anglo-French relations were at a low. To make matters worse, the French continued the development of a large fortified naval arsenal at Cherbourg, thereby menacing the English south coast harbors. There was thus an intense (and perhaps somewhat irrational) fear of a sudden invasion, a fear that had long been embedded in the British national psyche. This anxiety had surfaced intermittently throughout the past centuries with periodic concerns and spasmodic calls for national vigilance, rearmament, and construction of coastal fortification. With this background of concern, a Royal Commission was set up to reconsider the defenses of Britain. The members of the commission realized that refortifying the entire coastline was scarcely feasible, so they concentrated on defending the main naval bases. Britain's coastal defenses were subsequently completed following the report of the Royal Commission of 1860 with recommendation for new works at the royal dockyards of Portsmouth, Plymouth, Pembroke, Chatham, Portland, Dover, and Cork.

In the meantime, the new German system of "polygonal fortification" was rapidly imposing itself and was introduced in Britain.

Polygonal Fortification

Detached Forts

The 17th and 18th centuries had been the era of fortified cities enclosed within continuous bastioned enceintes, but the 19th century (particularly the second half) was the time of rings of autonomous forts. Montalembert (and German engineering officers) had foreseen the coming necessity for detached forts instead of continuous enceintes.

With the development of rifled artillery, the post–Napoleonic high masonry towers with their numerous embrasures proved too visible and vulnerable. The emphasis started to shift from the protection of forts against bodily assault to the protection of heavy guns from destruction by bombardment. The fort's artillery and most important elements (ammunitions stores and garrison quarters) had to be better protected. Thus, new forts had to be designed and established at a greater distance as range too had increased. Their profile had to be as low as possible and their surface wide in order to offer the enemy artillery open spaces and less concentrated targets. The various elements of the new forts had to be dispersed. These self-containing works were armed with heavy artillery covering a wide radius of fire. Each fort, soon armed with rifled cannons, was a powerful base of fire that could defend a large area and eventually could cover its neighbors with overlapping fields of fire. A belt of such forts (supplied by military railways) could defend a whole city and could replace a continuous enceinte. A line of them could protect a port, a coastal sector, a strategical passage like a valley or a pass in a mountainous site, or even a whole frontier. Defense in depth, made up of a series of parallel lines of fortresses strengthened by fortified complexes around manufacturing cities and communications centers, could turn a whole region or even a whole country into one impregnable stronghold.

At least, this was the theorical reasoning behind the American, British, German, French, Belgian, and Dutch military construction in the second half of the 19th century. Indeed, every industrialized European power subscribed to this strategic logic in some way. Besides, forts, together with armored steam battleships, size of army, heavy artillery, and soon railway mileage and colonial possessions abroad were symbols of national status. However, it should be noted that tradition and habitude were still strong with some military authorities, and the bastioned system was never totally abandoned particularly in Dutch and French fortifications. Until the 1870s, bastions and walls with thick breastworks used in combination with wide moats were still encountered.

Main Features of a Polygonal Fort

As already discussed, Montalembert and German engineers were the precursors of the so-called polygonal system that dominated the second half of the 19th-century military architecture.

All polygonal forts in the period c. 1840–1885 were different. Whether French, German, Dutch, Belgian, American, or British, no two of them were the same as they were always adapted to their local situation. The size, firepower, and number of garrisons also varied a lot depending on their intended task and on the terrain in which they were established. However, all forts of that period had common features and consisted of the same combinations of basic elements.

The outline of a polygonal fort was either a quadrilateral trapezoid or a large pentagonal (five-sided) gun battery—as it were a large bastion or a large redoubt. It was composed of a front displaying two faces forming an obtuse angle turned toward the enemy. The sides were called flanks. The rear part, called the gorge, was the back façade of the fort turned toward the friendly side. The gorge was usually given a straight or reentering layout for flanking the entrance. The ramparts were of massive section, made of thick masonry filled and covered with thick layers of piled earth. They included bombproof casemates sunk into the rear faces giving ample accommodation for the garrison and supply stores. All forts were autonomous, self-defensive, and surrounded by a deep and rather narrow ditch with vertical masonry scarp and counterscarp. Beyond the ditch, just like in bastioned fortification, there was a covered way on the counterscarp (for advanced defense, sortie, and patrol) and a wide, flat, and bare glacis denying any cover to attackers.

The ditch was no longer defended by bastions but by new flanking elements called caponiers. A caponier was a low vaulted combat chamber attached to the scarp and extending across the ditch. It was fitted with loopholes enabling flanking fire to be brought to bear on the width of the ditch. A typical polygonal fort often featured three caponiers: one double caponier was placed in the middle of the front covering both faces; and one or two simple caponiers placed at the shoulder of the fort, across the ditch at the point of junction of face and flank, each covering a side of the fort. In certain cases, under German (Prussian) influence, the caponier was developed and grew to two or even three stories in height in order to concentrate massive firepower in the ditch. It could also widen to contain its own supply magazine and barrack and even in extreme cases took on a rectangular form so as to include an inner yard or parade ground.

The entrance to the fort—the most vulnerable part of the fortress—was obviously placed in the gorge at the rear. The entrance featured a bridge across the ditch, with the last section in the form of a draw- or rolling bridge known as Guthrie bridge. Unhinged, the movable Guthrie bridge remained horizontal when retracted within the gates of a fort, operating in a similar way as a modern thrust bridge. Guthrie bridges can still be seen in forts of the Portsdown Hill line at Portsmouth, e.g., at Fort Nelson, as well as at Cambridge Battery and Rinella Battery in Malta.

The entrance was defended by a caponier or casemates placed in the rear scarp and controlled by a guardhouse placed in a soberly decorated gatehouse.

Each fort was a self-containing unit with everything available for a rather spartan life purposely intended for operating the armament. The garrison was housed in bombproof, heated, and ventilated barracks placed either in the gorge of the fort or on the terreplein in its middle, or under the ramparts bordering the faces. Barracks and all other buildings in the fort were made of strong masonry and covered with sturdy layers of piled earth several meters thick, which offered a good protection against enemy projectiles. Close to living quarters, facilities existed such as lavatories and washing rooms, a bakery, a kitchen, food stores, water supply (cisterns or a well), and an infirmary for small injury and light casualty. In addition, the commanding officer had a headquarters, an office for administrative affairs, and a telegraph exchange (later telephone) often placed in the keep described below. There was also often a prison cell to detain undisciplined soldiers. Each fort had one or more powder houses where explosives and ammunitions were stored. For safety reasons, these store places were placed as far away as possible from the living quarters, carefully guarded, and heavily constructed with heavy doors, strong masonry walls, and a roof covered with thick layers of earth. Each fort also included various casemated buildings including gun garages, expense magazines, workshops, and various stores all connected by vaulted corridors. There was also an armory or an arsenal where ammunitions were stored, shells and cartridges were filled, and where small weapons were stored, maintained, or repaired.

The garrison was composed of engineering troops for maintenance of fortification, artillerymen serving the guns, and riflemen for close-range defense. There could also be a small detachment of cavalrymen for liaison, reconnaissance, and patrol. All of them were trained and drilled to operate closely together.

In those days, forts had to possess numerous artillery pieces as the basic issue in coastal artillery was bringing sufficient firepower to beat on the target. Loading the muzzle-loading gun, running it out, firing, sponging, reloading, and running it back from where it had recoiled all took time, so breech loading was gradually introduced. The object of the coast gun was firmly defined—it had to smash through enemy armored ironclad warship with great velocity, powerful charges, and penetrating shot. Some of the pieces of artillery were mobile and mounted on wheeled carriages. They were intended for long-range firing at the enemy and, if necessary, for supporting a neighboring fort under attack. The guns fired *en barbette*, meaning that they were placed on a platform protected at the front by a parapet, a breastwork made of masonry filled with thick piled earth. The artillery emplacements were open (thus enabling toxic fumes to dissipate quickly) and located along the frontal and lateral ramparts. They were divided into compartments so that in case of a direct hit, the damage was limited. For this purpose, emplacements were fitted with traverses (raised earth banks placed across the front) to protect from enfilade and raking fire and parados at the back, which were traverses protecting from projectiles exploding behind them. Other guns were static and mounted on swiveling carriages inside a casemate. Since the Napoleonic era, a new kind of firing chamber had been introduced: the so-called Haxo casemate. Named after the inventor—the French general and military engineer François-Nicolas Haxo (1774–1838)—such gun emplacement combined good protection and good ventilation. The Haxo casemate was a vaulted firing chamber installed inside the rampart under a thick layer of dirt. It was fitted at the front with an opening, named embrasure, through which the gun was fired. The innovation was that the back of the casemate was fully open permitting smoke and toxic fumes of the discharge to clear rapidly. The open rear also allowed for daylight and fresh air to penetrate inside the chamber.

The infantry riflemen manned the caponiers in the ditches and were deployed along the walls in the open for close-range defense. They were protected by a thick breastwork furnished with a banquette or fire step, a platform on which they stood for firing; after shooting, they could step down and thus were completely protected by the parapet while reloading.

In the polygonal forts built in 1850–1885 in Britain, the Netherlands, Belgium, and Germany, there was quite often a *réduit* (aka redoubt or keep), a closed independent self-defensible work. In this stronghold within the fort, the garrison could withdraw when the rest of the fort was taken. Of various shape, size, and strength, the keep was always placed in the gorge. It could be compared to the medieval keep, the residential and strongest tower of the castle of the Middle Ages. The keep could be used as quarters for the garrison and often included the commander's headquarters. It was a small fort in reduction including facilities such as kitchen, troop chambers, rooms for officers, supply stores, infirmary, ammunition store and arsenal, observatory, own glacis and ditch, own gate, close-range defense combat emplacements, etc.

Opinion was divided regarding the last-stand keep. Some military engineers maintained that the keep stiffened resistance in depth as it allowed the garrison to withdraw to a position from which it might later sally out and eventually recapture the fort. Others considered it only encouraging the idea of retreat. In the French forts of the same period, the keep was always omitted for psychological reasons: the French high command considered that without a safe place to withdraw, the defenders had no other choice than to fight to the bitter end or to abandon the work and retreat at the right time. In the end, the French were correct. As a matter of fact, if the attacking party got this close, the fort and its keep were as good as lost. Depending on tactical condition, of course, it is well known that a withdrawal in good order at the right time is often a better move than a desperate fight till the last man. After 1885, attached and autonomous keeps were abandoned as relics of old-fashioned thinking.

The distance between the forts was various, depending on the range of artillery. In order to fill the eventual gaps that could occur in the defensive system and according to local situation, additional independent redoubts and batteries were built as reinforcement. So each fort with its interval annexes could defend a large zone as well as neighboring strongholds. Shapes, degree of protection, armament and garrison of annexes, redoubts, and batteries varied a lot according to the task they had to perform and depending on the local situation. Some were small independent enclosed

Part 4. Fortifications in Britain in the 19th Century

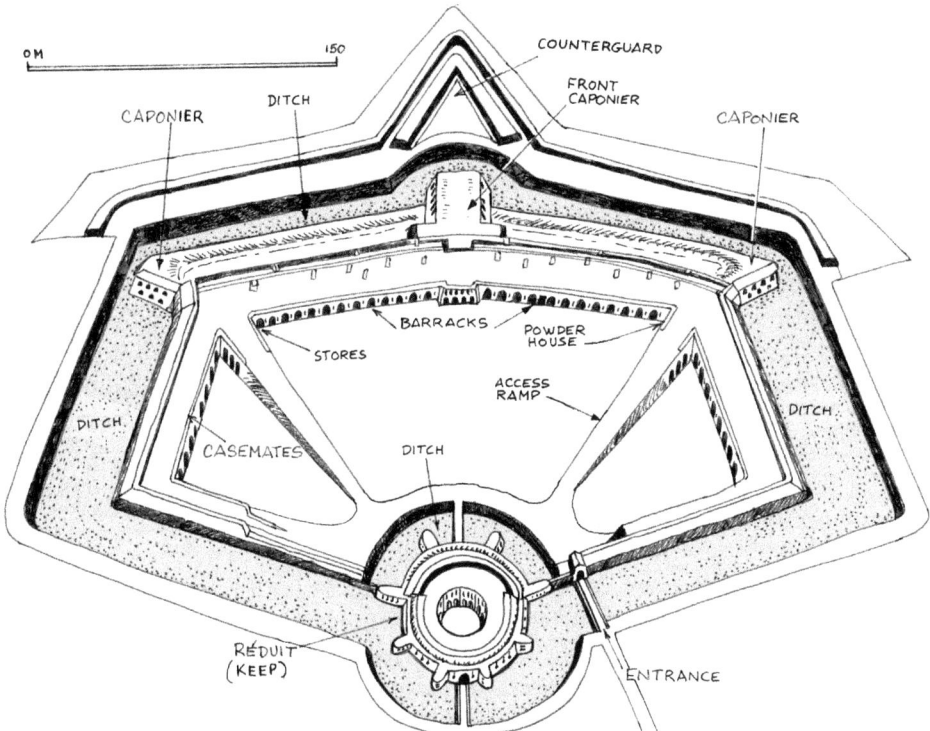

Brockhurst. Fort Brockhurst, located at Gosport (Hampshire), was built between 1858 and 1862 in order to protect the approach to the naval base of Portsmouth. It is a typical example of British polygonal fortification.

Keep at Fort Brockhurst. The keep (aka reduit) was a small, circular moated fort within the fortress placed in the rear.

pentagonal fortlets with a ditch defended by caponiers. Others were trapezium-shaped or redan-shaped. Still others were very simple earth entrenchment composed of a main front with no or little protection at the back. In time of crisis or war, forts, redoubts, and interval batteries were reinforced with fieldworks including trenches, infantry shelters, and various obstacles, notably ditches and later barbed-wire entanglements.

For the purpose of building and (when they were completed) supplying such large forts forming a wide entrenched camp, there were often military roads and spur tracks of (narrow gauge) railways.

Thus, within half a century, British fortification maintained the principles of detached works, but their physiognomy had greatly been changed from large free-standing casemated artillery towers to low-profile earth-covered redoubt-shaped forts.

Palmerston Forts

The so-called Palmerston Forts were a group of 76 forts and associated structures, built along the coast of Britain during the Victorian period on the recommendations of the 1860 Royal Commission on the Defence of the United Kingdom. The designation of the forts came from their association with Henry John Temple, 3rd Viscount Palmerston (1784–1865), who was prime minister at the time and advocated the project. The works were also known as Palmerston's Follies, as by the time they were completed, the French threat (if it had ever existed) had passed, largely due to the defeat of France in the Franco-Prussian War of 1870, the abdication of Napoleon III, and the reestablishment of the République Française. Furthermore, the technology of the guns had become outdated. Anyway, when it was launched in the early 1860s, the so-called Palmerston program was a large, ambitious, and expensive scheme following concerns about the strength of the French navy and the uncertainty about Napoleon III's foreign intentions. Strenuous debates in Parliament were held about whether the cost could be justified. Indeed, the Palmerston forts were the most costly and extensive system of fixed permanent defenses undertaken in Britain in peacetime. The defenses were built according to the polygonal system to defend a number of key areas of the British, Irish, and Channel Island coastline, in particular areas around military bases, including Alderney, Belfast, Berehaven, Bristol Channel, Chatham, river Clyde, Cork, Dover, Isle of Wight, Milford Haven, North Thames and East Anglia, North East England, Plymouth, Portland Harbour, Portsmouth, and Lough Swilly.

The vast majority of these fortifications were coastal forts and batteries, which generally had the appearance of pentagonal or curved casemated artillery structures designed to defend the seaward approaches to the most vital harbors.

In addition to holding enemy ironclad warships at bay, it was necessary to prevent attacks from the landside by invading forces that would have landed elsewhere. For this purpose, rings of mutually supporting polygonal forts were built around the important British ports, notably at Plymouth, Portsmouth, and Chatham.

Describing each and every one of the numerous 19th-century works, forts, redoubts, and batteries would take several books, so only a few typical examples are dealt with below.

Shoreham Redoubt

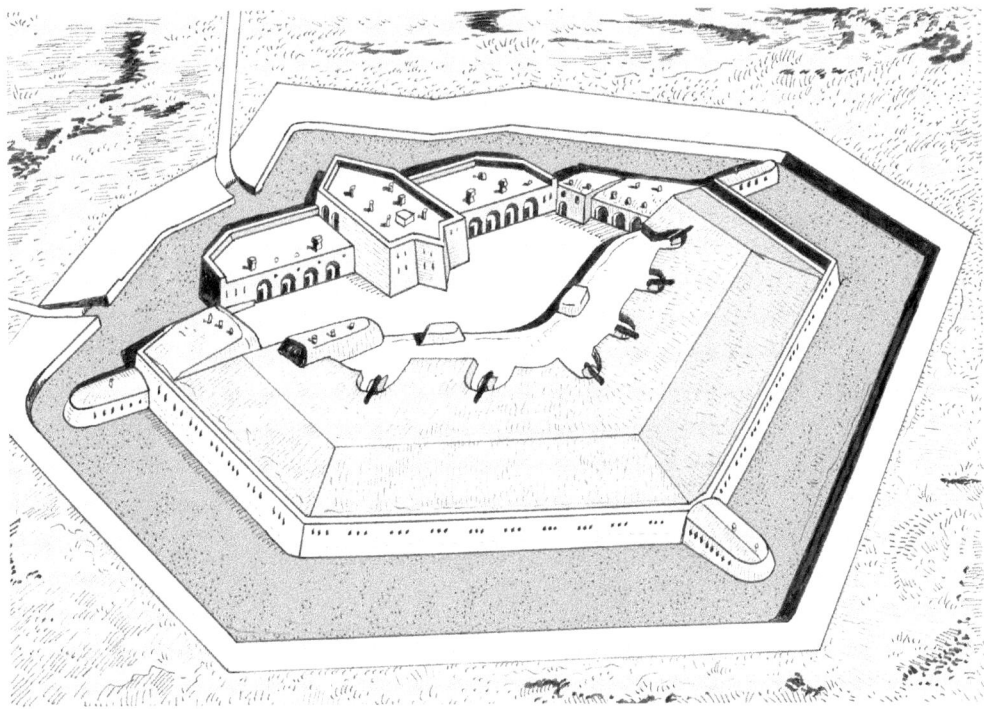

Shoreham Redoubt.

Shoreham Redoubt is located at the entrance to Shoreham harbor, at the mouth of the river Adur in West Sussex, England. It was planned during the 1850s in a period of great alarm about a possible French attack. Construction was completed in June 1857. The pentagonal fort consists of a gun platform 15 feet (4.6 meters) above sea level in the shape of a crescent with a bombproof barrack in the gorge. The fort is surrounded by a ditch defended by a Carnot wall and flanking caponier at each of the three angles. The fort was armed with six 6-pounder guns on traversing mounts. It was reshaped and modernized in 1873, 1886, and 1940. The redoubt is currently a museum.

Yaverland Battery

Located near Sandown on the Isle of Wight, the coastal battery was built between 1861 and 1864. Placed on a commanding position on the east cliffs, it was intended to defend Sandown Bay. Essentially, the battery was almost triangular in nature, with the widest side to the south. The narrow rear of the fort was protected by two caponiers and was the location of the store and the barracks with accommodation for the 2 officers and 57 men. Entry to the barracks was via a drawbridge over the ditch near the northeast corner of the fort. The parade ground included two bombproof magazines in the middle and two cartridge stores close to the guns. The fort was dismissed in 1956 and has become a holiday center.

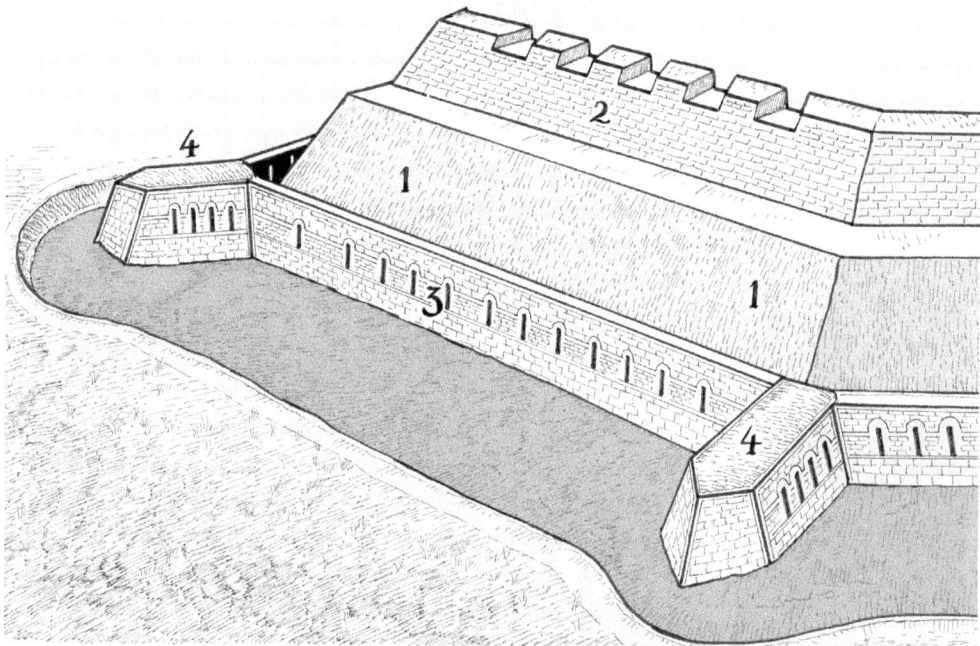

Yaverland Battery. The depicted south face included a sloping earth scarp (1) with a stone parapet (2) with embrasures protecting the artillery, a detached Carnot's wall (3), and flanking caponiers (4) projecting in the ditch.

Sandown Battery

Guarding a vulnerable beach at Sandown on the southeast coast of the Isle of Wight, the first Sandown Castle was built by order of King Henry VIII. Undermined by the sea, the castle was replaced with a square bastioned fort in the 1630s and later with a Victorian battery built in 1861–1863. It was one of three batteries recommended by the 1860 Royal Commission on the Defence of the United Kingdom to be built around Sandown Bay in order to prevent seaborne landings. The battery was rectilinear in plan, enclosed by a dry moat containing a loopholed brick Carnot wall, which was flanked by musketry caponiers. Entry to the battery was via a drawbridge and gate on the landward side. The fort included a cookhouse, a guardroom, several supply and ammunition stores, a smith's workshop, a coal reserve, an ablutions room, and a barrack for the garrison. The battery was initially armed with five 7-inch rifled breech-loader guns, but armament was modernized with more powerful guns in 1872, 1882, and 1893. Major rebuilding work began in 1900 with the construction of underground magazines and concrete emplacements for three 6-inch breech loaders. Sandown fort played a significant role in the D-Day landings (June 6, 1944), as it housed 16 pumps for the PLUTO (Pipe Line Under the Ocean) operation to Allies supplied with fuel. Each of the 16 pumps supplied 36,000 imperial gallons (1,000 barrels; 160,000 liters) of fuel per day at a pressure of 1,500 pounds per square inch. Following its abandonment by the military in the 1950s, the site now houses the Isle of Wight Zoo.

Part 4. Fortifications in Britain in the 19th Century

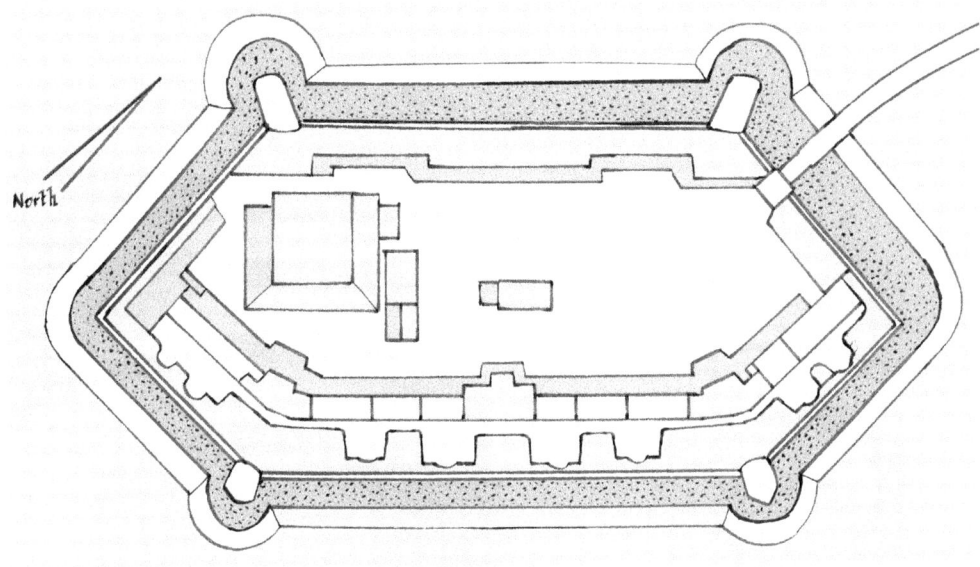

Plan Sandown Battery.

Fort Bovisand

Fort Bovisand, Plymouth, Devon. (1) Stone jetty and slip, (2) Staddon Height battery, (3) granite casemates.

Granite casemate at Fort Bovisand, Plymouth, Devon.

Fort Bovisand in Devon, England, was built on the mainland to defend the entrance of Plymouth Sound, at the narrows opposite the east end of Plymouth Breakwater. In 1816, a stone jetty and slip were built for boats from sailing warships anchored in Plymouth Sound to collect fresh water from the nearby reservoir. In 1845, the first fort at the site, named Staddon Height Battery, was started and still exists in the upper position of the present fort. As part of the recommendations of the Royal Commission on the Defence of the United Kingdom, the main part of the fort was constructed between 1861 and 1869. It consisted of a single-story semicircular coastal battery including 23 granite casemates facing the sea and originally armed with 22 nine-inch rifled muzzle loaders (RMLs), one 10-inch RML gun, and 180 men. In the 1880s, armament included 14 ten-inch and 9 nine-inch RML guns. During the same period, deep underground tunnels were dug to store artillery ammunition safe from enemy gunfire. The fort was rearmed in 1942, closed in 1956, leased in 1970, and today houses a diving school.

Part 4. Fortifications in Britain in the 19th Century 133

Fort Tregantle

Fort Tregantle is the main fort of the western defenses of Plymouth (the Antony Position), built between 1858 and 1866. Fort Tregantle had an irregular six-sided plan and included a dry ditch; a gatehouse and bridge; an oval keep with its own ditch; barrack blocks; open artillery batteries with earthen parapet; a central square or parade ground; and caponiers defending the ditch. It had a garrison of 1,000 men and was armed with 35 large guns. It was used as infantry barracks until 1891 and was vacated after 1918. However, in 1938, it was reopened and used as the Territorial Army Passive Air Defense School. During World War II, it was used first as the Army Gas School and from 1942 as US Army accommodation. After 1945, it remained in use by the British army and is still used today for Royal Marine training purposes. The fort is built of snecked rock-faced limestone rubble with stone dressings.

Crownhill Fort

The fort at Crownhill is one the best preserved of Lord Palmerston's ring of forts that surrounded Plymouth in Victorian times, protecting the Royal Dockyard at Devonport from attack and bombardment by land and by sea. Crownhill represented the cutting edge of fortress design during that period. However, Crownhill's guns were never put to the ultimate test, for advances in artillery soon overtook it, and fortresses of this type became obsolete. Unlike many of the mid–Victorian forts in Plymouth and elsewhere, Crownhill was retained by the British army for over a century. Therefore, it did

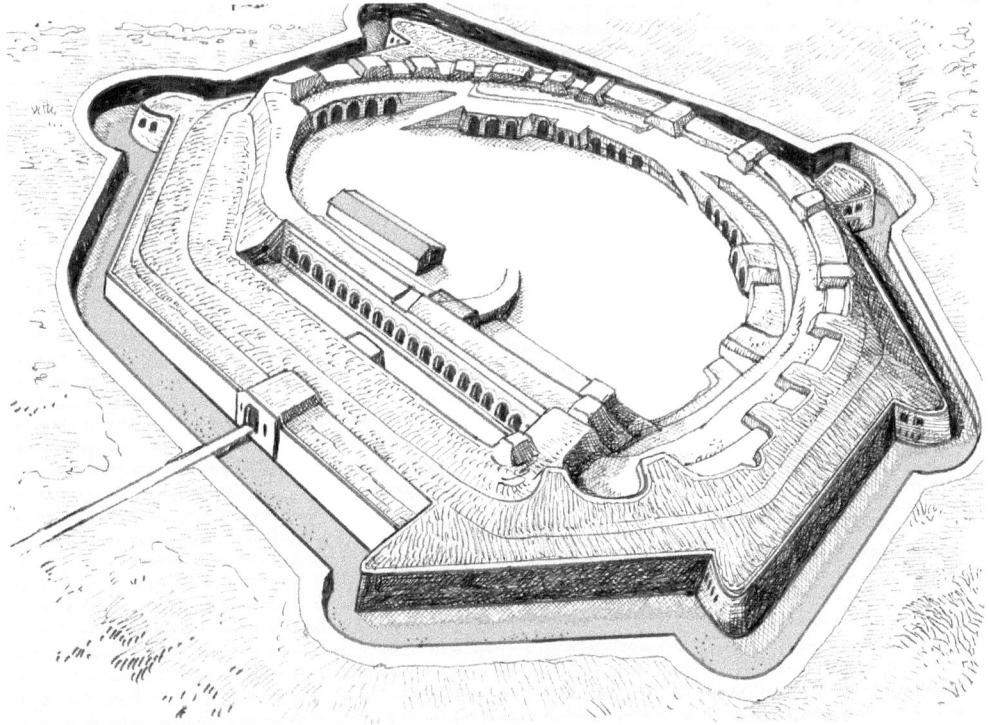

Crownhill Fort, Plymouth, Devon.

not suffer many irreversible alterations and still displays a good example of masonry polygonal style. It was acquired in 1987 by the independent Landmark Trust and is now open to the public.

Woodlands Fort

Woodlands Fort, located just west of Crownhill, was built in the 1860s. It is a polygonal fort surrounded by a dry ditch flanked by a caponier and a counterscarp gallery, which were accessed by tunnels from the barrack block. Woodlands Fort is now used as a community center and library and can be visited. The two-story casemated barracks survive in good condition. The terreplein is overgrown, but gun positions and two Haxo casemates can be seen as well as some small expense magazines.

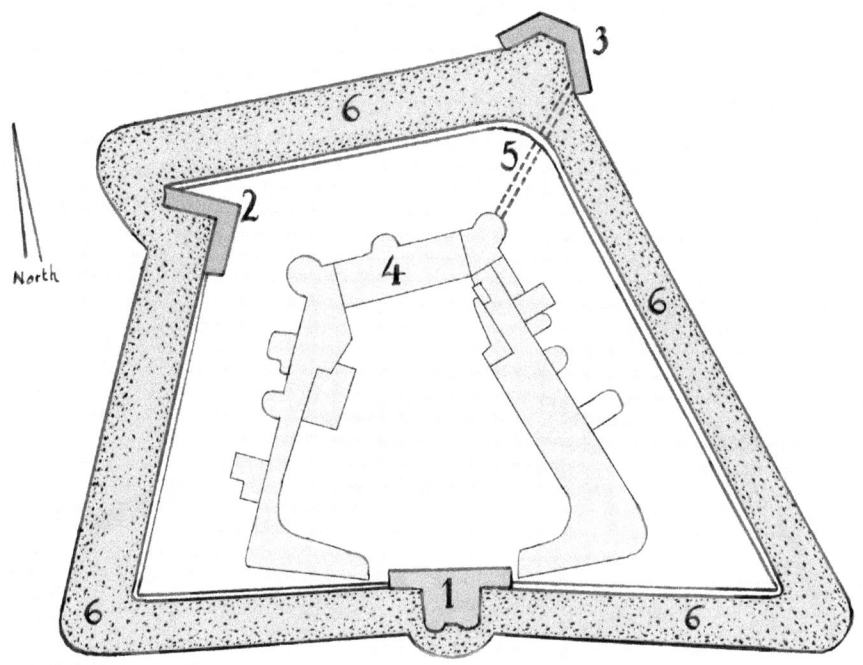

Woodlands Fort, Plymouth, Devon. The plan shows (1) entrance in the gorge with caponier flanking the southern ditch; (2) caponier (flanking the west ditch); (3) counterscarp gallery or coffer covering the north and east ditch; (4) bombproof barracks; (5) underground gallery leading to counterscarp coffer; (6) dry ditch.

Part 4. Fortifications in Britain in the 19th Century

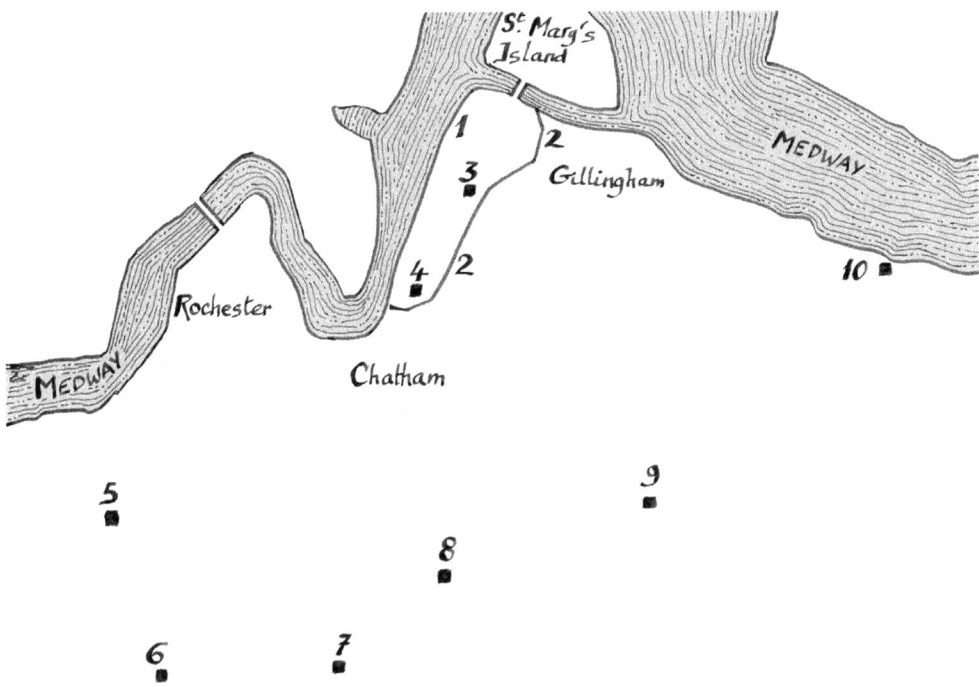

Map of Defense of Chatham Dockyard. The 18th century and Napoleonic defense of Chatham Dockyard (1) included the Cumberland Lines (2) with Townsend Redoubt (3) and Fort Amherst (4). Between 1860 and 1900, a new ring of five polygonal forts was established to defend the southern approaches on the landside with Fort Borstal (5), Fort Bridgewood (6), Fort Horsted (7), Fort Luton (8), and Fort Darland (9). The coastal defenses were reinforced with the addition of Twydall batteries (10) and the circular island forts on Darnet and Hoo. Today, Forts Horsted (the largest), Luton, and Borstal are still standing, while Forts Bridgewood and Darland have been demolished.

Fort Borstal

Built on high ground overlooking the Medway and the western approach to Chatham, Fort Borstal was constructed by convict labor between 1875 and 1883. The polygonal-shaped fort has a continuous rear gorge wall defended by loopholes. It is surrounded by a dry ditch defended by counterscarp galleries at the angles. A single rear caponier flanked the gorge of the work. Access to the fort included a roller bridge. The fort has two rows of casemates, the front two serving as a command area and the rear nine providing barrack accommodation. The gun positions on the ramparts were intended for movable armament served by handling rooms with expense magazines beneath. Fort Borstal and the other Chatham forts were used in training exercises, known as siege operations, notably in 1907. Observers watched these mimic battles and tried to draw conclusions on the best way to attack and defend fortifications with the latest weapons. The fort also served as barrack accommodation during World War I and had a 4.5-inch antiaircraft battery on its rampart during World War II. Disused from military service since 1957, the fort is listed as an ancient monument and is now used as a private residence.

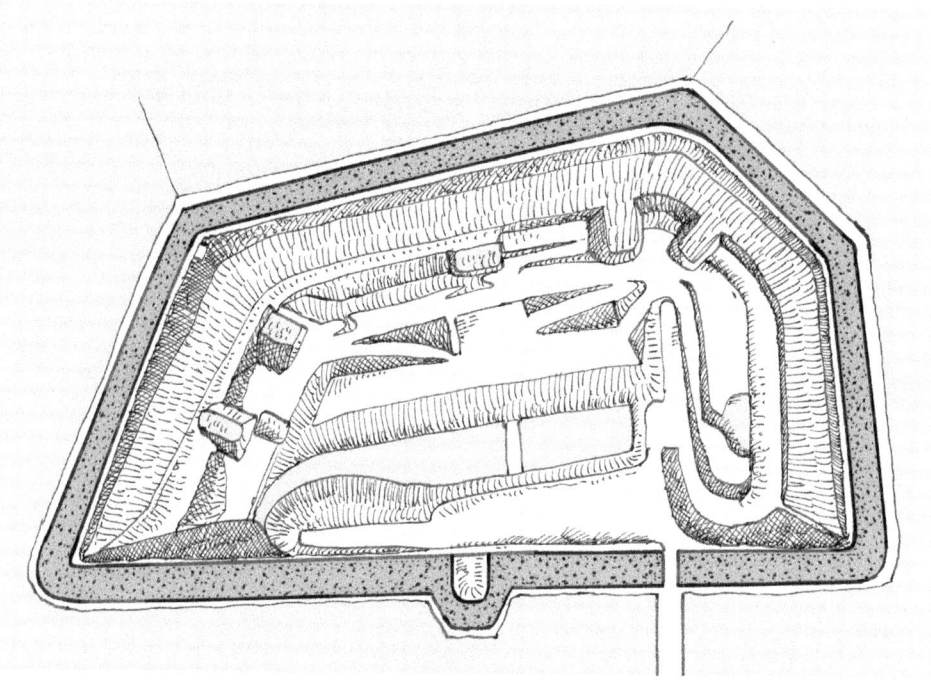

Fort Borstal.

Fort Horsted

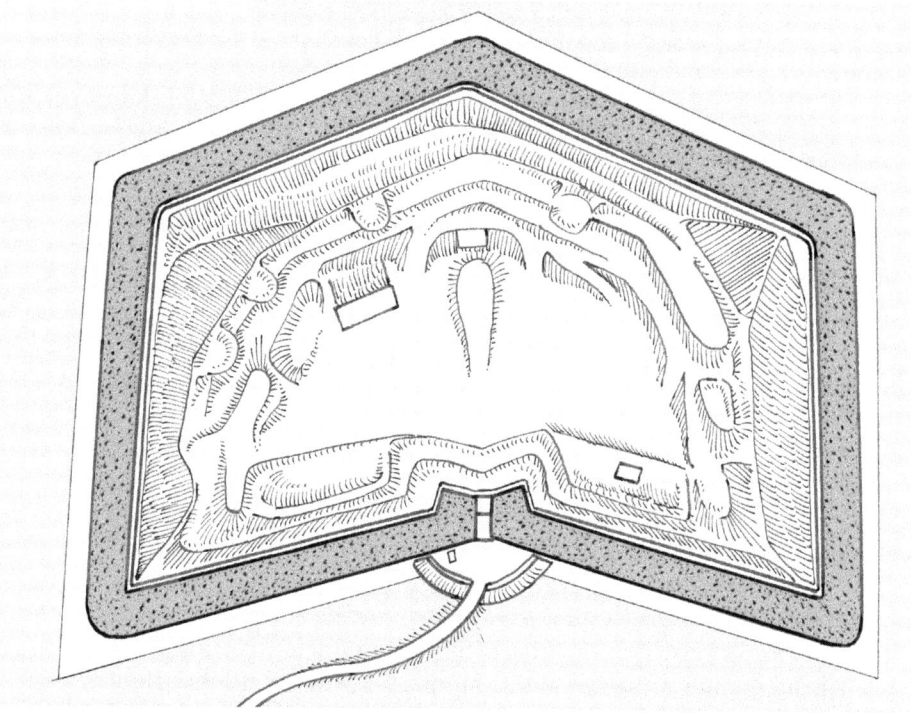

Plan of Fort Horsted, Medway.

Fort Horsted is located at the southern end of Maidstone Road, Chatham, and near Rochester Airport, about three miles from Chatham Dockyard. It was the largest of the five forts built to defend the southeastern approach to Chatham Dockyard. The construction started in 1880 and was completed by 1889. It was named after Hors who was an Anglo-Saxon king in the 5th century. It had the shape of a large pentagonal bastion and would have housed a garrison of about 400. Although it was never permanently manned and armed, in time of invasion it would have been armed with about 50 cannons and howitzers which were emplaced on the ramparts and in casemates. The front and the sides were defended by counterscarp galleries, which were accessible by steps and tunnels down and under the dry ditches. At the rear of the fort, there was the entrance with a movable drawbridge defended by embrasures. The fort included water tanks, a guardroom, magazines, stores, and barrack accommodation linked by underground galleries. At the beginning of the 20th century, the fort was used for the manufacture and storage of ammunition. In World War II, it served as ammunition depot and antiaircraft guns were installed. In the early 1960s, the fort was no longer needed by the military and was sold in 1963. Over the following years, the fort was owned by various private owners, and since the 2000s, it is a business center.

Hoo Fort

Hoo and Darnet forts are located on small and flat islands that lie either side of the inner navigable channel of the river Medway, north of Gillingham, Kent. Their construction was commenced in 1861, but in 1867, they were reduced from the planned three tiers to two after severe problems with subsidence, and they were commissioned in 1871. Accommodation casemates and magazines were on the ground floor with gun

Fort Hoo.

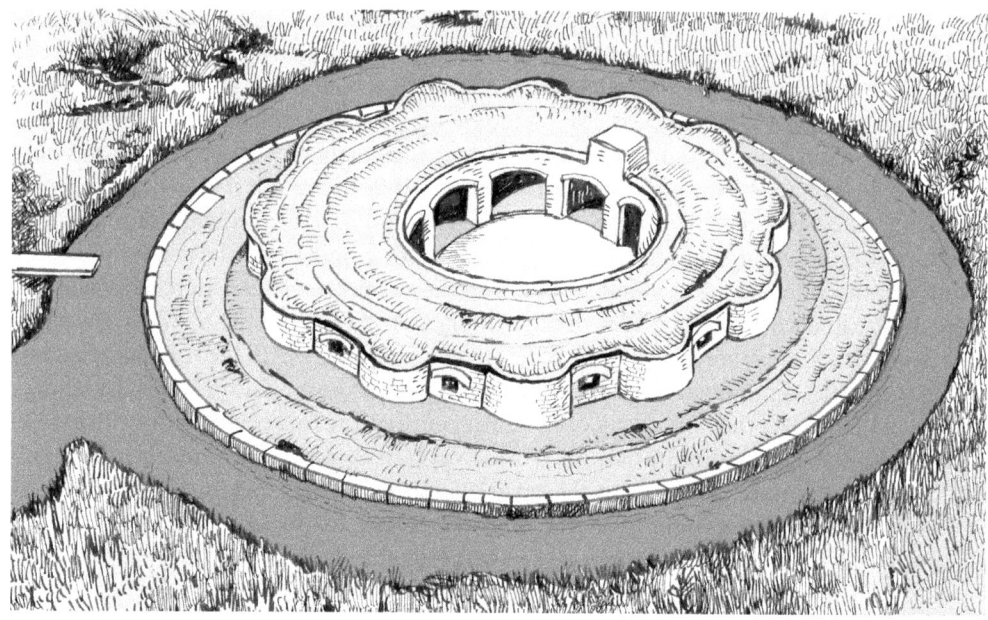

Fort Darnet. Forts Hoo and Darnet differed from the polygonal design. They were circular casemated works giving all-around defense.

casemates above, all casemates being built around a circular parade ground. The forts were disarmed before World War I but remained in care and maintenance until 1920. In 1930, experiments were carried out at Hoo Fort, and to a lesser extent at Darnet Fort, to ascertain the likely damage to underground magazines caused by accidental explosion of stored cordite. During World War II, both Hoo and Darney served as observation posts, with small brick structures built on their tops. Both forts remain in rather good condition. Hoo is still owned by the Ministry of Defence.

Fort Purbrook

Purbrook occupies the east end of the defensive line on Portsdown Hill at Portsmouth. It is 2,400 yards east of Fort Widley and built on a seven-sided tracé. It had

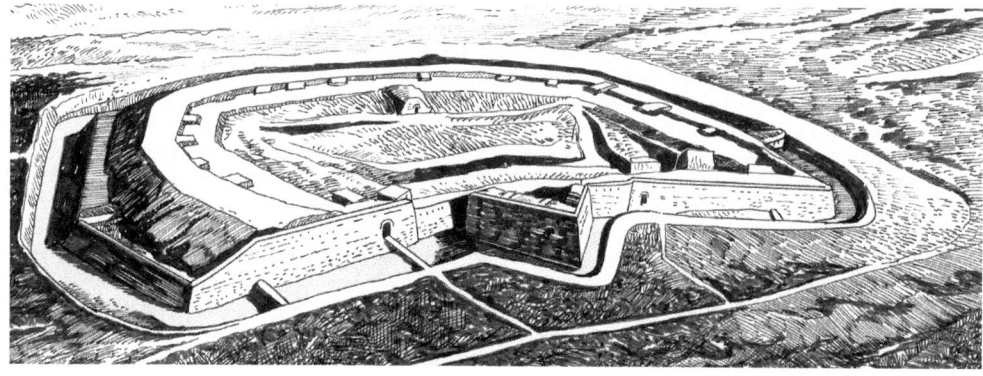

Fort Purbrook (Portsmouth).

a garrison of about 227 and was armed with 78 guns covering the northeastern approaches, and the double caponier featured a five-bay mortar battery behind. The vulnerability of the fort to its northeast and east led to the construction of two outposts: Crookhorn and Farlington Redoubts. Purbrook is now owned by Portsmouth City Council and is used for youth activities. It is largely intact and in a good state of conservation.

Portsdown Hill

Fort Nelson was built in the 1860s on Portsdown Hill above Portsmouth. The fortress is a classic and typical example of Palmerston's Follies. It forms part of a line along Portsdown Hill with Forts Widley and Purbrook. It was armed with ten 70-inch rifled breech loading guns and six rifled muzzle loaders. During World War II, it was used as an antiaircraft battery. Today, the fort is open to the public and is the home of the Palmerston Forts Society, a registered charity founded in 1984 that connects enthusiasts interested in 19th-century British military architecture.

Fort Fareham

After the Gosport Advanced Line of Fort Brockhurst, Fort Elson, Fort Rowner, Fort Grange, and Fort Gomer had been approved by the Royal Commission on the Defence of the United Kingdom, a decision was made to build an outer line of three more forts two miles in advance of the Gosport Advanced Line. Of these three projected forts, only

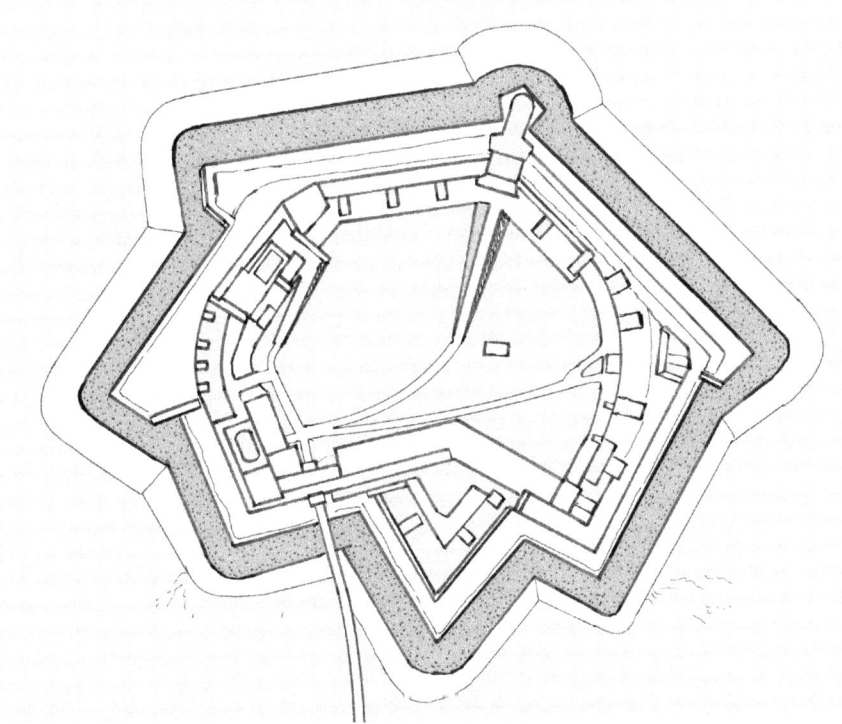

Fort Fareham, Hampshire.

Fort Fareham was built due to the need to cut costs. Fareham acted as a hinge between the forts on Portsdown Hill and those of the Gosport Advanced Line, filling the gap at Fareham. Fort Fareham was also to provide protection for the road linking Gosport to the Portsdown forts. Due to the need to protect the ground in front of the Gosport Advanced Line and because of its isolated position, Fort Fareham was constructed with a strong ability to defend itself from being overrun. Its position made an attack on the Gosport Lines very hazardous without it first being captured or disabled.

The construction of the fort was begun in 1861, and the fort was declared complete in 1868. In plan, Fort Fareham is a pentagon with a west-facing rampart for the main armament and a north-flanking rampart with armament to command the Fareham gap toward Fort Wallington. The ramparts to the south were given a greater measure of protection by hiding the guns into a complex range of Haxo casemates. This fort is the only one of the Palmerston forts to receive such a concentration of Haxo casemates. The fort is surrounded by a ditch that was intended to be wet (filled with water), but problems with obtaining and holding water in this meant that for much of its life, the northern portion remained dry. The ditch was protected by a large double caponier to the west with smaller semicaponiers at the north, southwest, and south salient angles. A redan was constructed at the rear for closing the gorge of the fort rather than a self-contained keep like those of the inner three forts of the Gosport Advanced Line. By 1902, most of the land front forts had their armament withdrawn, including Fort Fareham. The fort then continued to serve as a barrack right up to World War II. In 1941, Fort Fareham was used as brigade headquarters for local antiaircraft batteries. In 1965, it was sold by the Ministry of Defence to Fareham Urban Council.

Sea Forts

Where the harbor approaches were too wide and thus out of gun range from strongholds placed on land, sea forts were built in the 1860s at great difficulty either on small rocky islets (when available) or else at enormous expense on shoals or on the seabed itself. As can be easily imagined, a sea fort cost much more to build than a land fort, due to the difficulties involved in constructing foundations 20–30 feet underwater on sand banks or rocky seafloor. It was actually creating a kind of artificial island. Engineers first had to build cofferdams (walls of piling forming enclosures from which the water could be pumped out), and caissons (enormous cylinders containing air at high pressure to keep the water out) sunk into the seabed. Then the subsoil could be dug out and granite foundations be laid forming a ringwall about 16 meters thick and 70 meters in diameter enclosing a massive disc of cement-covered rubble. On top of this foundation was constructed a circular masonry fort protected by layers of iron plates varying in thickness from 4.6 centimeters to 6.4 centimeters. The seabed forts were not all of the same design or exactly of identical size, but they were virtually static warships comprising tiers of casemated gun emplacements and a top deck armed with heavy artillery, and some were planned to mount revolving armored turrets. All forts, of course, included observation and firing control devices as well as accommodation for artillerymen and supply and ammunition stores. Such technically advanced, difficult-to-construct, and expensive forts were built in the 1860s as coastal gun batteries for example off Royal Navy bases at Plymouth, at the harbor of Portland in Dorset, at Stack Rock Fort off the Spithead between the Isle of Wight and Portsmouth, and Fort Cunningham off the Royal Naval Dockyard in the British overseas Bermuda islands in the North Atlantic Ocean. Those sea forts represented an impressive technical tour de force. They were extremely expensive, but it was then thought there was no alternative of covering dangerous gaps in a defensive coastal system.

Interestingly, the French also built such a motionless oval stone casemated warship (Fort Boyard constructed between 1801 and 1857) located between Aix Island and Oleron Island in the Charente Maritime department. The American Fort Drum (completed in 1909) in Manila Bay defending the approaches to the U.S. naval base at Cavite (the Philippines) is another remarkable stationary concrete ironclad battleship bristling with weapons.

Spitbank Fort

Spitbank Fort is one of the four sea forts located in the Solent, near Portsmouth, England—the others being St. Helen's Fort, Horse Sand Fort, and No Man's Land Fort. Construction of Spitbank Fort started in 1861 and was completed in 1878. Spitbank's main purpose was to provide a further line of defense in the wide Spithead between the Isle of Wight and Gosport/Portsmouth. It is 162 feet in diameter at its base, with one floor and a basement revetted with metal armor plating on the seaward side. In 1898, the role of the fort was changed to defend against light craft, and the roof was fitted out with two 4.7-inch guns and searchlights. In the early 1900s, all but three original large guns were removed. Minor upgrades to the smaller guns and searchlights continued through the years and during the two world wars. The fort was declared surplus to requirements in 1962 and disposed of by the Ministry of Defence in 1982. The fort is now privately owned and used as a small luxury spa hotel.

Spitbank Fort (Portsmouth).

Horse Sands Fort

The Spithead forts were built in the Solent in the 1860s to protect Portsmouth from bombardment from the sea at the same time as the Gosport and Portsdown land forts were built. Another seabed fort was planned named Ryde Fort, but its construction was abandoned because of unstable foundations.

View of Horse Sands Fort.

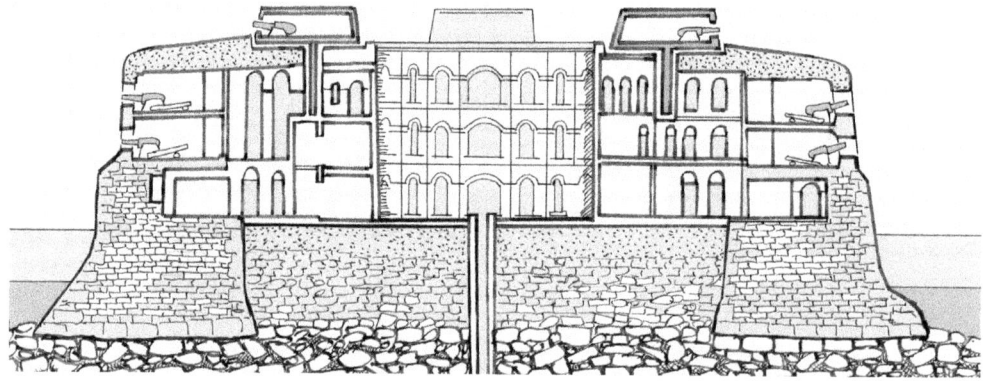

Cross-section of Horse Sands Fort. Each of the Spithead forts was intended to house a garrison of 30 men.

Horse Sands and No Man's Land are identical at 200 foot in diameter and fully armor plated. The other two are slightly smaller at 150 feet diameter with iron plating on the front only.

Defense of Dover

Western Heights

The defenses of Dover on Western Heights were subsequently completed following the report of the Royal Commission of 1860. During the 1860s, the defenses were strengthened and admittance was restricted to two gateways: Archcliffe Gate (demolished in 1963) and North Entrance. Other gun batteries were added around the perimeter of the high ground later in the 19th century—St. Martins, Citadel, South Front, and North Lines Right Batteries. Grand Shaft and South Front barracks provided additional accommodation to hold a total of around 4,000 soldiers. The northern side of the hill was totally relandscaped and the connecting lines between the Citadel and Drop Redoubt improved. The massive ditches (9–15 meters in depth) were upgraded and revetted with brick.

Fort Burgoyne

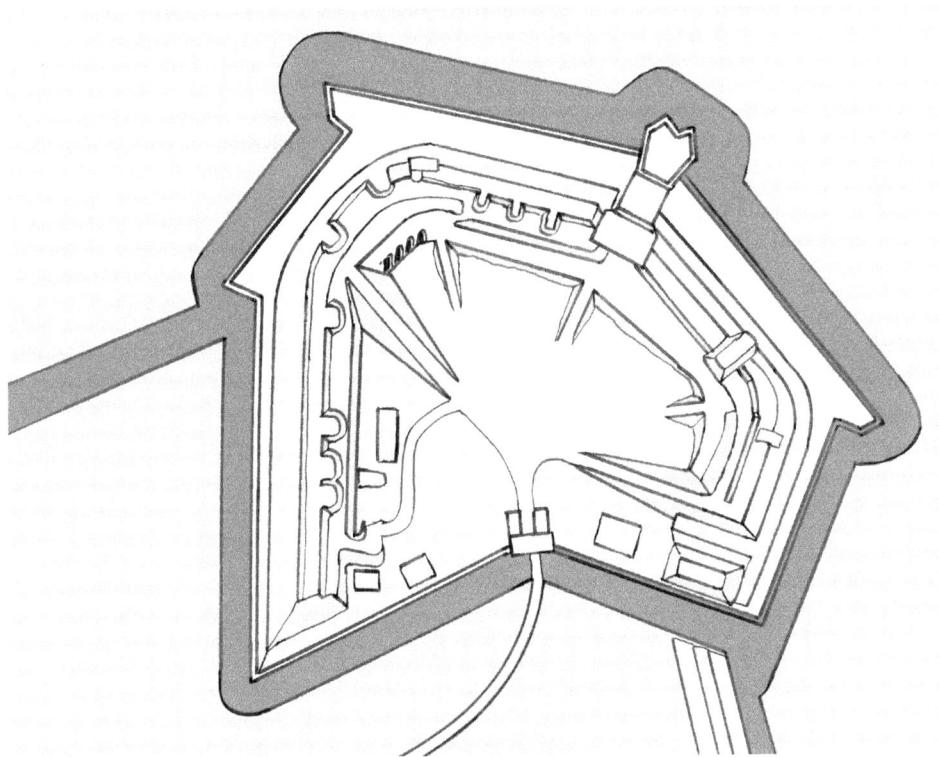

Fort Burgoyne, Dover, 1860. Fort Burgoyne, originally known as Castle Hill Fort, was built in the 1860s as one of the Palmerston forts around Dover in southeast England.

In addition, a fort was built on the high ground northeast of the strategic port, just north of Dover Castle, which was deemed a weak spot. Originally named Castle Hill Fort, this new stronghold was later known as Fort Burgoyne after the 19th-century general John Fox Burgoyne, inspector general of fortifications and son of John Burgoyne who fought in the American Revolutionary War. The new fort was a large polygonal structure surrounded by a ditch, flanked by three demicaponiers and a double caponier to the north. Two detached wing annex batteries to the east and west of the main fort were also constructed in spurs off the main ditch. The works formed a kind of entrenched camp. The main fort comprised a large central parade ground, to the north of which was a long row of casemates that provided the barrack accommodation for soldiers and lodging for officers. Above the casemates, on the terreplein, were Haxo casemates, which housed the guns. Until recently, the central part of the fort was still used by the army as Connaught Barracks, which has allowed the interior of the fort to be well preserved, but the ditch is very choked with vegetation. Today, Fort Burgoyne may be visited with permission.

Ferro-Concrete Fortification

Advances in Attack Weapons

In the long history of military architecture, there was a constant leapfrogging between the development of defensive fortifications and the invention of offensive projectiles capable of destroying them. While millions of pounds were swallowed up for the construction of polygonal fortifications all over Britain and in the empire (and everywhere in Europe), another unexpected revolution in artillery suddenly occurred in the second half of the 1880s. Then, new artillery with new materials and new techniques were introduced, notably the adoption and combined use of rifled artillery, breech loading (which greatly increased the rate of fire), and much more powerful explosives replacing the traditional black gunpowder.

As a result, attack weapons made a brutal and abrupt advance. Tests against masonry forts indicated that all fortresses built previously had become obsolete, and a new axiom was introduced: "Masonry, if seen, will be destroyed." Masonry forts were thus insufficiently resistant. Artillery and soldiers deployed on top of the open superstructures were now dangerously exposed. By then, indeed, a new artillery shell was introduced: the so-called shrapnel antipersonnel ammunition containing a large number of individual metal bullets, fired and exploding close to the target, and thereby releasing and scattering with enormous violence a sudden shower of deadly small metal projectiles in all directions. Small arms also had developed from smoothbore muskets with a rate of fire of a single ball every two minutes to breech bolt-operated rifles and repeater carbines, as well as automatic machine guns spitting out bullets at a rate of 500 rounds per minute. The good news was that defenders could also use repeater rifle, quick-firing guns shooting antipersonnel rounds, and machine guns. The unexpected increase in firepower meant that the garrison involved in the defense of a fort could now be drastically reduced, as a few servants of modern weapons had now as much firepower as a whole battalion of musketeers.

Concrete

Within a few decades, all previously designed and built fortifications had lost a lot of their efficiency. Again, as in the 1860s, there was a serious crisis, and a new solution was required. After 1885, military engineers set about adapting their works to meet the new projectiles. They were compelled to revise their designs for new defensive works and adapt existing ones to the new predicament.

Fortification entered the expensive ferro-concrete age.

Indeed, the answer was found in the use of a new material: high-strength concrete. Soon, it was found that reinforced concrete was still better for the construction of shelters, casemates, and other fortification structures. Reinforced concrete, as the name implies, is concrete featuring steel rods (aka rebars), reinforcement grids, or plates that have been incorporated to strengthen the concrete in tension. In the last decades of the 19th century and during the 20th century, the development of reinforced concrete allowed new fortifications to deal with the new explosives.

Obviously, transcontinental countries like Russia and the United States as well as the British Empire could not reconstruct all their defenses in all their extensive borders. Nevertheless, they tried creating national strongholds by constructing fortresses intended to obstruct the strategic approaches to vital areas. No nation, even the richest and the most advanced, could now afford a complete reconstruction of all their defenses. One had to be selective, and in most cases, one took the decision to building new ferro-concrete fortresses only at the most strategic bases and vital places. In Britain, that meant the coastal forts defending the Royal Navy ports, coaling stations, as well as dockyards and bases. Betterments included adding concrete layers to cover some particularly vulnerable parts such as magazines, ammunition stores, personnel quarters, observatories, as well as casemates and firing emplacements under the new material.

Revetments were enormously strengthened and designed so that their weight resisted overturning. Concrete roofs were made from six- to ten-foot thick, and in many cases, the surface of the concrete was left bare so as to expose a hard surface to the shell without any earth tamping. Another approach to reinforcement consisted of combining materials for making a sort of "second skin." Placed above existing masonry chambers and brick-made stores, for example, it was composed of a three-meter-thick layer of earth, then a one-meter-thick concrete burster layer (extending well beyond the underground chamber to guard against oblique shots), then a one-meter-thick layer of sand. The projectile would be slowed down by the earth, stopped and burst by the concrete, and the sand layer would absorb the explosion wave and the whole protection would prevent the destruction of the substructure.

Counterscarp Coffer and Profile

Another upgrade concerned the masonry caponiers for close-range defense of the ditch, which proved extremely exposed and vulnerable to the new high-explosive projectiles, too. They were discarded, demolished, and replaced with concrete casemates, named counterscarp coffers, fitted with embrasures and (often) own ditches, placed under the protective cover of the counterscarp at the outer angles of the ditch. The coffers were reached from the fort by underground galleries and tunnels passing beneath the ditch. In the last decade of the 19th century, counterscarp coffers were armed with modern repeater rifles, machine guns, and light quick-firing guns.

After 1885, some strongholds and forts were given a so-called Twydall profile—for example, at Woodland Redoubt and Grange Redoubt near Chatham as well as at Beacon Hill Battery (Essex, 1889), Penlee Battery (Cornwall, 1889–1892), Steynewood

Battery (Isle of Wight, 1889–1894), North Weald Redoubt (Essex, 1890), Fort Farningham (Surrey, 1890), Fort Macaulay (Victoria, British Columbia, 1894–1897), and Culver Battery (Isle of Wight, 1904–1906). The Twydall profile included a scarp in the form of a continuous gentle slope instead of a vertical wall. The main advantage of the Twydall scarp arrangement was that defenders could totally see what was going on in the ditch from their position on top of the wall. Beyond that, there was a steep earth or masonry counterscarp. By then, the ditches were enclosed by security grilles or grating known as Dacoit fences. Those fences were strong and about three meters high, consisting of frameworks of aligned spiked metal bars that were difficult and dangerous to climb.

After 1900, the covered way and the glacis around a fort was defended by a new obstacle: fences of thick and impassable network of barbed wires and, later, antipersonnel mines.

In 1871, the British army adopted the breech-loading Martini-Henry rifle, which had a rate of fire of 12 rounds per minute and a deadly effective range of 400 yards (370 meters). In 1886, the crew-served Maxim heavy machine gun was introduced; it was water cooled, recoil operated, used .303 ammunition, had a rate of fire of 600 rounds per minute, and an effective range of about 2,187 yards (2,000 meters). The combination of wire obstacles with rapid and dense fire reestablished the superiority of defense—a domination that culminated in the stalemate of trench warfare in World War I.

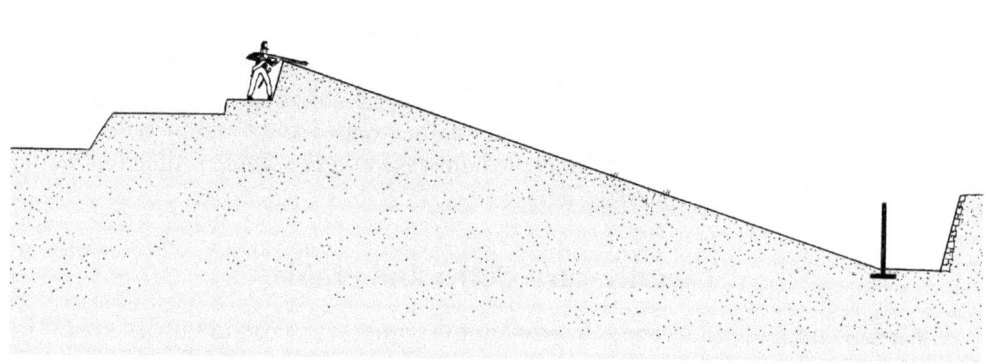

Twydall profile.

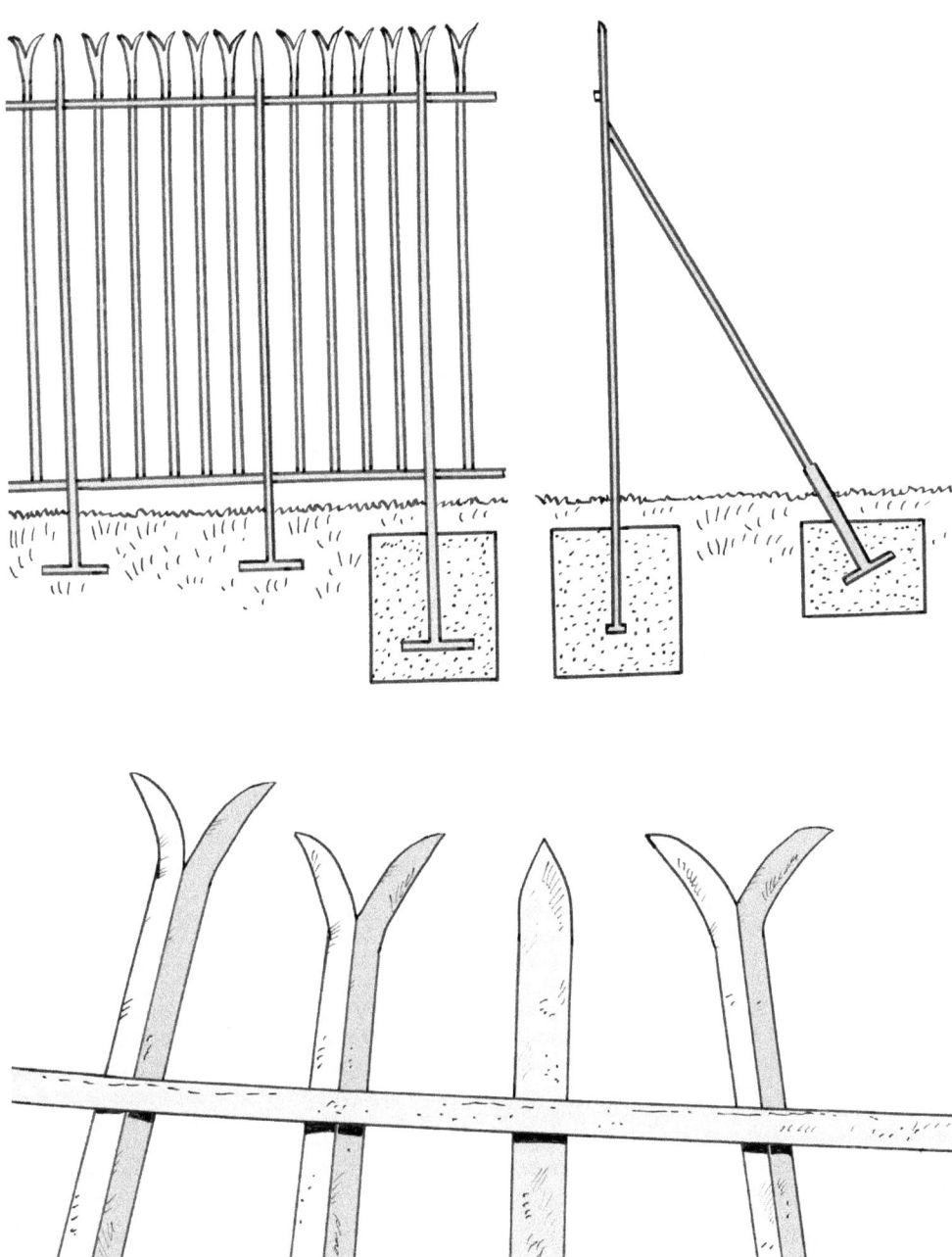

Dacoit fence.

Armored Fortification

In the second half of the 19th century, iron was used in civilian architecture for the constructing of bridges and viaducts, factories, train stations, tracks for railways, and exhibition halls, so it was only logical that it should be employed in military architecture, too. In the 1860s, experiments with iron plating armor had been carried out in Britain and during the US Civil War, notably on steam gunships (e.g., *Monitor, Merrimac*). The concept of armored turret for land fortification was also pioneered by the Prussian lieutenant colonel Maximilian Schumann and developed in Germany by the engineer Hermann Gruson of Krupp's Steel Company and by the French engineer Henri Mougin from the Saint-Chamont Company. In the 1870s and 1880s, advances in the metallurgical industries allowed for new fortifications combining concrete and armor in innovative ways. Systems of laminated armor using rolled iron, cast iron, and steel were introduced and used to cover casemates and provide protection. Since the primary target of coast fortifications was the ship, it followed that the armament of forts kept close step with that of warships. An important step was the development and generalization of armored turrets for both fortifications on land and as naval armament on ships. Armored turrets proved to deliver effective tools to naval architects and fortification engineers in the everlasting race between attack and defense. The so-called Coles turret, experimented in the 1860s and 1870s by Captain Cowper Philipps Coles (1819–1870), could now be produced and was soon adopted in the British navy and British-built turret ships. Coles got his idea from improvised artillery barges built for Black Sea service during the Crimean War. At first, Coles's turret was a simple armored shield mounted on a turntable, then it took the form of an enclosed circular box, and eventually became a fully revolving turret. The armored turret resting on rollers allowing a wide traverse with an all-round arc of fire of 360° and well covered with solid laminating armor plating gave a good protection to crew, gun, and ammunition.

Breech Loading

The generalization of breech loading made it easier to protect the crew now that they did not need access to the muzzle, so armored turret and cupola mountings could proliferate. Besides, to exploit the advantages of breech loading, a method was invented to control the recoil with a hydraulic brake, which absorbed recoil, and a cylinder with compressed air that pushed back the barrel forward to its original position. There were, however, experts who held the opinion that armored turrets were too delicate, arguing that a chance hit might easily jam their movement or derange their pivot. Others objected that the true strength of fortification did not depend on great machinery works intricately pieced together at vast expense but on organization, communications, and invisibility by good concealment.

Eclipsing Armored Turret

Numerous experiments were carried out, and it was ultimately often decided that the risk was worth being taken. The armored turret (armed with one, two, or even three guns) was widely adopted by various nations, for both ironclad warships and for fixed permanent fortification. For land and coastal fortifications, fully rotating armored turrets were placed on top of underground structures, which contained the rotating system, ammunitions stores, and crew quarters. Only the top of the turret was visible. The lower edge was usually protected by a small glacis in the shape of a thick and flat concrete plate or an armored sheet.

Armored turrets could avoid direct-fire damage to their gun embrasures only by facing away from the direction of fire. So a further refined degree of protection was introduced leading to the design of the so-called eclipsing turret. Encased into a concrete pit, the turret could retract in the face of attack, leaving only its domed top surface exposed. The eclipsing or retractable/"disappearing" armored turret was raised vertically by a complex system of counterweights and levers for firing and lowered for reloading or when not in operation. As a result, the crews and guns were perfectly protected, and the only visible feature from a distance was a grassy mount with the upturned metal saucer turret roof flush with the concrete surface. For a while, umbrella-shaped cupolas hiding retractable turrets became fashionable in military architecture. Retractable cupolas and eclipsing turrets could also be used to house searchlights and machine guns. These devices would remain retracted until an assault on the fort was attempted and were then suddenly unmasked for firing.

The retractable system had, however, a number of drawbacks. Due to the enormously high cost involved, forts saw their firepower vastly reduced from that considered advisable in previous years. Moreover, as a result of the enormous weights involved, armored eclipsing turrets could be armed only with comparatively light guns, which could engage only one target at a time. The use of retractable eclipsing turrets was thus restricted to relatively light guns with caliber generally not exceeding 7.5 centimeters (3 inches). Any heavier weapon and the retracting cupola had to be lighter in weight—hence, with thinner armor—or else it would be too heavy to operate. The use of disappearing turrets demanded massive counterweights and complicated hydraulic systems to move up and down, and those mechanisms often appeared to be temperamental and prone to jamming even without the hazard of enemy shellfire. When firing (or fired at), they presented serious problems of noise, condensation, and ventilation for the gunners. Although they were fitted with small observation slits, the occupants could not see a lot and could suffer from claustrophobia. Eclipsing turrets were immensely expensive to purchase and maintain and rather complex to operate.

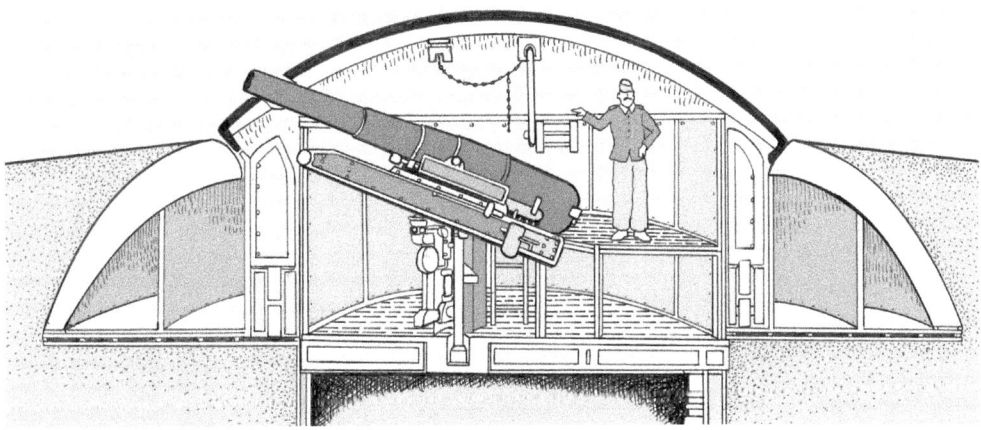

Armored turret cross-section.

The worst objections to the cupola were the military disadvantages of isolation and immobility and the multiplication of mechanical arrangements. For a successful round from a disappearing cupola, the elevating and traversing arrangements, the elevating and loading gear of the gun, and the telephone communication must all be in good order.

So the eclipsing armored turret was not the panacea after all. Artillery firing *en barbette* (over a parapet) or placed in casemates did not disappear and was still often used due to the lower cost. After all, the object of fortification is not to obtain a resisting power without limit but to put the men and guns of a work in an advantageous position to defend themselves as long as possible against a superior force.

Disappearing Gun

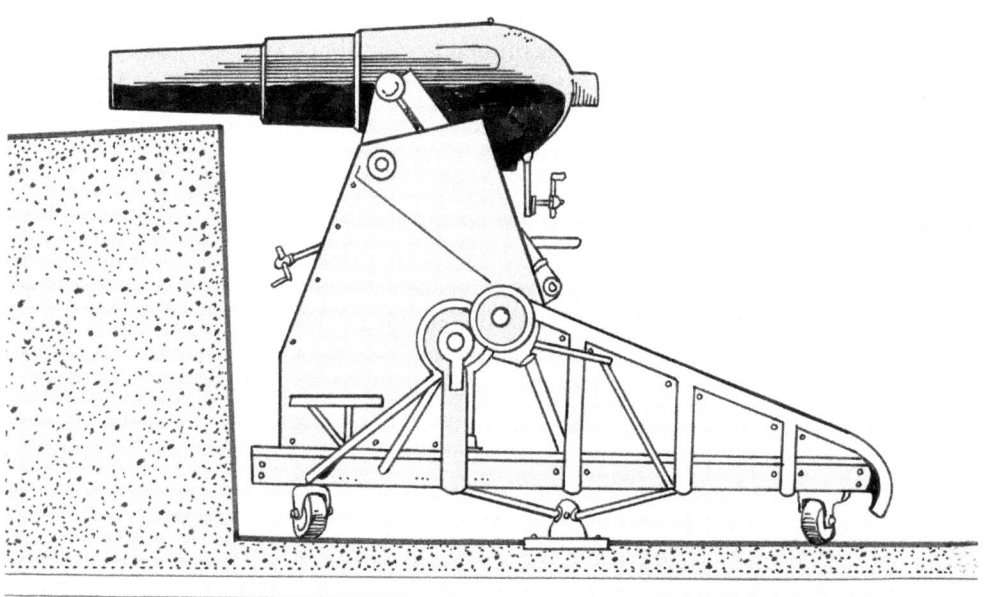

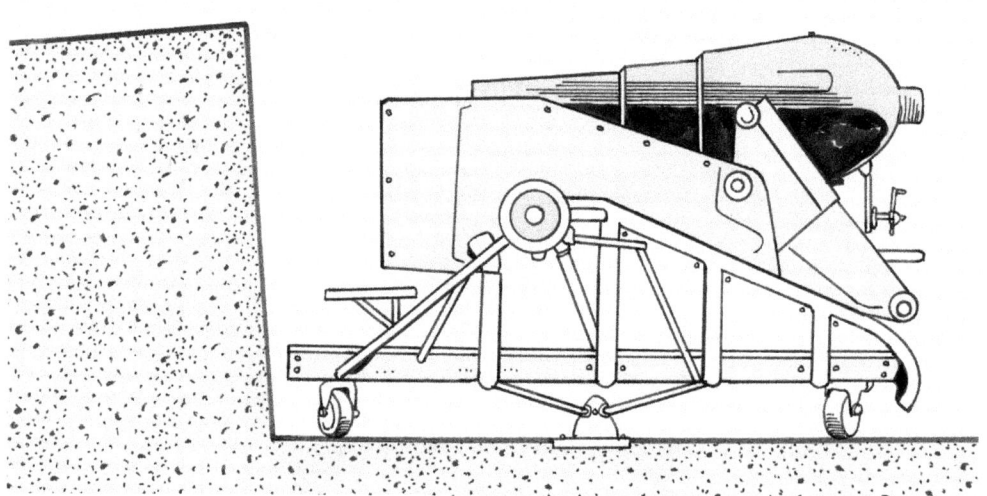

Disappearing gun. Top: firing position; bottom: loading and rest position.

Ironclad ships were constantly improved with more accurate and more powerful cannons, so a new protective and quite ingenious system called disappearing gun was invented with the aim of minimizing the visibility and exposition of coastal artillery. The so-called disappearing gun was a type of heavy (mainly coastal) artillery where the gun was mounted on a mobile curved ratchet cradle allowing retracting and recoiling down into a protected pit or behind a wall after firing. The system was invented in the 1860s by Captain (later Sir) Alexander Moncrieff. The so-called Elswick mounting (named after a ward of the city of Newcastle-upon-Tyne, England) was developed by the Armstrong Tyneside Manufacture. The disappearing gun used the energy generated by the recoil of a gun to lift a counterweight in order to get from upper firing position into lower loading position. In another system, the movement of the weapon was employed to compress air by means of a pneumatic piston. After reloading, the same energy was then used to briefly raise the gun back in its upper position for firing the next shot. The American generals William Crozier and Adelbert Buffington further refined the concept in the late 1880s by incorporating hydropneumatic recoil control to assist the counterweight action. The advantages of the "disappearing gun" system—revealing itself only when actually firing—were protection of the gun and crew, concealment, and cover from enemy fire, especially during reloading. Disappearing guns became highly popular in the British Empire and in the United States—often employed in coastal artillery. In the United States, the Endicott Board formed in 1885 shifted the focus from the structure of the fort to the disappearing weapons contained within it. American coastal forts became simple and rather low-lying concrete emplacements or pits totally blending with the environment with the guns visible for only a few seconds when firing and otherwise totally obscured from attacking warships. However, the concept of disappearing gun fell out of use in the 20th century due to the following disadvantages: high cost involved, technical complexity, limited rate of fire, restricted elevation (maximum 20°), and vulnerability to air attack when bombers, dive bombers, and ground attack aircraft were introduced and developed in the 1930s.

Another form of protection used in coastal defense (in the United States notably) was to sink high-angled mortars into deep concrete-lined pits. The ammunition magazines were on the same level as the floor of the pit and protected by enormously thick traverses consisting of earth layers and concrete placed between the mortar positions.

Underground Forts

So, at the close of the 19th century, the impact of technology was complete. The number of first-class strategical fortresses had been reduced to a few Royal Navy bases and dockyards. Tactically, traditional forts and enceinte had become obsolete, while new ferro-concrete fortresses were developed. By then, a modern fort was like a land warship or an underground submarine. From then on, it was divided into two distinct parts. All open positions including parapets, parados, and traverses as well as infantry breastworks and artillery emplacements were removed or covered with concrete or armored metal plates. Concealment was essential, and where the lie of the ground did not help, it must be obtained from earth parapets or plantations. So from a distance—once the scars of construction had been concealed by grass and bushes—only a few features were visible on top of a fort including the low, armored (retractable) artillery turrets, a few armored observation cupolas, the entrance at the back, and the occasional ventilation chimneys and pipes, airshafts, and flanking rifle ports that could easily be camouflaged. All other structures were safely buried underground: ammunition magazines, stores, workshops, headquarters, telephone exchange, infirmary, power plants to move the turrets, as well as barracks and facilities for the garrison. Indeed, everything was subterranean, illuminated by electricity produced by plant powered by internal combustion engines, which also drove all the machinery necessary for living in and operating the fort—notably, the ventilation and drainage systems—and the power-driven hoists supplying guns with rounds from underground stores. Electric searchlights were used in all important works and batteries. They could be placed in disappearing cupolas and (although quite expensive, rather vulnerable and fragile, and useless in thick foggy weather) were of great value for discovering enemy engineering working parties at night and lighting up the foreground during an attack.

Passages, posterns, tunnels and galleries linked these facilities, and fortress troops became troglodytes. Gradually, the quality of concrete improved, poured in one batch instead of layers forming internal lines of weakness liable to fracture under stress. Concrete was now systematically reinforced and internally bonded with steel bars. Steel replaced iron. Wireless telegraph and telephone were introduced as communications became of the first importance, not merely to facilitate the movement of the enormous stores of ammunition and matériels required in the fighting line but also that defenders may fully utilize the advantage of acting on interior lines. Military railways and roads ran from the center of the place to the different sectors of defense, all round, and in rear of the line of forts.

Soon, the invention of the internal combustion engine would revolutionize transport and the implements of war themselves in the form of trucks, armored cars, self-propelled artillery, tanks, and airplanes. With the development of repeater magazine rifles, machine guns, and barbed wire in the last decade of the 19th century, a modern fort could be efficiently defended against an infantry attack. Good examples of this subterranean ferro-concrete fortification were the fortresses built in the 1890s by the Belgian general Henri Brialmont around Liège in Belgium. They were designed to withstand the .83-inch (21 millimeter) field howitzers, which were all that could be deployed in mobile warfare at that time. Another example would be the ring of Dutch concrete forts built around Amsterdam between 1885 and 1907. Yet another example would be the German *Festen* (fortified groups) constructed between 1890 and 1918 around Strasburg, Metz, and Thionville (today in France).

By then, military architecture had thus entered a new age, and needless to say, that was a very expensive era as indeed such sophisticated forts with all their refinements cost a fortune to build and to maintain. The gigantic cost and the increasing complexity of equipment demanded new garrisons, not only riflemen soldiers but also skilled experts in specialist technicalities. In step with the general advance in armament, there was an equal improvement in fire control, notably the determination of range of a moving target. So new specialist gunners appeared operating new optical instruments known as range finders relying on trigonometric relationship between height, depression angle, and distance.

Mass production, accumulation and maintenance of reserve stockpiles of ammunitions, equipment, and supplies, and the development of communication systems pressed hard on the national defense budget. At the dawn of the 20th century, permanent fortification had reached its zenith. Billions had been transmuted into concrete, steel, and masonry, but while the fortress designers were busy, artillerymen were active, too, and the 20th century became another challenging time. The speed at which new fortifications were designed, introduced, updated, and rendered obsolete even before completion was phenomenal.

In all industrialized countries (namely, Germany, France, Britain, Italy, and the United States), weapon-designing bureaus, ammunition factories, and firearm manufacturers advertised their wares. International expositions were held where the marvels of contemporary weapon technology were exhibited. The displayed products were sold to their own country but could also be purchased by anyone prepared to pay big money. What was seen as "progress" was in full bloom, and "advancement in civilization" could cynically be identified with man's growing capacity to kill ever increasing numbers of people by single and less intensive efforts.

Radical capitalism greatly prospered, tycoons and financiers thrived, large companies expanded, and industrial empires grew tremendously. More than ever, producing weapons and building fortifications were lucrative businesses.

Defense of London

Armored and Artillery Trains

While gigantic funds were consumed for the modernization of land and coastal defenses, attention was also given to the protection of London.

In the 1860s and 1870s, William Bridges Adams (1797–1872), an authority on railroad, and several military engineers (e.g., Lieutenant Arthur Walker, Colonel E.R. Wethered) suggested the utilization of armored trains and heavy artillery mounted on railroad cars for defending the shores of Great Britain against invasion instead of expensive static fortifications. It must be noted that by 1850, the United Kingdom railroad system was in a leading position and already counted 6,621 miles of tracks. Thus, armored fortresses rolling on tracks would constitute a strong line of mobile defense along the whole coast as they could quickly bring troops and artillery to defend threatened sectors and places where enemy forces could be landing. More especially for the defense of London, the employment of armored trains was also advocated as the most efficacious and the most economical line of defense the city could have. A military strategic circular railroad line would form a complete defensive cordon around the capital at a distance of 15 miles from the center. For unknown reasons, these suggestions were not retained, and military funds were employed for the improvement of static coastal permanent defenses.

London Defense Scheme

The London Defense Positions were late 19th-century earthworks installed in the southeast of England, intended to protect London from enemy landing on the south coast.

The 1859 Royal Commission on the Defence of the United Kingdom report on Britain's defenses believed that London was practically undefendable. However, the

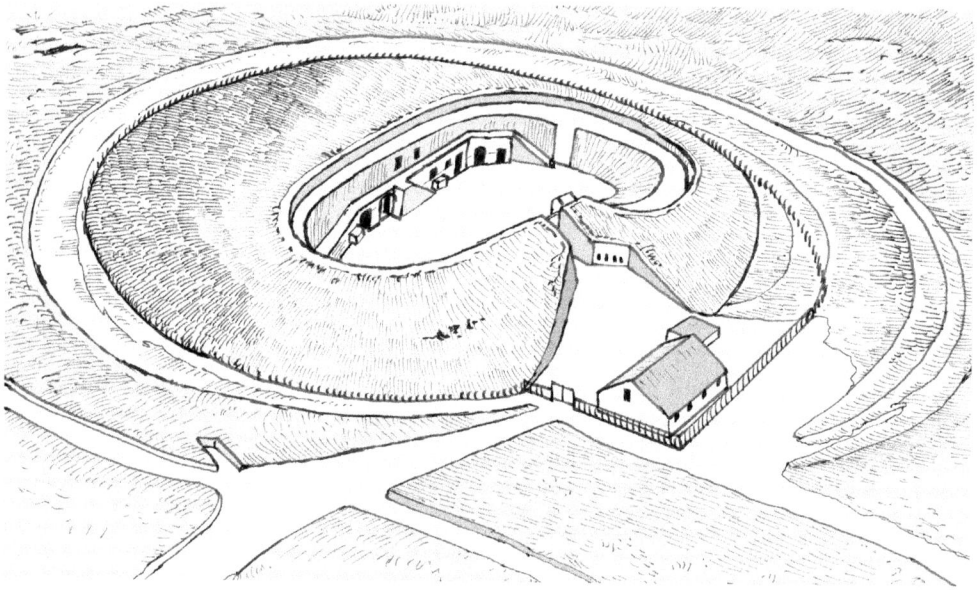

Fort Henley Grove. The fort, located at Guildford, Surrey, was built in the late 19th century for the protection of London. Fort Henley was never used as a stronghold in the conventional sense but served as a magazine, as a ground to train volunteers, and as a conscription meeting place and later as an ordnance store.

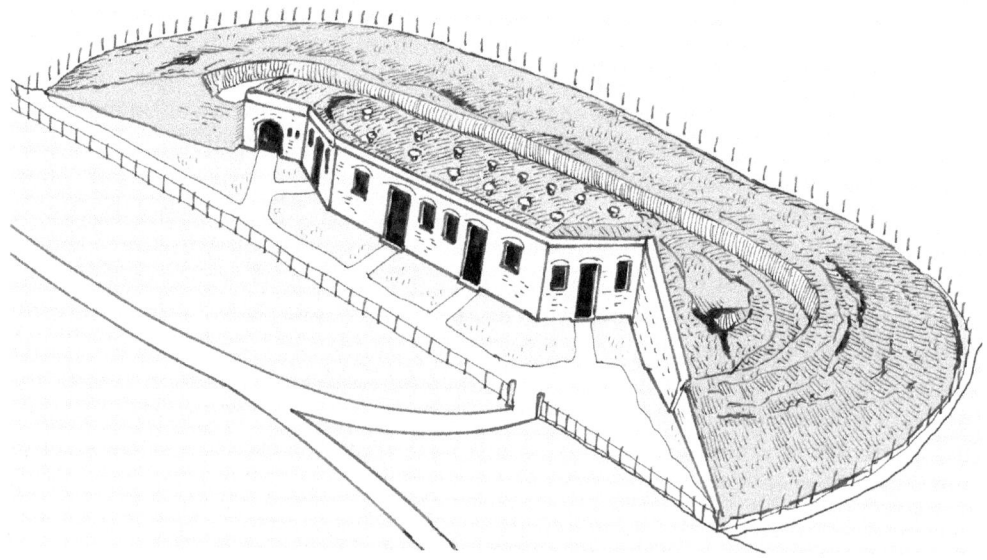

Boxhill mobilization center.

commission proposed the construction of a fort at Shooters Hill to defend the Royal Arsenal at Woolwich, but this was never carried out.

Following a number of other proposals by senior military figures based on simple earthworks for infantry and movable armaments, the London Defense Scheme was announced in Parliament in March 1889. The London Defense Positions were to be earthworks refurbished and garrisoned in time of war but backed up with permanent works, which acted as stores and magazines at five-miles (eight-kilometer) intervals. They were built along a 70-mile (113-kilometer) stretch of the North Downs from Guildford to the Darenth Valley. Several sites were chosen—for example, at Pewley Hill, Henley Grove, Denbies, Box Hill, Betchworth, Reigate, East Merstham, Fosterdown, Woldingham, Betsoms Hill, Halstead, Farningham, and (to the north of London) at Warley and North Weald. The design of each site varied, but they were never very elaborate, just magazine and storehouses for the mobilization of troops, with limited defenses. These sites' purpose—in addition to holding ammunition and other supplies—was to act as strong points in an almost continuous line of field fortifications. The trench lines joining the defense positions could be rapidly excavated on the outbreak of an invasion. These positions were quickly viewed as obsolete, and the whole scheme was abandoned in 1907.

PART 5

Fortifications in the British Empire in the 19th Century

Colonialism

The Industrial Revolution in the second half of the 19th century provided the European powers tools, weapons, machines, and instruments, as well as a belief of technical, racial, and intellectual superiority giving rise to an aggressive and racist spirit that assured them world dominance. The consequence of this infatuation was a scramble to conquer and establish permanent colonies overseas in "underdeveloped" regions of the world. The primary aim was obtaining cheaply raw materials, create economical markets, and incidentally "civilize" native populations and parenthetically "save their souls" by bringing them a "true religion." Non-White people were regarded with contempt as backward, savage, retarded, and ignorant and therefore inferior and treated with condescension as no more than serviceable labor force for rough exploitation. In this context, the European (and the American) White man was convinced he had the right and even the duty to dominate other people—if need be by force. By then, his superior technology and weapons enabled him for the most part to score more or less easy victories in Africa and Asia over native armies many times their own size. Colonial conflicts also offered opportunities for lethal experimentation with new weapons.

Imperialism, a policy of extending a country's power and influence through colonization, use of military force, or other means, was not new. It had existed since the antiquity, notably in Rome. It had been revived at the time of the great discoveries in the Renaissance, but in the late 19th century, it dramatically developed particularly in Africa and Asia. Many people in the late 19th century viewed colonial acquisitions as a true indication of having achieved nationhood. Public opinion eventually was generally favorable to conquering prestigious and exotic African, Asian, and Pacific colonies. These aspirations were nurtured by numerous scientific institutions, geographical associations, and pro-colonial lobbies. Imperialism was also fired up by the military, particularly the naval authorities who wanted to have coaling bases for their steamships. Another aspect of colonization was converting natives to the "one and only true" religion. The Protestant aspect of Christianity was considered by many British people and leaders as part of the larger battle against the pope and the Roman Catholic nations of continental Europe. Ever since the Reformation, religion represented not only a spiritual difference between the Catholic and Protestant faiths but was part of a far larger cultural, economic, and political rivalry between eager competitors. Portugal, Spain, and France were the Catholic "papist" nations that developed successful commercial empires before Anglican Britain and Calvinist Netherlands were able to do so. Religion gave an excellent excuse for this commercial competition to turn into political rivalry and military hostility. The very success of the Protestant nations in challenging the Catholic hegemony in the New World and the East Indies seemed to confirm that God might be on the Protestants' side after all—although this did ignore the fact that the English and Dutch sectarian Christian co-religionists were just as frequently found at the throats of each other. So racism, colonial conquests, religious rivalry, and imperialist economical interests created new and dangerous tensions between antagonist European powers.

In the late 19th century, European interests were advanced and defended by armies and fleets until almost every part of the earth was conquered. The result of European predatory colonialism was that their influence and power extended over the whole globe. By 1900, large parts of North and South America, Africa, Oceania, and Asia had been explored, mapped, estimated, divided, conquered, occupied, and exploited. This had been made possible by technologically superior powers, the combination of good conscience, monopoly of modern firearms, professional qualities of drill and discipline, and the blessing of God. At the same time, steamships, telegraph, and railways opened up the interior of remote lands in Africa and Asia and gave European colonial armies a mobility that compensated for the smallness of their size.

There were very few people and very few territory whose natural resources were not developed and ruthlessly exploited by European and American conquerors and no native society whose culture was unaffected by the so-called Western civilization.

The British Empire

The loss of Britain's 13 American colonies after the war of 1776–1783 was an unpleasant and humiliating blow for Britain. However, that was partly compensated by the spectacular growth of Upper Canada (now Ontario) after the emigration of loyalists from what had become the United States. Indeed, the 19th century marked the full bloom of the British Empire. New settlements were established in Australia from 1788. The Napoleonic Wars and the collapse of France provided further additions to the empire. The Treaty of Amiens (1802) made Trinidad and Ceylon (now Sri Lanka) officially British. By the Treaty of Paris (1814), France ceded Tobago, Mauritius, Saint Lucia, and Malta. Malacca joined the empire in 1795, and Singapore was acquired in 1819. Canadian settlements in Alberta, Manitoba, and British Columbia extended British influence to the Pacific. In India, the British conquered several territories including Agra and Oudh and the Central Provinces, East Bengal, and Assam. New Zealand became officially British in 1840, after which British control was extended to Fiji, Tonga, Papua, and other small islands in the Pacific Ocean. In 1877, the British High Commission for the Western Pacific Islands was created. In the wake of the Indian mutiny (1857), the British Crown assumed the East India Company's governmental authority in India. Britain's acquisition of Burma (Myanmar) was completed in 1886, while its conquest of Punjab (1849) and of Balochistan (1854–76) provided substantial new territories in the Indian subcontinent itself. The completion of the Suez Canal (1869) provided Britain with a much shorter sea route to India. Britain responded to this opportunity by expanding a refueling port at Aden, establishing a protectorate in

Somaliland (now Somalia), and extending influence in the sheikhdoms of southern Arabia and the Persian Gulf. Cyprus was occupied in 1878 and, together with Gibraltar and Malta, became an important stopover in the chain of communication to India through the Mediterranean Sea. Elsewhere, British influence in the Far East expanded with the development of the Straits Settlements and the Federated Malay States, and in the 1880s, protectorates were formed over Brunei and Sarawak. Hong Kong became British in 1841, and an informal empire operated in China by way of British treaty ports and the great trading city of Shanghai.

The greatest 19th-century extension of British power, however, took place in Africa. Britain was the acknowledged ruling force in the officially Turk-Ottoman Egypt from 1882 and in Sudan from 1899. In the second half of the century, the Royal Niger Company began to extend British influence in Nigeria and the Gold Coast (now Ghana). The enclaved Gambia in French Senegal also became a British possession. The Imperial British East Africa Company operated in what are now Kenya and Uganda, and the British South Africa Company operated in what are now Zimbabwe (formerly Southern Rhodesia), Zambia (formerly Northern Rhodesia), and Malawi. Britain's victory in the South African Boer War (1899–1902) enabled annexation of the Transvaal and the Orange Free State in 1902 and to creating the Union of South Africa in 1910. The resulting chain of British territories stretching from South Africa northward to Egypt realized an enthusiastic British public's idea of an African empire extending "from the Cape in the South to Cairo in the North." By the end of the 19th century, the British Empire comprised nearly one-quarter of the world's land surface and more than one-quarter of its total population.

Colonial Fortifications

The British possessions had to be protected as much against local opponents as against European rivals. With the extension of imperial conquest, Britain's defense commitments stretched across the whole world. In order to secure their communication routes, the British aggressively fought for a mastery of the seas and oceans so that its navy could pass unthreatened almost anywhere. They developed a system of fortified intermediate calls, naval arsenals, and coaling stations at convenient distances defending the sea routes to their colonies. They managed to secure a lifeline along which supplies, a great volume of trade, both imported and exported, and urgent troop reinforcement in time of crisis could be quickly shipped.

Of course, imperial fortifications followed the evolution and the designs of those built in the motherland according to the quickly changing patterns of 19th-century evolution and required enormous funds.

Originally, individual settlements had to provide for their own defense, and primitive urban enceintes and improvised forts were the first structures built in the oversea colonies. They were constructed for protection against European rivals for land and influence and also to guard settlers, administrators, and traders against those natives who dare react with hostility to the foreign intrusion into their homelands. Fortifications were also intended to impress and deter the local inhabitants and acted as a perpetual reminder of the foreign White man's domination. In time of crisis, they were used as places of refuge. Although many forts were only temporarily built and although many were eventually destroyed, abandoned to deteriorate, and swallowed up by nature and vegetation, the sites of many of these early forts and trading places often decided for the location of settlements that later became permanent villages, towns, ports, and cities. These early forts were usually quite simple in design, mainly small trading posts and garrison places from which soldiers and private company militiamen could operate. They were usually built with perishable earthwork, timber and logs, mud or adobe. The actual construction of colonial fortifications was subject to the following main influences: the current trend in military architecture back in Europe, the local style of building, the presence of skilled military engineers and personnel, and the availability of materials, funds, and local serviceable manpower.

As time went by, larger and more significant fortified places were constructed with more durable material as colonial forts and fortified posts served as permanent bases for the military, administrative, commercial, and religious exploitation of the colony. The builders often attempted in their designs to reproduce examples familiar to them from their homes so far as the materials available to them permitted. No real systematic pattern of defensive construction was adhered to because elaborate European formalism was of little practical value to guard a sparse population against raiders, pirates, and rivals. However, most of these forts were inevitably built on the traditional bastioned system developed in Europe but in a most rudimentary fashion. Although simple field fortifications were often enough against poorly armed natives, several Palmerston polygonal forts were constructed following the 1860 Royal Commission report. Eventually, some colonial forts abroad could be as intricate and as sophisticated as the fortresses built at home, although there was always improvisation and simplification in the wildest and most remote places.

Fort Albert

The British began intentional settlement of Bermuda in the 1610s and gradually established fortifications. The principal forts at Dockyard, St. Catherine, Victoria, Albert, George, and Cunningham were all built in response to the independence of the United States. If that country had remained within the British Empire, it is likely that no forts would have been built in Bermuda after the 1760s and Bermuda would have remained "in obscurity." Both Forts Albert and Victoria are located in Retreat Hill at St. George's, Bermuda and are close to each other. They were built to protect the northeastern coastline of Bermuda. Fort Albert, named after Queen Victoria's husband, Prince Albert, was completed in 1842. Fort Victoria, named after Queen Victoria, was also completed around the same time. Fort Albert was of pentagonal shape and surrounded by moats. It had provisions for housing officers and soldiers and was armed with three different types of artillery pieces placed in protected platforms: cannons for low long-range targets, howitzers for higher short range, and mortars for very high, short-plunging fire. In 1865, the fort's artillery was modernized with four 10-inch rifled muzzle loaders.

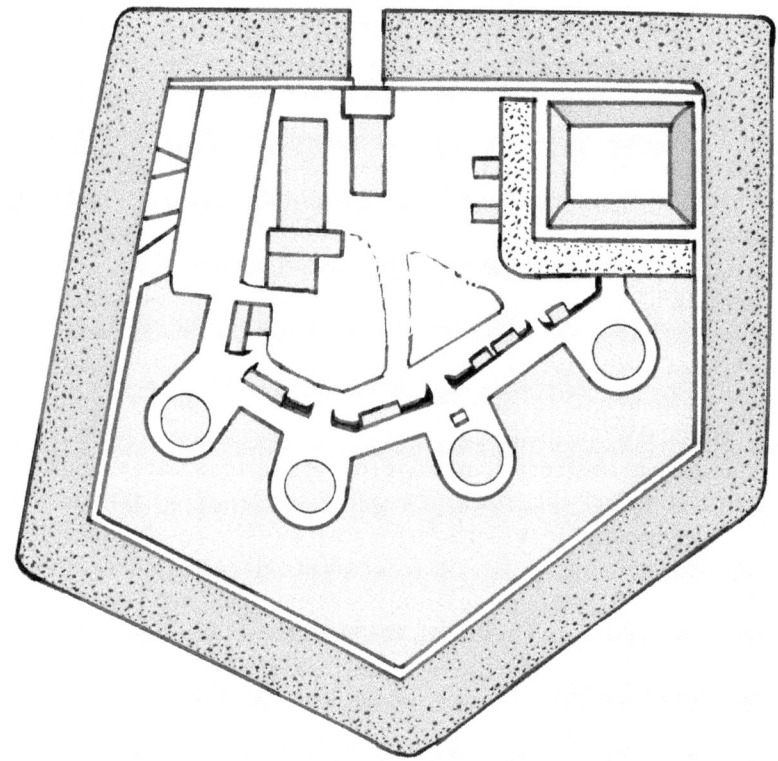

Plan of Fort Albert at Retreat Hill, Bermuda.

Fort Cockbourne

The tiny isolated volcanic Ascension Island located in the Atlantic Ocean became a British colony in 1815. Fort Cockbourne (in 1830 renamed Fort Thornton) was a gun battery built in 1822 on a rock promontory and was intended to defend the entrance to the harbor of the capital, Georgetown (named after King George III). It has been significantly modified over time, notably in 1830 and 1880 and during World War I and World War II. Ascension Island and the neighboring Saint Helena and Tristan da Cunha were garrisoned by the British Admiralty from 1815 to 1922. They were important safe havens and coaling stations to the Royal Navy in the middle of the Atlantic Ocean. During World War II, they were naval and air stations, especially providing antisubmarine warfare bases in the Battle of the Atlantic. Other fortifications were built on Ascension Island to complement and support Fort Cockburne/Thornton: Fort Hayes to the west, Fort Warren to the immediate east, a battery on the slopes of Cross Hill, possibly below Bate's Cottage, another immediately west of the cottage, and eventually Fort Bedford to the west.

Part 5. Fortifications in the British Empire in the 19th Century

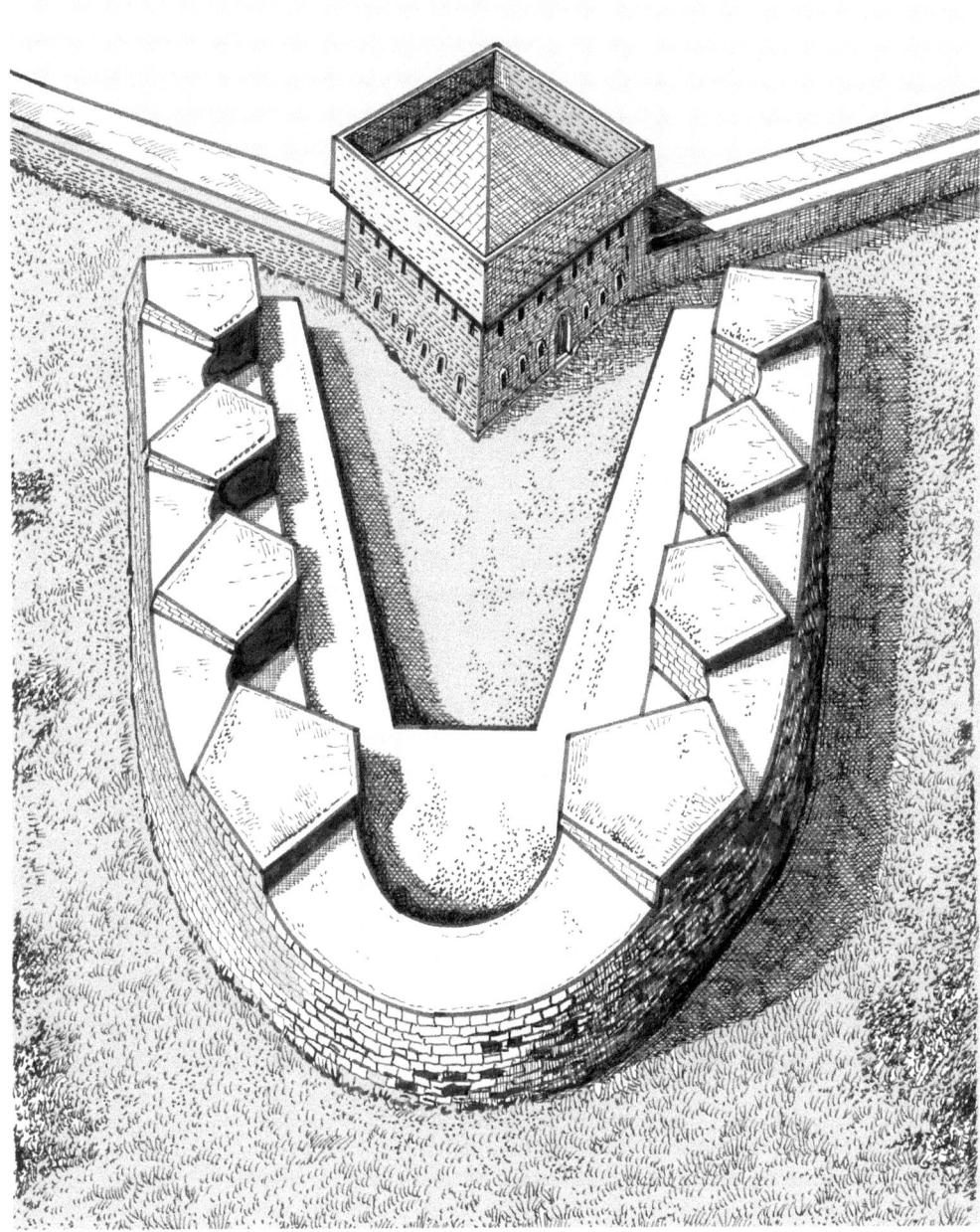

Fort Cockbourne, Georgetown, Ascension Island, c. 1830.

Martello Towers

Martello tower, Alcaufar. The tower is located at Sant Lluis in Menorca (Balearic Islands, Spain). Menorca was invaded by the British in 1798 during the French Revolutionary Wars, but it was finally and permanently repossessed by Spain by the terms of the Treaty of Amiens in 1802.

In the 19th century, British fortifications were upgraded following the general trend and using more elaborate and stronger masonry structures (later concrete) in order to resist attack from European rival steamships and rifled artillery. Martello towers were constructed in the empire, notably in Antigua and the Virgin Islands (Fort Recovery), Barbuda, Bermuda (e.g., Ferry Point), Jamaica (Fort Nugent), Sierra Leone (Freetown), Australia (Sydney Harbor), and South Africa (Cape Town and Fort Beaufort), for example. An impressive scheme of Martello towers was also built in Canada against the French, notably at the naval base of Halifax, Nova Scotia, in 1796 and 1798, as well as at

St. John's, Newfoundland, at Quebec City, and at Kingston, Ontario. At Minorca in the Balearic Islands between 1798 and 1802, the British built a number of towers like the Santandria tower with a curious continuous ring of machicolation. Several towers were built on the island of Ceylon (present-day Sri Lanka) at Hambantota, with an unusual design at Trincomalee: a tower with a square plan constructed in 1806.

British Fortifications in Canada

The British were persistent in their efforts to dislodge and oust the French from North America. This was achieved in 1763, and by then, the now-British Province of Quebec was divided into Upper and Lower Canada in 1791 and reunified in 1841. In 1867, the Province of Canada was joined with two other British colonies—New Brunswick and Nova Scotia, forming a self-governing entity named Canada.

Fort Frederick

Tower at Fort Frederick, Kingston, Ontario, Canada. Note the presence of a conical timber snow roof covering the top platform and bulbous caponiers (casemated structures) projecting from the main body to provide flanking fire and defend the dry ditch around the tower.

Fort Frederick was built on the south end of Point Frederick, the site of the Kingston Royal Naval Dockyard. The original fort, consisting of bastioned earthworks, was built during the War of 1812 for protection against naval attack. Four large stone Martello towers were built to strengthen Kingston's defenses in 1846 during the Oregon Boundary crisis between the United States and Britain. The towers were meant to protect the shipyard and the entrances to the Rideau Canal and St. Lawrence River from possible U.S. aggression. Fort Frederick was one of these and was built on the site of the original fort. Fort Frederick was abandoned in 1870 and today has become a museum.

Halifax

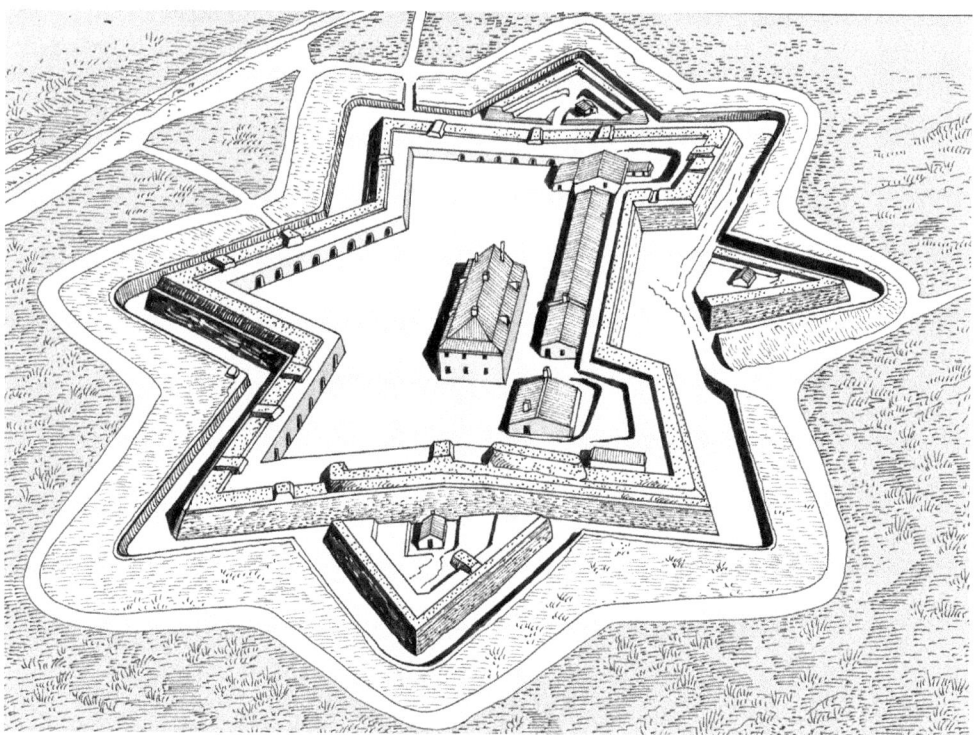

Citadel of Halifax.

Halifax, the capital of the province of Nova Scotia in Canada, was founded in 1749 by the British as a new settlement on the shores of Chebuco harbor. To ward against Wabakani and Mikmaq natives and French Acadians' attacks, the settlement and the harbor (later enlarged to a strategic Royal Navy dockyard) included a stockade enceinte, coastal batteries, and timber palisade forts built in the 1750s at Citadel Hill, Fort Sackville, Dartmouth, and Lawrencetown. The present-day citadel, aka Fort George (named after King George III), was designed by Colonel Gustav Nicholls, and its construction started in 1828. When it was finally completed in 1856, it was obsolete due to the introduction of new rifled cannons. The citadel of Halifax is a rare and late example of tenaille fortification with regular projecting V-shaped walls and three ravelins in the ditch.

Quebec

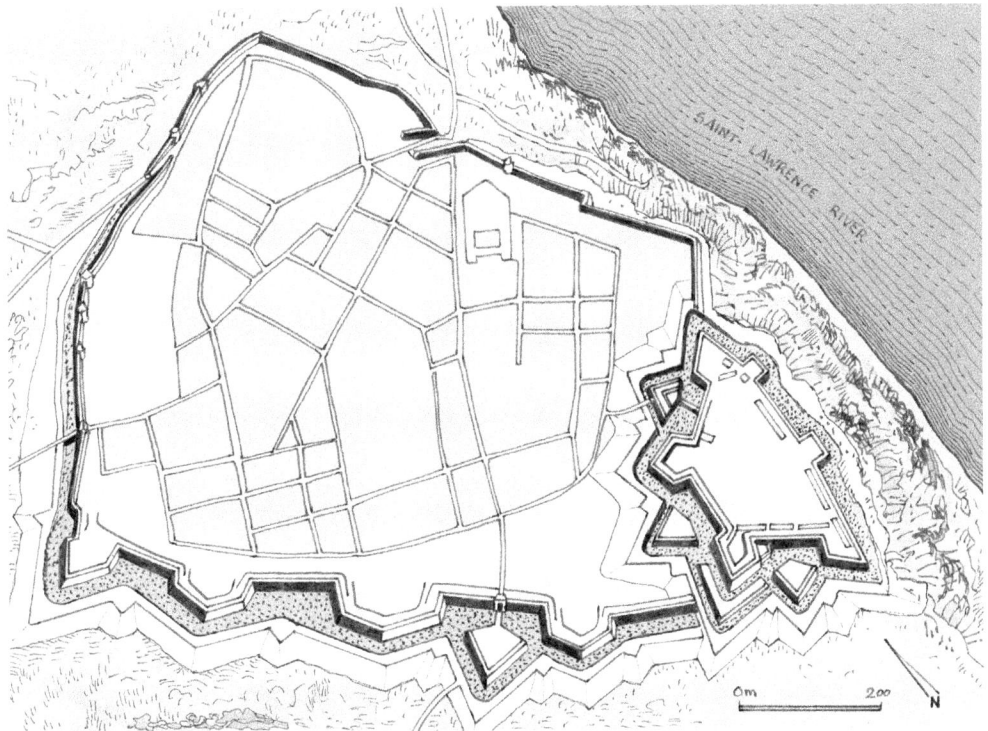

Quebec (Canada).

The Saint-Lawrence Bay was explored in 1534 by the Frenchman Jacques Cartier, and the territory was annexed and baptized Nouvelle-France. Samuel de Champlain developed colonization and founded Quebec city in 1608. Situated on a 106-meter-high promontory dominating the Saint-Lawrence and Saint-Charles Rivers, Quebec was the fortified capital of the Belle Province. The early bastioned enceinte, built in the 17th century, was designed by Louis de Buade, lord of Frontenac. Modifications were later carried out by the French military engineer Jacques Levasseur de Néré in 1701, and further works were added by the engineer Gaspard-Joseph Chaussegros de Léry in 1745. The town was besieged and taken by British troops under General James Wolfe's command in 1759. This decisive battle marked the end of New France and the birth of British Canada. The British expanded the city's fortifications and constructed a citadel.

Quebec Citadel

The citadel of Quebec, located on the highest point of Cap Diamant, was built between 1820 and 1850 by the British engineer Elias Walker Durnford. It incorporated a section of the French enceinte of 1745 and displays an irregular bastioned layout with bastions, straight curtain walls, ditches, and demilunes all constructed with locally quarried sandstone. Within its walls are several buildings, notably the officer's barrack that was one of the official residences of the governor-general of Canada. The citadel was intended to secure Quebec City against American aggressions and serve as a

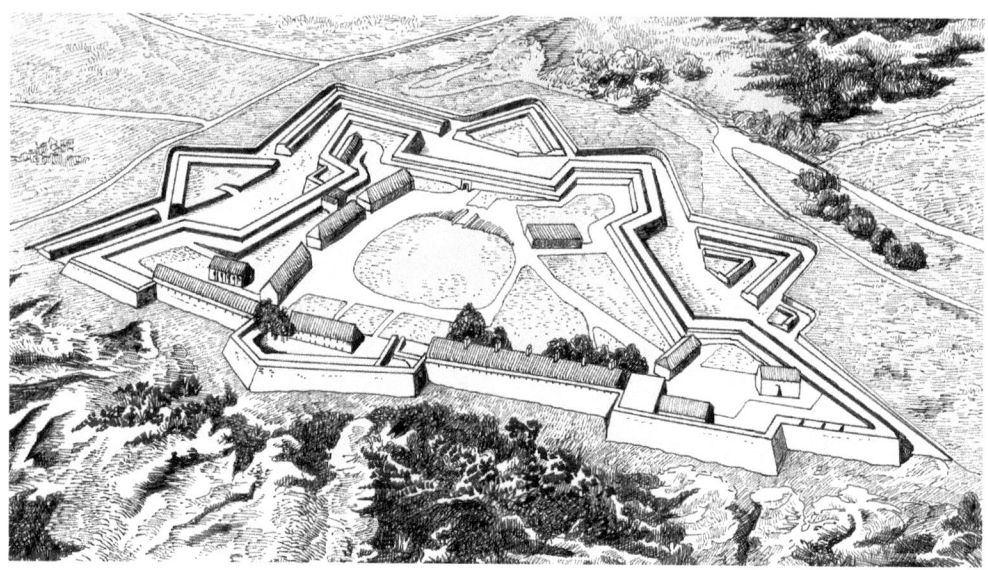

Quebec Citadel.

refuge for the British garrison in the event of attack or rebellion by the newly conquered French-speaking people. It was occupied by Britain until 1871. Today, the citadel houses the famous French-speaking Royal 22nd Regiment of the Canadian Forces.

Fort Henry

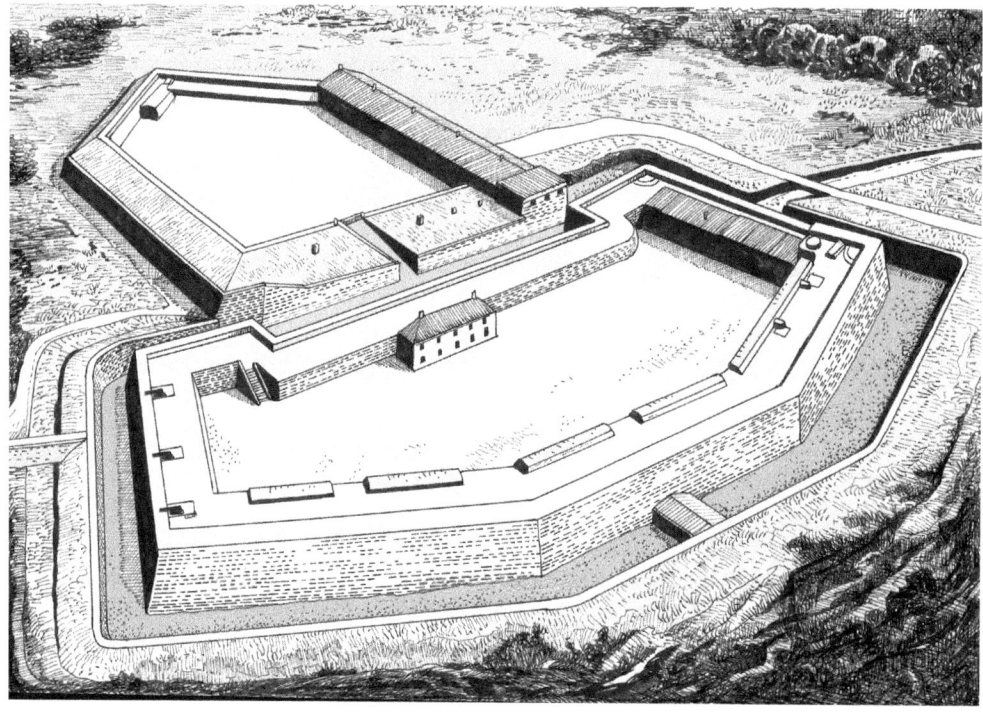

Fort Henry, Kingston, Ontario, Canada.

Located on a strategically elevated point near the mouth of the Cataraqui River where it flows into the St. Lawrence River at the east end of Lake Ontario, the first fort was established in 1812–1814 in order to protect the Kingston Royal Naval Dockyard. This first fort was demolished and another was built in 1832–1836. It was temporarily used as a prison in 1837–1838 and remained occupied by a Canadian militia until 1891 when good relations with the United States made redundant the need for defenses along the border. The fort was then abandoned and fell into ruin. During World War II, it was refurbished and served as an Axis prisoner of war camp. It is now open to the public.

British Fortifications in Gibraltar

In the 19th century, and more particularly after 1869 when the Suez Canal was opened (thus significantly shortening the journey by sea to India), British attention was focused on holding key positions on a permanent basis: Gibraltar, controlling the entrance to the Mediterranean Sea; Malta, which dominated the narrows between Sicily and the North African coast and effectively blocked the passage from the eastern to the western Mediterranean; Corfu, the key island that locked up the Adriatic Sea; and Cyprus, another convenient intermediate call acquired in 1878. These islands were of crucial importance along the shipping route to the British Empire in East Africa, India, and the Far East. Since Minorca, Corsica, and Sicily were out of control, the possession of Gibraltar, Malta, Corfu, and Cyprus became vital.

The rocky site of Gibraltar in southern Spain had been British since 1704. The existing Moorish and Spanish fortifications had been modernized, and the main strength lay in the sheer character of the rock enclosed by the sea on three of its sides. The rock was honeycombed with underground facilities, galleries, and tunnels, which had been dug after the siege of 1782. During the 19th century, cutting tunnels and excavating casemates in the north face of the rock continued unabated. Eventually, the Rock of Gibraltar was converted into a strongly defended fortress, particularly in the 1890s when huge hydraulically operated 100-ton guns were installed. Gibraltar was indeed a labyrinthine maze featuring vaulted gun casemates, bombproof barracks, power plants, workshops, sheltering caves, a hospital, and supply store places connected by means of tunnels and underground galleries. In addition, there were large fueling stations, a dockyard with dry docks, a shipping jetty, a wharf, a slipway, an internal railways network, and various naval installations and facilities. Rubble excavated from tunneling the rock was used to create landfill. In order to hold enemies at bay, breakwaters and outworks at sea were established.

British Fortifications in Malta

The small island of Malta in the Mediterranean Sea displays numerous, impressive, well-preserved, and highly interesting vestiges of fortifications. The diversity and density of fortifications in Malta is quite amazing ranging from early 16th-century artillery forts, compact bastioned urban enceintes, citadels, coastal watchtowers, and coastal forts (constructed by the Knights Hospitaller between 1530 and 1798) to 19th- and 20th-century defenses, coastal forts and batteries, and concrete bunkers (built by

Great Britain from the 1800s to 1964). The small but strategically important archipelago of Malta, Gozo, and Comino had been occupied by numerous foreign rulers including the Phoenicians, Romans, Greeks, Arabs, Normans, Sicilians, Swabians, Aragonese, the Knights Hospitaller, and the French. After the Napoleonic Wars, Britain assumed a protective role over the Maltese islands, which it maintained for a century and a half. Britain took over a bewildering assemblage of existing fortifications, notably coastal batteries and forts and rings of bastioned fortifications extablished by the Knights of St. John of the Hospital of Jerusalem for the defence of Valletta, the Three Cities (Vittoriosa, Senglea, and Cospicua/Bormla), and the Grand Harbour. The British modernized the existing fortifications and improved armament and barrack accommodation. They also increased the defenses of the landward approach to the harbor, arsenal, and naval dockyard at Valletta. In the 1860s, they began to build a series of new coastal forts. With the opening of the Suez Canal in 1869, Malta's importance increased, and the fortifications were modernized again with new coastal defenses notably Fort Delimara (1876), Fort Tombell (1872), Fort Leonardo (1873), Fort Rinella (1878), and Wolseley Battery (1897), just to name a few. The Maltese forts display a curious trait of British engineering. Unlike French work of the period, they were not standardized but exhibited a great variety of shapes varying from nearly square to fan shaped, lozenge shaped, regular hexagonal, rectangular, and quite irregular.

The British army also established the so-called Victoria Line across the island. Finally, by the end of the 19th century, a degree of standardization was introduced in the ordnance of defense, and barbette batteries mounting breach-loading guns, singly or in pairs, firing smokeless powder, safer to load from concealed positions, and laid from a distance were installed in all the main batteries at the naval stations.

By then, the process of renewal had reached its peak with the introduction of these standardized breach-loading guns and supporting defenses in the form of mines, electric searchlights, and motor torpedo boats. In relation to the size of the places occupied, the expenditure by British standards was vast and, some might say, wasted, for Malta was not attacked—at least until 1940. The problem of usefulness versus high costs always remains an open discussion, of course. But in general, it was believed that fortifications were easier and cheaper to maintain than a large standing army. Besides, Malta and other stations like Gibraltar provided the Royal Navy with a safe refuge and anchorages in which it could refit, rest, and restock with provisions and refuel.

Fort San Lucian

Fort San Lucian, also known as Saint Lucian Tower or Fort Rohan, is a large watchtower located above the shore of Marsaxlokk Bay on the headland between Marsaxlokk and Birżebbuġa (Malta). The original tower was built by the Order of Saint John between 1610 and 1611. A semicircular artillery battery with an arrow-shaped blockhouse was added to the tower in 1715. It was upgraded into a fort between 1792 and 1795, when the tower and the battery were enclosed with a ditch. This was designed by the engineer Antoine Étienne de Tousard, and the stronghold was renamed Fort Rohan after the reigning grandmaster, Emmanuel de Rohan-Polduc. In the 1870s, the British rebuilt the fort in the polygonal style. On the seaward side, a low coastal battery was installed with three large casemates facing out across Marsaxlokk Bay toward Fort Delimara. The fort was decommissioned in 1885 but was used as a Royal Air Force bomb depot between World

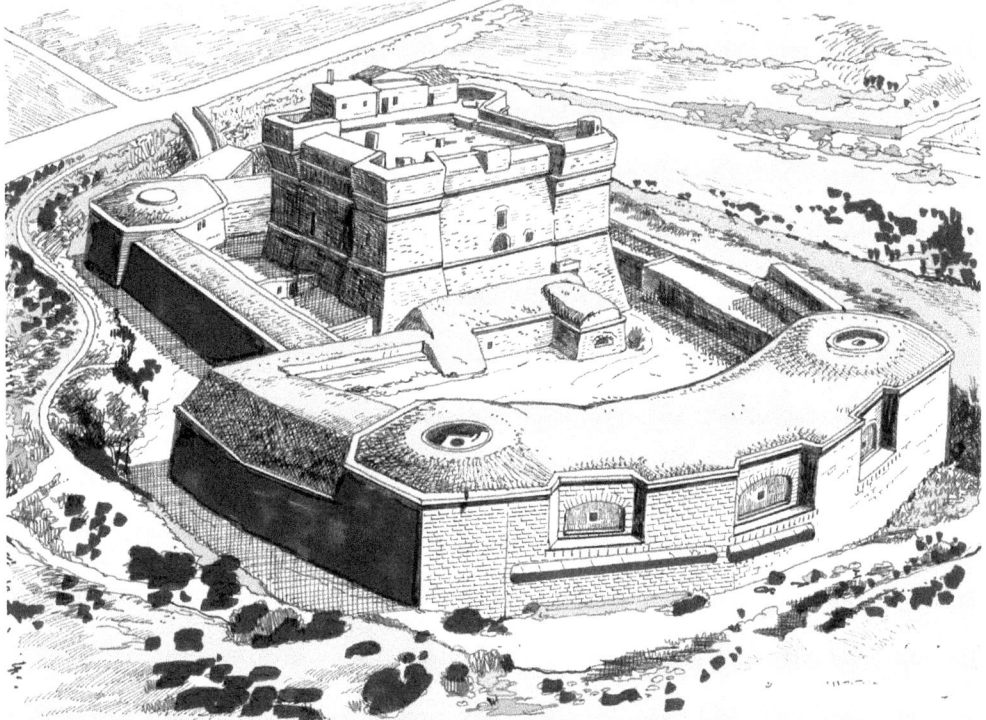

Fort San Lucian.

War II and the 1960s. It was handed to the government of Malta upon independence in 1964. Today, the tower and fort are used by the Malta Aquaculture Research Centre.

Victoria Line

The Victoria Line, named after the commemoration of Queen Victoria's Diamond Jubilee, was originally known as the North West Front and sometimes unofficially known as the Great Wall of Malta. Work on the Victoria Line began in 1875. It followed the Great Fault—a natural barrier of high grounds, which runs across the island in the east-west direction cutting off the northern half of the main island. Spanning 12 kilometers across the width of Malta and exploiting the defensive advantages of geography, it divided the bare north of the island from the more heavily populated south. Built in local limestone, the Victoria Line was completed in 1899. It provided an advanced defensive position for the dockyard and the capital city, Valletta. The line includes a ditch and a continuous infantry wall with riflemen firing slits. At regular intervals, the wall is reinforced with fortlets, defensive towers, flanking gun batteries, mortar emplacements, and forts—notably Binġemma (built in 1874), Madliena (1878), and Mosta (1878). The line remained in military use until 1907 and today is partially intact.

Fort Mosta

The construction of Fort Mosta commenced in 1878. The polygonal fort was designated as the central pivot of the land defenses of the North West Fault, which later

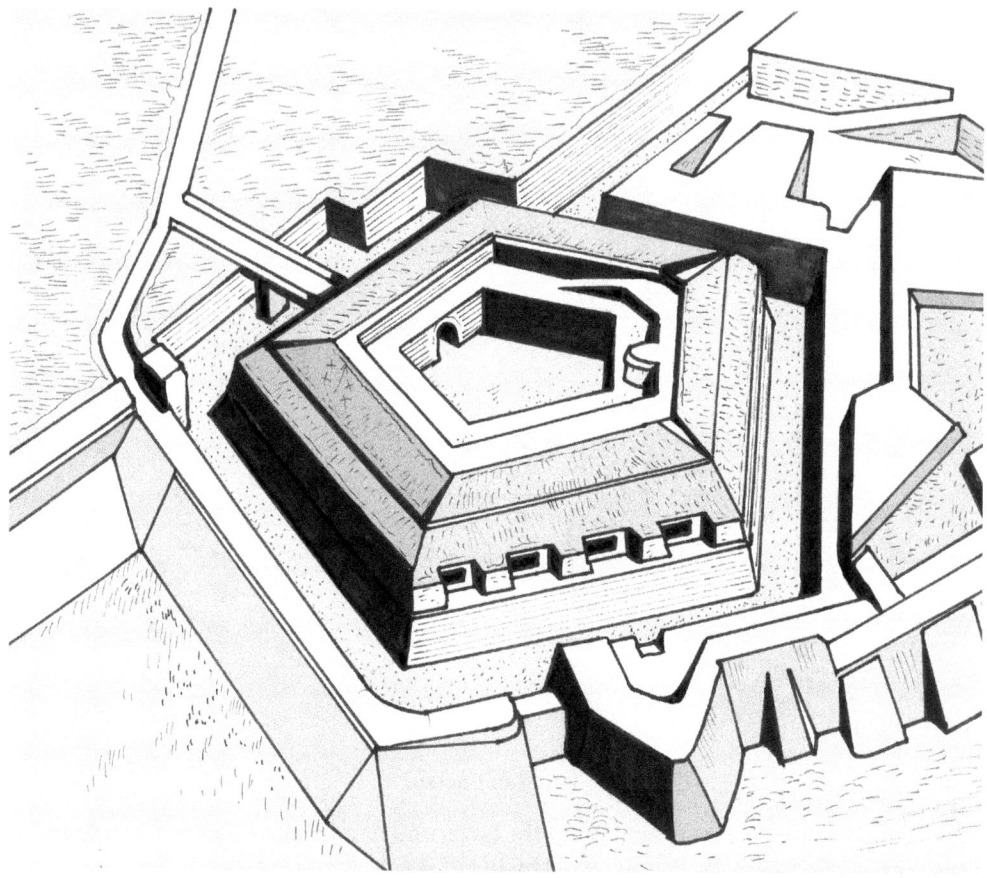

Fort Mosta, Malta.

developed into the Victoria Line. Fort Mosta comprised a squat pentagonal casemated keep with a central courtyard surrounded by shellproof store places and barracks that provided accommodation for the garrison. Around it, there was a narrow-hewn rock dug ditch with vertical walls defended by caponiers and counterscarp galleries. Beyond that, there was a fortified perimeter with open and casemated gun emplacements and howitzer platforms, as well as underground magazines and artillery crew shelters. In the 1880s, two large six-inch (100-millimeter) guns on hydropneumatic Elswick disappearing carriages were installed. Given the small size of the island, the heavy long-range artillery mounted in Fort Mosta was capable of firing at both sea and land.

Fort Leonardo

Fort Leonardo is located at Żabbar between the villages of San Leonardo and Żonqor above the shore east of Valletta's Grand Harbour. The coastal fort, built between 1872 and 1878 by the British, displays a triangular layout divided into an inner core forming one corner and a broad platform with gun emplacements facing the sea. Fort Leonardo remained in use by the British military until the 1970s.

Part 5. Fortifications in the British Empire in the 19th Century

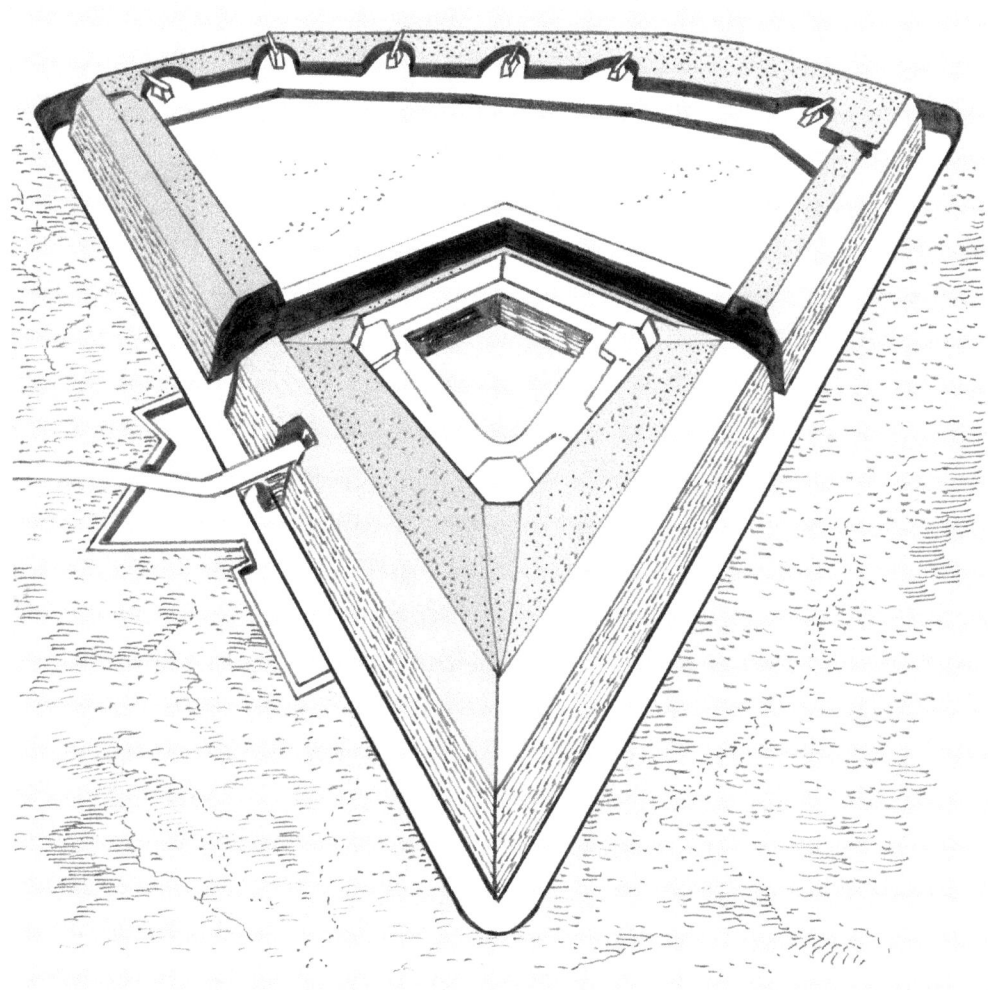

Fort Leonardo.

Fort Rinella

Located at Kalkara, Fort Rinella was built between 1878 and 1886 and formed a part of a series of four coastal batteries intended to defend the eastern entrance of Grand Harbour at Valletta. The battery had a polygonal layout with a narrow and deep dry ditch enfiladed by three caponiers and a counterscarp gallery. The fort included two levels: the lower one for the magazines, coal store, and the loading chambers; the upper one for the garrison accommodation and machinery chambers. The main armament (installed in 1882 on the top open platform) consisted of a huge 100-ton Armstrong gun mounted *en barbette* on a wrought-iron sliding carriage high enough to fire over the top of the parapet of its open platform. The rather monstruous gun was rifled. It was powered, traversed, depressed, and muzzle-loaded by a steam-generated hydraulic installation. The gun had an overall length of 32 feet and an unprecedent caliber of

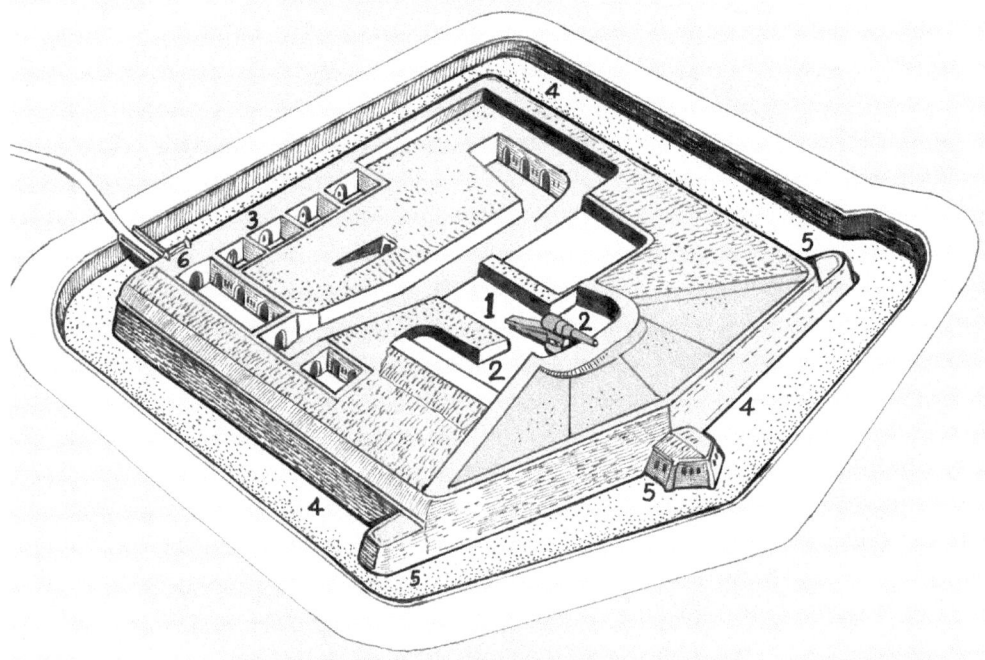

Fort Rinella. (1) 100-ton gun, (2) loading chambers, (3) underground barracks, (4) ditch, (5) caponier, (6) entrance at the rear.

17.72 inches. It could fire a ten-ton shell up to three miles and could penetrate 21 inches of wrought-iron armor. A similar battery was placed at Gibraltar. The enormous 100-ton guns were deemed necessary by that time because the Italian navy was equipped with all-metal ironclad warships armed with powerful guns. After 1906, the mighty 100-ton Armstrong guns were phased out, and Fort Rinella after 20 years of purely deterrent service was dismissed. Following decades of neglect, Fort Rinella has been restored and is now open to the public.

British Fortifications in India

By the early 19th century, India was effectively under British control, although there remained a patchwork of states, many nominally independent and governed by their own rulers, the maharajas (or similarly titled princes) and nawabs. While these "princely states" administered their own territories, a system of central government was developed. Trade and profit continued to be the main focus of British rule in India, resulting in wide exploitation and far-reaching changes. Iron and coal mining were developed, and tea, coffee, and cotton became key crops. A start was made on the vast rail network, and irrigation projects were undertaken. Just like in other colonies, important towns (e.g., Delhi, Calcutta, Madras) were fortified. All harbors (e.g., Surat, Bombay, Pondicherry), which were crucial for maintaining warships of the Royal Navy, were defended too with coastal batteries and forts and provided with facilities and installations in which ships could be refitted, repaired, restocked with provisions, and refueled.

Delhi

In 1803, during the Second Anglo-Maratha War, the forces of the British East India Company defeated the Maratha forces in the Battle of Delhi, ending the Maratha rule over the city. As a result, Delhi came under the control of the British East India Company. Delhi passed into the direct control of the British government in 1857 after the First War of Indian Independence. The city received significant damage during the 1857 siege. Afterward, the last titular Mogul emperor, Bahadur Shah Zafar II, was exiled to Rangoon, and the remaining Mogul territories were annexed as a part of British India.

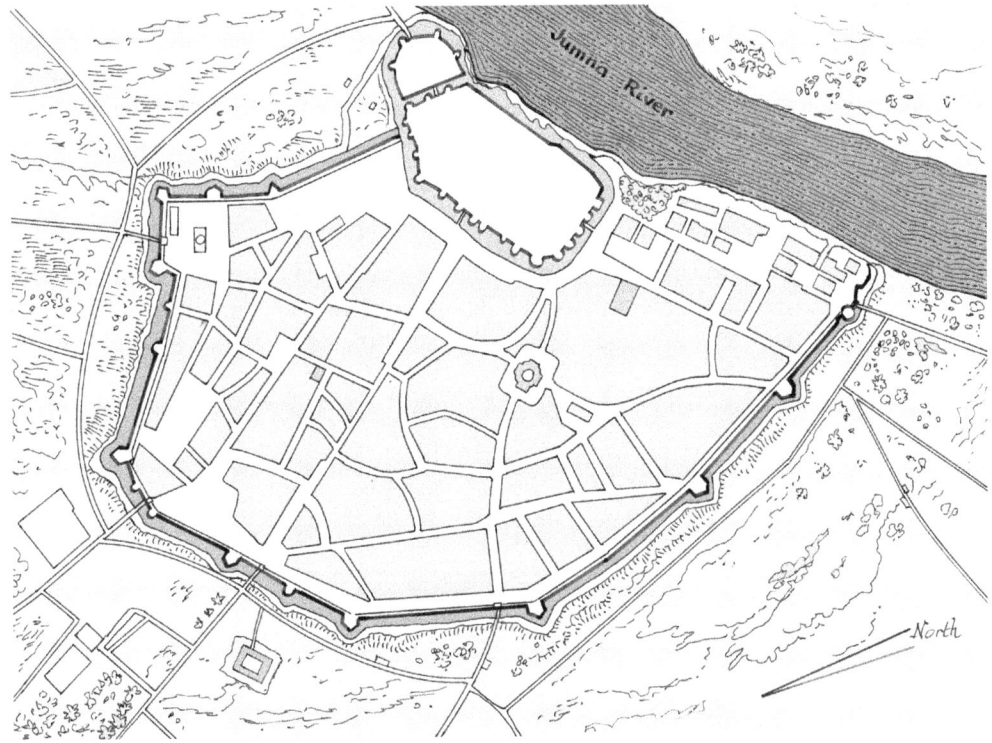

Delhi.

British Fortifications in New Zealand and Australia

Created in 1788, the establishment of New South Wales became the colony of New Zealand in its own right in 1841. New Zealand was rather isolated in the Pacific Ocean and thus far from most conflicts, but the islands' long coastline and dependence on seaborne trade had periodically spurred the British to fortify the islands' harbors. The introduction of steam-powered ships and the aftermath of the Crimean War (1853–1856) sped up the construction of coastal defenses in New Zealand. In the 1880s, Russia was threatening British interests in North India, and a military naval intervention in the South Pacific was not excluded. This threat led to the building of elaborate coastal forts. It was decided to construct fortifications and purchase gunboats, which would protect the harbors at Auckland, Wellington, Lyttelton, and Port Chalmers. The fortifications were built from British designs but adapted to New Zealand conditions. The first program of coastal fortifications included batteries (some with disappearing guns) and command posts. By 1885, work started in earnest on the construction of coastal forts defending the main harbors of Auckland.

The fort on Mount Victoria is the highest of Devonport's volcanic cones at 87 meters and was selected for the observation and control post for local coastal defenses. In 1885, four 64-pounder muzzle loading guns from HMS *Nelson* were emplaced on a terrace on the northern face of Mount Victoria, Takarunga. In 1898, it was decided to place one 8-inch Armstrong Elswick rifled breach-loading gun on a hydraulic-pneumatic disappearing carriage. Fort Takapuna, built between 1886 and 1889, was part of a chain of new defenses around Auckland Harbour. Other forts were built at North Head, Bastion Point, and Point Resolution (above the Parnell Baths). Fort Takapuna housed two 6-inch disappearing guns, which controlled the approaches to the Rangitoto Channel. These guns were mounted on two circular gun pits in the underground part of the fort.

Australia was also fortified notably with Fort Queenscliff, located on the Bellarine Peninsula in Victoria. The fort was constructed in the 1860s, and an open battery was established on Shortland's Bluff to defend the entrance to Port Phillip. The fort, which underwent major redevelopment in the late 1870s and 1880s, became the headquarters

Fort Queenscliff.

for an extensive chain of forts around Port Phillip Heads. Since 1982, Fort Queenscliff is a museum authorized by the Department of Defence.

British Fortifications in South Africa

Zulu War, 1879

The so-called Zulu War, fought in 1879, was caused by the invasion of the Zulu Kingdom by Britain. The British were interested in Zululand for several reasons, including their desire for the Zulu population to provide labor in the diamond fields of Southern Africa and their plan to create a South Africa federation under their control in the region (thereby submitting sovereign local African states). The offensive was more difficult than expected, though, as Zulu resistance proved remarkably strong. A final route at Ulundi in July 1879 marked the definitive British victory. During the Zulu campaign, the building of fortifications was one of the major Royal Engineers tasks. The structures included border posts (along the exposed Natal-Zulu frontier) as well as defended camps, fortified villages (aka laager) with defensive stone buildings for White civilian refugees, farmers, settlers, and traders. Other fortifications included wagon barricades, small earth and stockade entrenchments, temporary field fortifications and trenches, artillery field batteries, redoubts, defended supply depots, and camps like Fort Helpmekaar, Fort Bengough, Fort Pinetown, and Fort Eshowe. Fort Tenedos, for instance, was a typical example of field fortification used during that campaign. Built in early

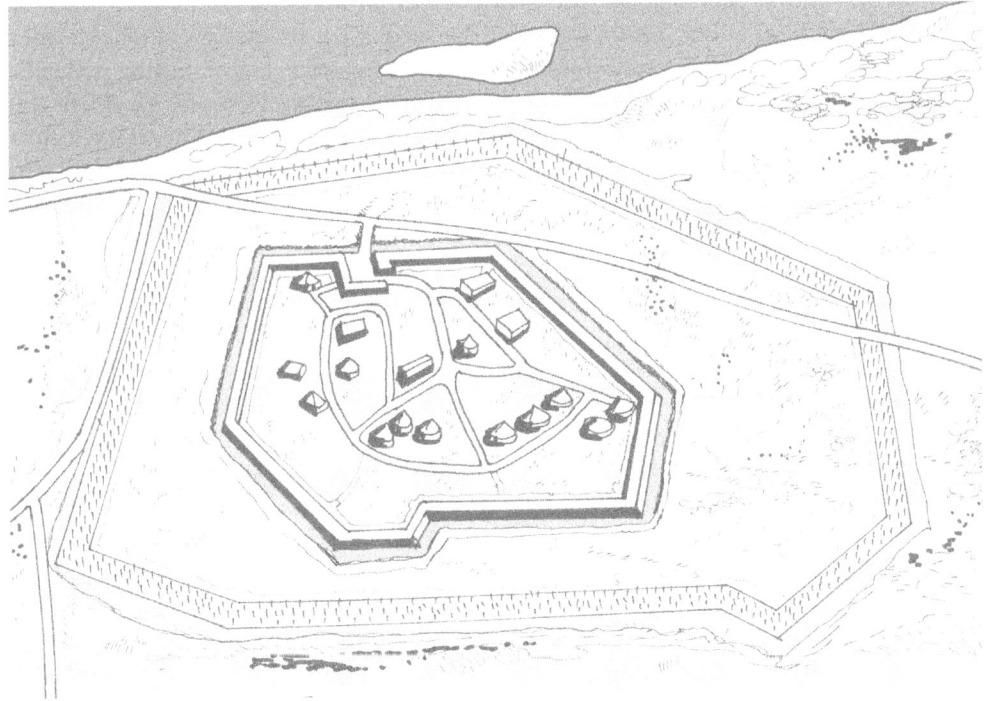

Fort Tenedos.

1879 by Captain Warren Wynne, it was established on the bank of the river Tugela. The fort included simple earth ramparts (walls and a ditch) with tents accommodating troops. It was surrounded by entanglements of barbed wires and man-traps called trous-de-loup (aka wolf pits), which were excavations about three feet square and three feet deep with sharpened stakes at the bottom. On a steep hill on the opposite bank of the river, there was another kind of simple redoubt designated Fort Pearson. Both forts connected by ferry and served as supply stores and anchors for British offensive operations in Zululand.

British Fortifications in the Boer Wars

Two armed conflicts (known as Boer Wars) were fought between the British Empire and the two independent Boer republics, the Orange Free State and the South African Republic (Transvaal Republic), founded by Dutch settlers known as Boeren ("peasants" in Dutch). The First Boer War (1880–1881), also known as the Transvaal War, was a relatively brief operation in which Dutch Boer settlers successfully resisted a British attempt to annex the Transvaal.

The Second Boer War (1899–1902), by contrast, was a conflict involving large numbers of troops. Unlike many colonial conflicts, the Second Boer War lasted three years and was very bloody. The weaker Boers often used guerrilla tactics involving the protracted harassment of British troops and sabotaging lines of communication in order to deprive them of essential supplies. To guard against Boer raiders, entrenched posts were placed at every strategically important place, and constant patrolling was maintained between these posts. Blockhouses (fortified small forts) were erected along the railway lines at distances of about 2,000 yards. For the most part, the design of blockhouses was standard consisting of a rectangular masonry three-story-high building with rifle holes and two small flanking turrets placed at opposite corners. Designed by chief royal engineer Major General Sir Elliot Wood (1844–1931), these stone blockhouses were quite expensive and time consuming to build, so another model was designed by Major General Sir Spring Robert Rice (1858–1929) in 1901. This alternative standardized new fortlet was a circular blockhouse that consisted of a mass-produced circular drum made of corrugated iron with a sloped roof and a base reinforced with stones or sandbags. Blockhouses designed by Wood and Rice were standard models, but other designs were used. Some resembled large granaries, others modern-day electricity "substations," while others appeared as fanciful little castles. Sometimes blockhouses where established by using existing structures along the railroad tracks. Although most blockhouses were eventually dismantled, a number still remain in silent testimony of the bitter Boer War.

All of the 8,000 blockhouses built during the South African War were intended to do the same job: keep the ever-mobile mounted Boers from accessing British supply lines. Therefore, most of the blockhouses were erected near essential railroad routes to protect vital supply lines.

Each blockhouse had a garrison of about ten men, and each was surrounded by barbed-wire entanglements, which, together with various kinds of alarm fences, were also placed between the blockhouses themselves in order both to impede the approach of the enemy and to warn the garrison thereof.

The British also resorted to the harsh practice of "pacification" and ruthless

British blockhouse designed by General Wood.

anti-insurgency tactics with inhumane treatment of elusive South African Boers using a new system: the concentration camps. This policy consisted of burning farms and crops, killing cattle, and rounding up and isolating the Boer civilian population into camps, in fact holding the whole civilian community as hostage. The wives and children of Boer guerrillas were sent to these camps with poor hygiene and too little food, resulting in high mortality rates. The death and suffering of innocent civilians finally broke the Boer guerrillas' will and forced them to surrender. Peace was signed in May 1902 with the conversion of the Boer republics into British colonies (with a promise of limited self-government). These colonies later formed part of the Union of South Africa.

PART 6

British Fortifications in World War I

The First Phase of the War

By the turn of the 20th century, the winds of war were sweeping through Europe again, and Britain deemed it prudent to prepare for the protection of homeland and empire. Thus, the Committee of Imperial Defence was created in 1904 and given the task of advising the prime minister on the best means and methods of defense. Nonetheless, little changed in Britain's defensive policy, and the maintenance of sea supremacy by the Royal Navy remained the basis for the system of defense.

On July 25, 1909, the Frenchman Louis Blériot (1872–1936) made a spectacular flight across the English Channel in a heavier-than-air monoplane that he had himself designed and built. Blériot made history, and his spectacular 33.3-kilometer (20.7-mile) flight from Calais to Dover (lasting 36 minutes) made a sensation and attracted much public attention. It also focused the minds of the military on the potential of aircraft as a future weapon of war. The advent of airplanes was followed by astonishing, remarkable, and quick development that brought a new, disturbing, and dangerous element to the defenses of Britain. Henceforth, attacks could come not only from the sea but from the air, too.

In August 1914, the fateful interplay of alliances brought all the leading European powers into war. All nations quickly mobilized their forces, and everywhere the departure of troops to the front was accompanied with scenes of enthusiastic joy and a mood of patriotic exaltation. Military "experts" prophesied that this war would last only a few months. Nobody realized that it would not end until four ghastly years later.

At first, it was a war of movement with large offensives and counterattacks in Belgium and northern France.

During that period, defensive fortified positions were established on the English coasts. They included artillery batteries equipped with modern weapons as large as the 9.2-inch guns, emplacements for land-launched torpedoes, infantry positions, and barbed-wire obstacles near potential landing sites. The other key British positions in Europe requiring special attention included Gibraltar and the island of Malta—two important outposts guarding the route to the Suez Canal, which connected the Mediterranean Sea to the Red Sea and the Indian Ocean and the important British colony of India.

Unexpectedly, as all attacks foundered, the western fronts stabilized in an enormous arc from southern Belgium near the North Sea to the Vosges and the Alps near the

Swiss border. After four months of bitter fighting, losses were staggering. Confidence in a quick victory evaporated, and the war grounded to a stalemate. The opposing armies dug themselves in for the longest and bloodiest war of attrition of history, a gigantic duel, and a trial of each side's endurance and resources.

Trench Warfare

Field Fortification

At the end of 1914, it was clear that a profound change had come over the situation on the western front. In long lines of field fortifications extending from the North Sea to the Vosges, opposing armies watched and fought, unable to make more than the slightest dint in the rigid contours of the front.

Field fortifications are defensive earthworks thrown up at short notice in wartime in order to defend previously neglected sites, to secure territory captured from the enemy, to hold a position temporarily before resuming advance, or to give temporary shelter to troops in campaign or on the battlefield. Field and siege works had, whenever possible, always adhered to the basic tenets of permanent fortifications, being carefully (be it sometimes hastily) erected with an eye to enfilade fire, flanking, protection, arc of fire, and to all the other factors governing the sophisticated permanent fortress design. By the end of the 19th century, field fortification had become a precisely defined, codified, and standardized army activity performed by the troops under supervision of specially trained engineering corps. Each army developed techniques and issued practical manuals. Field works began with individual foxholes, shelter pits, and small excavations dug by soldiers who threw earth in front of them to make a slight protective screen. The connection of individual pits moved to deeper and larger rifle positions for a few soldiers with fire steps and parapets (breastworks). Finally, collective and continuous trenches were established, whose depth, strength, and length were graded according to time available for excavation. These were completed with the establishment of loopholes, obstacles, and a variety of other devises intended to strengthen the vicinity of the position.

World War I Trench Warfare

Trench warfare was not new, as it had become a necessity since the introduction of firearms in the Renaissance, but the field fortification systems established during World War I were the most elaborate in history.

At first trenches were primitive excavations, which protected only against low-velocity small arms because military authorities on all sides expected resuming the war of movement at any moment. When it became evident that this was to be the battle line and when heavy artillery began to be used, some improvements began to be made, and troops disappeared below ground. The first requirement was to deepen and align trenches safely. Indeed, a simple straight excavation was a dangerous place, since one enemy at one end could kill every occupant for a very long way by enfilading fire. Trenches were therefore given a zigzag or meandering layout, they were made narrow and bent back and forth, and they were compartmented with traverses (thick wall established across trenches), which cut the position into short sections and so confined shell bursts and enfilading fire to short and irregular lengths known as fire bays. Trenches were usually fitted with a banquette (fire step), a small platform allowing soldiers armed with rifles to see and fire above the top of the parapet (so-called barbette fire). Screens of sandbags were raised on the breastwork in order to protect soldiers against small arms projectiles as well as grenade and shell splinters. To allow a soldier to see out of the trench without exposing his head, loopholes could be built into the parapet. A loophole might simply be a gap in the sandbags, or it might be fitted with a steel plate. Another means to see over the parapet was the trench periscope—in its simplest form, just a stick with two angled pieces of mirror at the top and bottom.

At intervals, dugouts were scooped out of the back of the trench to provide the crudest living accommodations. Front combat trenches had to be relieved at intervals by fresh troops, the wounded had to be evacuated, and food, supply, and ammunitions had to be safely brought to the front. So the next step was to connect the front to the rear. For this purpose, communication saps were dug perpendicularly from the front trenches in the direction of sheltered spots where they could safely debouch into open country at rear points invisible to the enemy. Communication saps, too, were narrow, deep, and zigzag shaped in order to prevent enfilading fire. Finally, as it appeared that the tenancy of trenches would become a long-term affair, some facilities for the occupants were provided; deeper underground dugouts and larger supply shelters were excavated with ample earth cover. They were gradually lined with board and strengthened with planks, beams, and courses of sandbags. Inside these crude chambers, men and equipment would be protected against artillery bombardments.

So quite amazingly, spade-excavated trenches and dugouts, strung with barbed wire, defended by machine guns, and backed by massed field artillery, soon proved as impregnable as any fortress of steel and concrete.

Needless to say, the changing seasonal climate and the various nature of the ground brought extreme difficulties. In many sectors (particularly in Flanders where the water table was a constant problem, or elsewhere because of rainfall), it was necessary to provide pumps and drainage systems to keep the water at bay, particularly after heavy artillery shelling had ruined the natural drainage of the area. In some extremely wet ground, entrenching was not totally possible, and the problem of water was so serious that it could only be countered by building upward, by raising high and thick parapets. On the contrary, in Picardy, very neatly designed trenches could be cut to any depth in the solid chalky ground. In many places, soft ground meant that excavations were bound to collapse. In this case, it was necessary to support and revet their walls with corrugated iron, expanded metal (XPM, formed by cutting slits into sheets of metal and then stretching them), woven wood hurdles, piled-up sandbags, planks and beams, and any other materials available. The floor of most trenches was lined with planking (aka duckboards) that allowed soldiers to move more easily in the mud. The proliferation and increased power of mortar and artillery fire rendered necessary stronger and deeper overhead protection for dugouts and shelters. Soon, wooden or steel beams, armored plates, and later concrete lining had to be employed for ceilings and corridors. Where dead angles could occur ("blind spots" below and beyond which a section of ground could not be seen and thus could not be fired at), semipermanent observatories and blockhouses (aka pillboxes) armed with machine guns with overlapping fields of fire and mortar nests were installed.

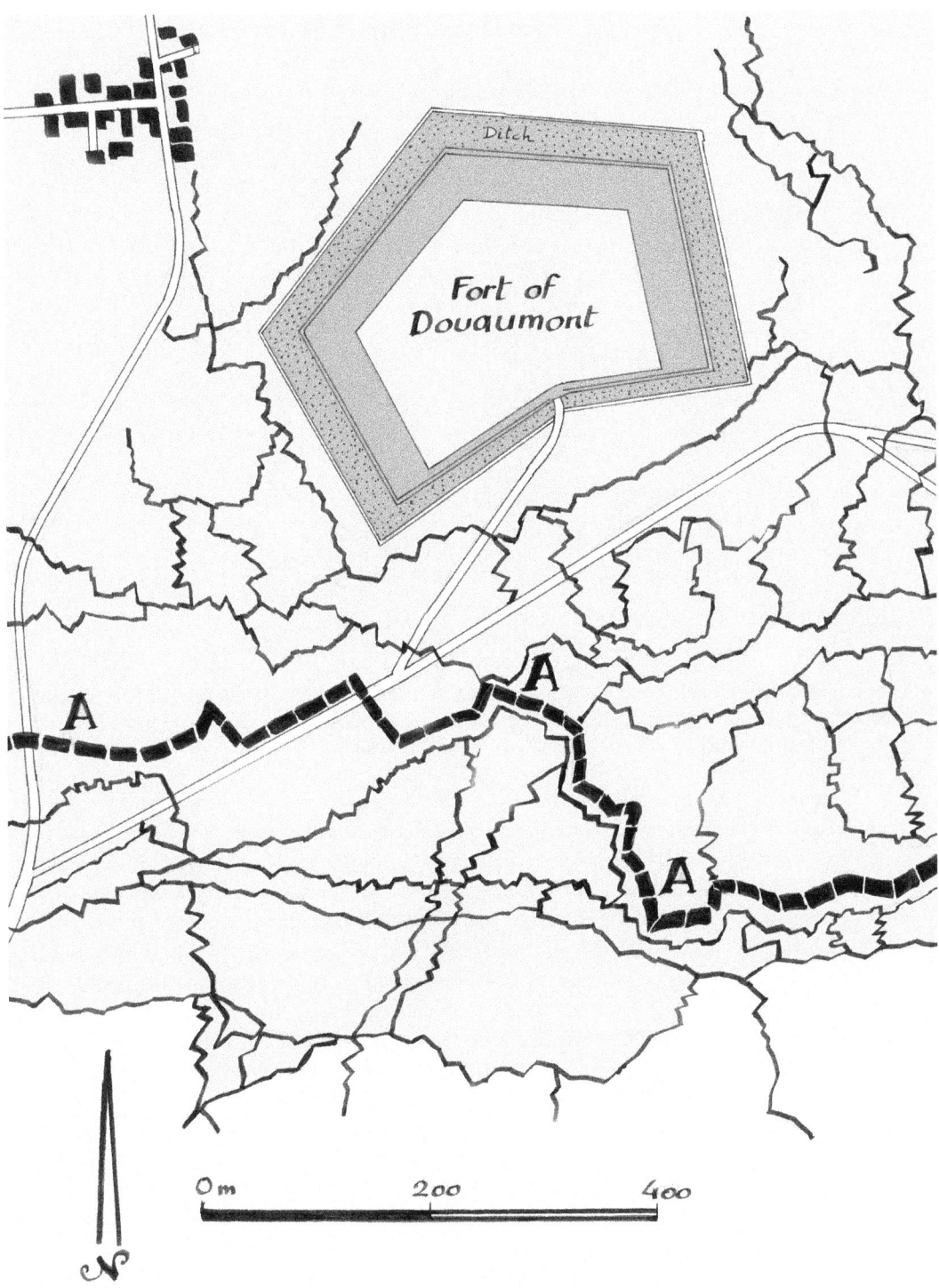

Trench systems around the French fort of Douaumont near Verdun. The thick black hachured line AA represents the front line.

No Man's Land and Barbed Wires

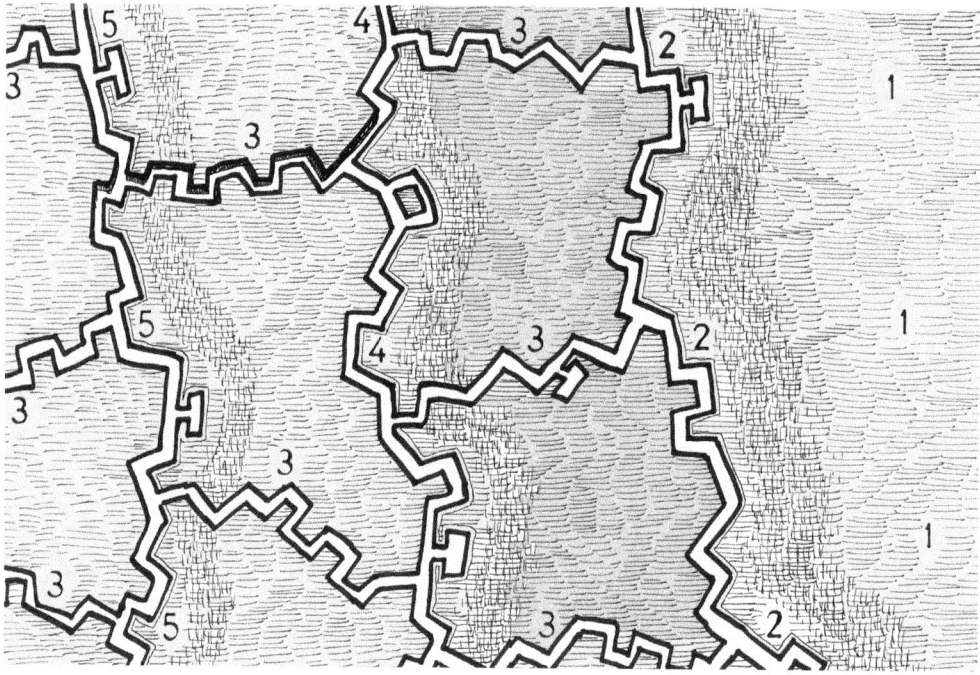

Trench network, World War I. Trench warfare was a combination of field fortifications, reinforced by continuous barbed wire entanglements. (1) No man's land, (2) front trench line, (3) communication saps, (4) support trench, (5) bombing trench.

The area separating the two opposing front lines was known as no man's land. To keep the enemy at bay, heavy entanglements of barbed wires were planted in front of the fore trench lines. Barbed wires were invented and patented in 1874 by the American Joseph Glidden, originally as a harmless farm-fencing device. It was first adapted to military purpose by the end of the 19th century, but its full significance was not appreciated until the Boer Wars in South Africa and World War I. Barbed wires are artificial metal thorns made of sharpened wire barbs twisted around a central strand. They could/can be used in different ways, and it should be emphasized that a wire entanglement was not simply a few strands of the type of wire one sees today around a meadow. Barbed wires were stretched on iron pickets (1–1.5 meters high, firmly hammered in the ground). Several lines were generally established and these were made impenetrable. Between rows, other wires in the form of coiled rows could be set making an entangled impassable network about 1.5–1.8 meters high (5 or 6 feet) and 3–6 meters thick (10–29 feet). Intended to stop or hinder infantry progression or channel attack to defended areas, barbed wire fully exploited the lethality of automatic weapons by holding their targets in the line of fire and by keeping them away from the gun and its crew.

Obviously, installing and repairing the wires was a very dangerous business, as any activity or movement in no man's land was likely to attract fire from snipers or a burst of machine gun fire. Activities such as work on the wires, patrol, and search for dead bodies were thus generally done at night. In World War I, screw pickets were used for the installation of wire obstacles; these were metal rods with eyelets for holding strands

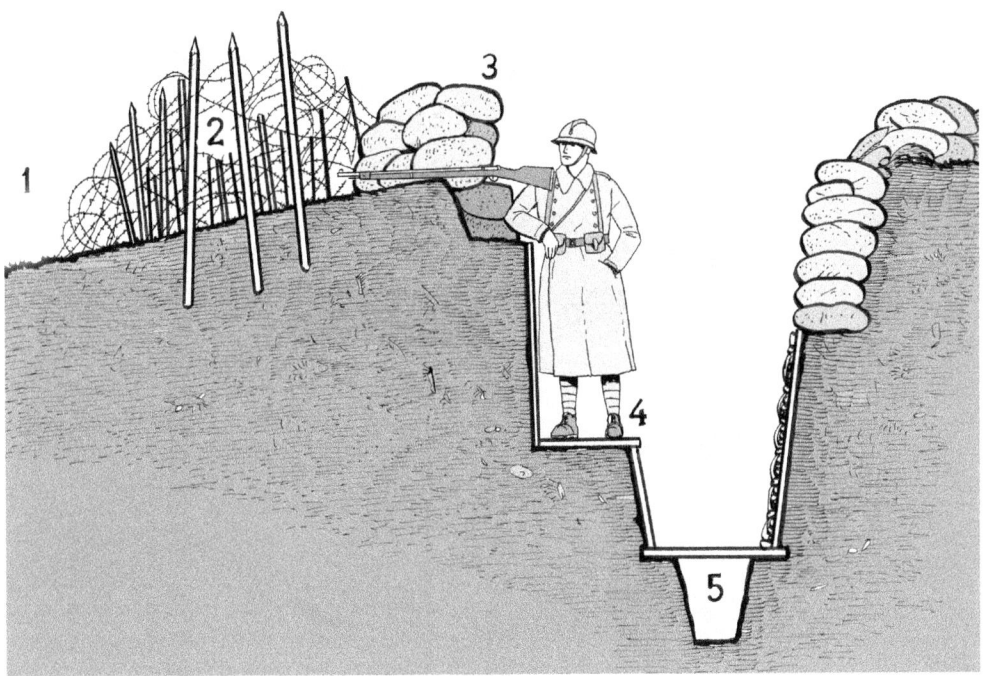

Cross-section trench, World War I. (1) No man's land, (2) barbed wires, (3) breastwork made of piled sandbags, (4) firing step, (5) drainage system.

of wire and a corkscrew-like end that could literally be screwed into the ground rather than hammered so that wiring parties could work at night and not reveal their position to enemy soldiers by the sound of hammers.

Here and there, gaps (covered by machine gun fire) were established at convenient distance to allow raiding and scouting parties to sally forth in no man's land and return. Barbed wires could be crushed and destroyed by heavy artillery bombardment or discreetly breached by a party using wire cutters being sent out at night to carve breaches so that an attacking force could pass through.

Trench Lines

As the war dragged on, the defense in depth became of great importance. From a simple beginning, the trench systems proliferated into labyrinths of incredible complexity, making sound and solid front lines on both sides. Behind the first delaying trench and outposts, some 100 or so yards, came a support trench, often so sited that the occupants could bring fire to bear over the heads of the soldiers of the fore trench and thus add firepower to break up any assault. Soon, a third line was dug, about 20 yards behind the support line, called bombing trench. Should the first trenches be conquered, counterattack parties deployed in the third trench could throw grenades at the intruders and retake the lost positions. All these combat trenches were, of course, linked together by zigzagging saps or sinuous communications trenches provided with recesses, first aid stations, dugouts, and underground chambers for troops ready for instant call as

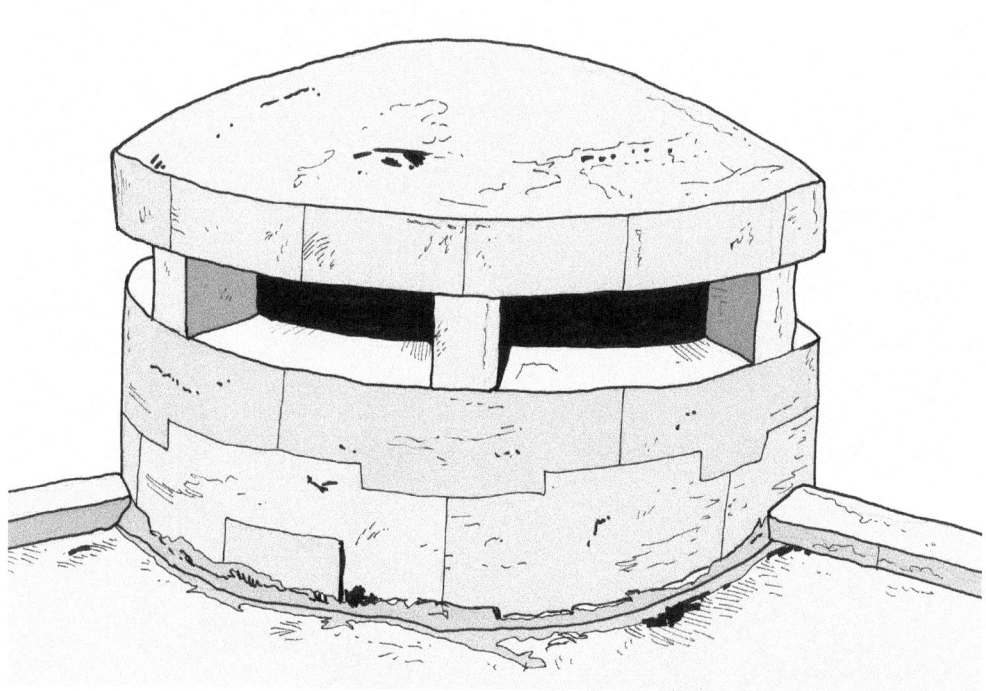

Moir pillbox.

reinforcement. Interval defenses included concrete pillboxes and small fortlets in the form of T-head or D-head branches armed with machine guns with calculated and mutually supporting angles of fire to give all-round defense should enemy troops break through the front line. For this purpose, the British designed a small pillbox. Patented by Sir Ernest Moir, minister of armaments, this was a small World War I British prefab cylindrical design based on the use of interlocking concrete blocks for extra strength. The pillbox was armed with a Vickers heavy machine gun served by two soldiers. It had a diameter of about 220 centimeters, and the concrete blocks were 20 centimeters thick. Its roof included a domed armored plate covered with a layer of concrete. The pillbox was accessible via an underground passage. It was designed for quick construction and enabled troops to establish firing strongpoints. It had an ingenious rotating steel shutter through which the machine gun could be fired, but its main disadvantage was the absence of a system to ventilate the toxic gases caused by firing the weapon. About 1,500 units seem to have been built in 1918 in Belgium and northern France. Remains can be seen at Yper in Belgium.

The patterns of trench lines could be repeated if needed to extremely complex systems covering a large area that allowed defense in depth. In some areas, the defensive zone stretched up to 6.5 kilometers (4 miles) back from the first sentry. At the rear of the complex, there were local command posts with observatories and telephone, ammunitions and supply dumps, field kitchens, field hospitals, quarters for reinforcement troops, and other facilities, many buried underground. Farther back were the heavy support artillery batteries with their command posts and fire control stations, and even farther back was regional headquarters. Of course, wherever possible, trench lines followed natural obstacles offered by the terrain—e.g., impassable woodlands, steep hills, wide rivers, canals, and swamps.

Daily Life in the Trenches

Life at the front was not always hectic with constant shelling and offensive and counterattacks. Going over the top as part of a raiding party or to take part in a major frontal offensive did not happen every day, fortunately. The amount of military activity varied a lot, and there were times when combatants went for weeks or even months without seeing any action at all. The soldiers' time was divided into watches, stand-by, rest, and duties (e.g., weapons inspections, bringing supplies, digging latrines, laying barbed wires). During the day, snipers and observers made movement perilous, so the trenches were mostly quiet. Consequently, trenches were busiest at night, when cover of darkness allowed the movement of troops and supplies, the maintenance and expansion of the barbed wire and trench system, patrolling no man's land, and reconnaissance of the enemy's defenses. Sentries in listening posts out in no man's land would try to detect enemy patrols and working parties or indications that an attack was being prepared. Much time was spent repairing or making new trenches and dugouts. Life was a struggle to dry, drain, and dig what the weather and enemy bombardments were continually destroying. Even at rest, soldiers always had to be vigilant as one never knew what the enemy was up to. Daily life (or better said, survival) in the trenches was both dangerous and boring and often grim, miserable, and always uncomfortable and precarious, with few moments of humor and comradeship, as well as periods of tragedy, weariness, discouragement, and despair. Fresh water, both for drinking and washing, was brought from the rear and always in short supply, so maintaining personal cleanliness was a difficult if not impossible affair. Unshaved faces gave the nickname of *Poilus* ("hairy ones") for French frontline combatants. Lice and fleas infested men's hair and clothing. Rats and other vermin, thriving on the scraps of food, litter, excrements, and the remains of decomposing bodies, contaminated stocks of food and spread disease. Wine and liquor were officially regulated; in fact, they were freely available, and drunkenness was often a serious problem. Food, too, was brought from field kitchens at the rear, and when it reached the front, it was often cold. Quantity and quality of the daily ration depended on circumstances. For four years, millions of men were submitted to seasonal changes in open country. In autumn and spring, rainfall was at its highest, transforming trenches and dugouts into morasses of clinging mud, and many men suffered from a form of foot rot known as trench foot. Winter in northeastern France can be extremely cold and wet with periods of snow and frost. Trench soldiers were thus exposed to serious frostbite that could lead to gangrene and result in amputation of the affected limbs. Summer can be very hot, and thirst, discomfort, and swarms of flies were other scourges. Spring and summer were also the dreaded periods of large-scale attacks, which decimated whole units. A remarkable and disgraceful fact is that a high proportion of senior officers were totally ignorant of the conditions in which front soldiers were surviving and fighting.

Morale in frontline armies varied a lot. It depended on many factors such as local conditions, quality of the leadership, the level of esprit de corps, intensity of combat and casualty level, the progress of the war, and most important, news and packages from home. Discipline was harsh. Lesser misdeeds (drunkenness, falling asleep on duty, petty theft, insult or light violence against comrades or officers) were punished on the spot, but serious offenses (cowardice, desertion, striking or murder of comrades or officers, insubordination, and mutiny) were dealt by merciless martial courts.

A sector of the front would be allocated to an army corps, usually comprising three divisions. Two divisions would occupy adjacent sections of the front and the third would be at rest to the rear. This breakdown of duty would continue down through the army structure so that within each frontline division, typically comprising three infantry brigades (regiments for the Germans), two brigades would occupy the front and the third would be in reserve. So soldiers generally spent a week on the front line, a week in reserve, and various lengths of time behind the front for rest, where they could at last wash, shave, change clothing, and sleep before searching relaxation. But it was not exceptional that due to circumstances, they had to stay in combat positions for a month or more without relief. It could also happen that soldiers were taken at the rear, not for relaxation but for drills and additional military training. Even so, the tedious, brutal, and harsh trench life, the forced sharing of poor sanitary conditions, boredom, loneliness, homesickness, acute suffering and constant fear, intense danger, and petty pleasures were often occasions of forging a sense of strong comradeship—a feeling of fervent and long-lasting brotherhood that stayed alive in postwar veteran associations.

As the conflict progressed and trench warfare conditions set in, the patriotic exhortations and the enthusiastic spirit of the early stage of the war no longer echoed the feelings of most of the serving troops. Flag waving became increasingly out of touch with the mood and motivation of the frontline soldiers, but amazingly, soldiers' morale did not collapse. Despite enormous losses, miserable conditions of survival, and atrocious fighting, the spirit dwindled and broke down in some units for only a short period at the time of the 1917 mutinies, but on the whole, morale and confidence in victory remained incredibly high.

Tactics and Weapons

New Weapons

Fought by highly industrialized and rich nations, World War I saw the improvement and perfection of previous army equipment and armament and the introduction of new ones. The specificity and requirements of trench warfare caused the appearance of new tactical weapons—some revolutionary—all in tremendous increase of number and lethality.

Science, metallurgy, and industry became tightly intertwined with the technology of war. The civilian was now responsible for providing the industrial means of war, and workshops and manufactures became as vital parts of the struggle as the battlefield. Workers and combatants were now fundamentally dependent on each other. Science and industry put into the hands of top military authorities of each great nation involved in the war machine of far greater power than any known before. Ironically, none of the improvements or special or revolutionary weapons were successful in bringing to an end the stalemate of the western front with all its useless slaughter. The men who led the armies of all World War I belligerents were unable to cope with a war, which produced such tactical innovations. Matters were simply beyond their competence. They underestimated the resilience of their opponents and failed to come to terms with the realities of siege warfare in which frontal attack was no longer appropriate. They had to deal with problems for which their experience and training had not equipped them, and few military leaders displayed much imagination in tackling these new issues. Their narrow strategic doctrine knew no horizons beyond the immediate needs of the battle. Their stupidity, apathy, and stubbornness failed to develop effective tactics, and they persisted in their errors long after soldiers and civilians had been aware of their futility. The French hero and commander in chief General Ferdinand Foch, supposedly one of the more enlightened of military leaders, is reported to have said, "*Aviation is good sport for civilians, but for the army it is useless*," epitomizing the prevailing attitude to an innovation that was to play a leading role in warfare. "Brass Hat" (senior officer) was not a term of respect in Great Britain. The rate of technological advance—exceeding military authorities' ability to understand it—resulted in enormous butcheries on the battlefields. All this further points to the senselessness and futility that seems so much to characterize World War I.

The dreadful trench warfare lasted three years before mobility was restored to the battlefield in the spring of 1918. This was eventually caused by the following main tactical factors: the employment of new weapons (guns and mortars, machine guns, flame throwers, hand grenades, poisonous gas); the introduction of armored tracked combat vehicles (aka tanks); the adoption of smaller but numerous infiltrating offensives instead of massing frontal charge in one "big push"; the development of assault tactics attempting to outflank enemy position; and the constitution of ample reserve of men and matériel, standing by for exploiting advance and breakthrough without delay and able to switch from one sector to another.

Mining

Mining was intensively used in trench warfare and used as an attack weapon. This was an adaptation of an old siege technique employed in previous centuries. An underground tunnel was driven toward the enemy position, explosives placed under it, and the gallery was blocked with earth to contain the blast of the explosion. At a suitable moment (in coordination with an infantry assault), the charge would be exploded, thus destroying the trenches above it. For example, a spectacular mine attack was use at Messines Ridge in West Flanders, Belgium, in June 1917, when 19 deep mines had been dug more than 100 feet below the surface and filled with no less than 1 million pounds of explosives by the British engineering corps. After they exploded, troops merely walked in to take possession of a huge smoking crater.

The success of mining depended on hard labor, patience, and secrecy, and opposants did not wait idle. Special *écoutes* were established. These were listening posts manned by alert and sharp-eared engineers equipped with geophones (detecting devices that recorded vibrations in the earth). When enemy mining activity was revealed and located, a countermeasure was taken in the form of a countermine, a subterranean gallery cut in the direction of the approaching attacking mine. The countermine tunnel was then packed with explosives and fired while the enemy miners were at work in hopes of blowing them and their mine. It could also happen that both mine and countermine struck each other, either inadvertently or on purpose. Then attackers and defenders fought a terrifying and creepy hand-to-hand battle in subterranean darkness. The grimness, danger, and immense labor entailed in such underground warfare can easily be imagined.

Aviation

Despite Marshal Foch's opinion about aviation, airplanes started to play an important role. When World War I broke out, most belligerents had a number of primitive and fragile aircrafts used for reconnaissance. Originally, planes had low performances, haphazard design, and flimsy structures, and for one airplane to shoot down another was unknown in 1914. Airplanes were considered to be the eyes of the ground forces and were confined to the reconnaissance and observation role. A technical race soon started for air supremacy, and each country sought for superiority in speed, ceiling and climb rate, range, strength, versatility, firepower, and maneuverability. The results were fine and reliable flying machines, which were larger, had greater speed, and were armed with machine guns. Soon, tacticians used aircrafts in a more daring role, and the value of harassing by means of bombing from aircrafts was increasingly exploited. At first, the Germans used zeppelins (huge hydrogen-filled, rigid, dirigible flying ships developed by Count Ferdinand von Zeppelin) for bombing missions. The hopelessly vulnerable airships were gradually replaced by larger bomber airplanes with increasing weight-carrying capacity, e.g., German Gotha and British Handley-Page. Although both civilian and military casualties due to aerial bombing were relatively light, these terrifying manifestations of a wholly new type of warfare made a deep impression on public opinion and caused a flurry of alarm. Their military impact was minor, even trivial compared to World War II destructions and casualties caused by large-scale strategic bombing. But air attacks had a corrosive effect on the morale of the civilians who experienced the terror of being struck in the heart of their supposedly safe cities far behind the front line.

By 1916, airplanes began to participate in the land battle by providing close support to ground forces, but the full impact and the revolutionary potential of air warfare, however, came a generation later.

Tanks

The weapon that finally helped break barbed wires and defensive machine gun fire was the tank. Developed in England and France more or less concurrently and with little official support, the tank was an armored tracked combat vehicle designed for the specific purpose of making a path by force for infantry frontal attack against entrenched positions. Employed as an instrument of rupture, they were armored pillboxes on caterpillar designed to resist machine gun fire, roll over difficult terrain, smash barbed wires, cross trenches, and support the advancing infantry with their weapons. Tanks—regarded by many senior officers as only useless gadgets—were first introduced into battle in September 1916, with mitigated result, as many of the new combat vehicles suffered mechanical breakdown. The most spectacular penetration through German lines (six miles) was achieved at Cambrai in November 1917, when a large number of tanks were engaged in a surprise assault without preliminary artillery bombardment. The Cambrai tank mass attack was a major tactical landmark, but military authorities remained skeptical. They failed to appreciate the lesson and did not fully understand the revolutionary tactical value of armored fighting vehicles. Because of inadequate quantities, failure to adopt appropriate tactics for their use, and the lack of sufficient reserves to exploit transitory advantages when the weapons were introduced, the tanks of 1918 did not bring an end to the war. It must be added that in spite of their terrifying aspect, they could not be decisive war machines because of technical imperfections, slow speed, poor mobility, short range, insufficient armor, and poor armament. Besides, the environment inside the crammed vehicle was extremely unpleasant and unhealthy. There was terrible noise inside, and as ventilation was inadequate, the atmosphere was soon poisoned with fuel and oil vapors, carbon monoxide from the engine, and other toxic cordite fumes from firing the weapons. Temperatures inside the tank could reach 50°C (122°F). It was so hot and noisy that crews became stunned after a few hours inside. The atmosphere was so unhealthy inside the vehicle that some crews could lose consciousness. As with the airplane, the early tank of 1917–1918 still had far to go before reaching its tremendous potential. Tanks and planes played only a minor part in the tactics of World War I land warfare, but they were weapons to be reckoned with in the future. They became battle-winning machines later in World War II.

Logistics

World War I frontline landscape.

Industry and technology also provided the means for new and improved logistics systems that not only supplied weapons and ammunitions but also distributed to the millions of front soldiers food, clothing, and a wide range of other goods and services (notably medical attention and evacuation of the wounded). Military railroad made possible to provide huge armies with enormous supplies for long periods of time, bringing everything needed to broad fronts from great distances. In operational zones, horse transport was still standard both for infantry and artillery. Horse-drawn wagons and carts played a central role for supplying distribution points. However and although it was at first a capricious and unreliable device, the internal combustion engine began to appear. Soon, trucks and cars were powered by this modern and revolutionary invention for both freight and personnel, while automobile caterpillar tractors began to be used for moving heavy artillery in difficult country terrain.

Finally, it must be remembered that impressive advances in medicine were of considerable account in preventing disease and epidemics. Owing to tremendous medical and chirurgical development, wounded soldiers in World War I (provided they were attended at the right time, which was not always possible) were far more likely to survive serious battlefield wounds than any of their predecessors.

Allied Victory

The year 1917 was marked by two major events destined to exercise a far-reaching influence on the history of World War I and the fate of the world: the entry of the United States into the war and the Russian Revolution.

The United States entered the war on April 2, 1917. In June, the first U.S. troops arrived in France and by the end of 1918 would total about 1 million men. Ultimately, the military and financial involvement of the United States in the war proved to be decisive.

In March 1917, the discredited czar of Russia was compelled to abdicate by a general revolt caused by hunger, misery, economical ruin, military disasters, and already 4 million Russian casualties. After complicated political moves, turmoil, and complex events, the leaders of the Bolshevik October 1917 Revolution abandoned the entente with bourgeois France and Britain. They sought peace with Germany, and this was signed on March 3, 1918, at Brest-Litovsk. This marked the withdrawal of Russia from the war. So the German army, freed on the eastern front, was able to launch major offensives in the west.

Last German Offensive

After the Russian Revolution, the German divisions were promptly shifted from the east to the west during the spring of 1918. The Germans committed all their forces for a last great push that would bring a quick and final victory. However, lacking reserve, the German offensive failed to make a decisive breakthrough while a last attack launched in Champagne ended in complete failure. The French and British armies with decisive assistance provided by 1 million well-equipped American soldiers succeeded in halting the German thrust. The German armies were now exhausted, and it was now the Allies' turn to attack on all fronts. Successful counteroffensives were launched in July 1918, and under repeated attacks, the German armies were forced to a general retreat. The tide had now turned, and pressing home their advantage, the Allies steadily continued their advance in the summer. At the end of September 1918, the German armies were at the point of collapse. The country found itself under threat of invasion, a revolution broke out, the German commander in chief Erich Ludendorff resigned, Kaiser Wilhelm II was forced to abdicate, and a hastily formed transitory civilian government sought an armistice, which was signed on November 11, 1918.

World War I was over at last.

It had been one of the bloodiest conflicts in history, and indeed, the only impressive results were the casualties.

According to the historian Lawrence James in his book *The Rise and Fall of the British Empire* (Abacus, 1994), the casualties suffered by Britain were 702,000 dead and 1.67 million wounded.

The Allies had won, Germany was defeated, but the war did not solve European problems or achieve lasting peace. World War I was not the war to end all wars, as many politicians proclaimed. Four years of hatred, suffering, misery, confusion, injustice, slaughter, destruction, and propaganda left numerous open wounds. This was reflected in the Treaty of Versailles, a vindictive peace imposed on Germany in 1919. This ill-fated agreement, and later the economic recession of 1929, would be the cause of a second, yet even more terrible, global conflict 20 years later.

PART 7

British Fortifications in World War II

Evolution of Fortifications in the Interwar

Militarily, the lessons learned from World War I were numerous. The full potential of aviation and armored vehicles had not been fully exploited during World War I, but in the interwar years 1918–1939, military airplanes and tanks underwent important development. They gradually became standard and essential attack weapons in modern armies, allowing for rapid offensives with great mobile firepower. The post–1918 period heralded a change in patterns of warfare and, as a reaction, a new phase in the design and use of permanent fortifications. Obviously, the era of the large autonomous forts inherited from the 19th century was over. Forts had proven to be only large and obvious targets and only deathtraps, as this system of permanent fortification had been discredited at Liège in 1914. One of the last ferro-concrete forts ever built (Fort Eben-Emael in Belgium constructed in 1931) was captured in a matter of hours in May 1940 by a commando of German assault engineers brought on top of the fort's superstructures by gliders.

To face and oppose planes and tanks as well as machine guns and artillery fire that had become mobile, extremely accurate, lethal, and destructive, a new concept appeared. Fighting during World War I had been dominated by trench warfare, by field fortifications stiffened by wire entanglements and glacis swept by artillery and machine gun cross fire. Even if fluid situations did develop, they could be brought to a full halt by the simplest of earthworks. So the new fortifications emerging from World War I experience were constituted of wide defensive zones with dispersal and concealment as main features. Those "prepared battlefields" included carefully designed, well-sited but dispersed purpose-built and permanently constructed trenches, and small concrete units (called bunkers) including shelters for troops, ammunitions and supplies storage places, observation and command posts, artillery casemates, and pillboxes for automatic weapons. These units were scattered, partly underground, and camouflaged. They featured ventilation devices and filters to counter the use of toxic combat gases, and their vicinity was systematically defended by antipersonnel and antitank obstacles and mines. Spreading over a large area or established along a line of defense, a prepared complex of trenches and small concrete strongholds was less vulnerable than a large fort and also comparatively cheaper to build. Scattered bunkers either in a linear or in a cluster disposition made possible a defense in depth. Indeed, wherever possible, all bunkers were positioned in order that each of them was always covered by fire coming from

Pillbox built in northern France by the British Expeditionary Force in 1939 during the "Phoney War."

a neighboring one. A system of concrete bunkers had great flexibility: the elimination of a single work did not mean that the rest of the line/group was put out of action, and it was always possible to add new reinforcement to any existing position. Gaps in the linear system were filled with temporary field fortification including prepared infantry trenches, shelters, dugouts, machine gun nests and mortar pits, barbed-wire fences, and obstacles of all kinds. Besides, in order to respond to sudden tactical developments, mobile reserve troops supported by armored vehicles and aviation could deal with infiltrated enemy forces.

That was the reasoning during the interwar period (1918–1939). As a result, permanent lines of such concrete defensive works were erected all over Europe. In the late 1930s as the threat of war increased, working on those linear complexes was greatly accelerated. Nazi Germany developed an aggressive army with tanks, paratroopers, and infantry supporting bombers and dive bombers and also constructed fortifications along the border with France, the so-called Westwall (aka Siegfried Line).

Czechoslovakia established a strong line of defense against Germany called the Benes Line. The same happened in Yugoslavia, the Soviet Union, Poland, and Spain. Even neutral countries like the Netherlands, Belgium, and Switzerland, built defensive lines. The most sophisticated and most impressive of these concrete defense lines was without doubt the French Maginot Line—a strong line of elaborate, expensive, subterranean forts established at the borders of France. Every continental country in Europe had great faith and expectation in belts of concrete fortifications. They were all fated to be disappointed.

Operation Sea Lion

As was the case in other Western European countries, the British were long reluctant to recognize the real nature of Hitler's racist, criminal, and bellicose regime. The Conservative UK prime minister from 1937 to 1940, Neville Chamberlain (1869–1940), pursued a controversial policy of "appeasement" in order to avoid war. He bargained and made compromises with an increasingly aggressive Nazi Germany. Chamberlain's efforts were in vain, and war broke out in September 1939 when Nazi Germany invaded Poland.

After defeating Poland and after a winter (1939–1940) of "Phoney War," the Germans launched a series of successful offensives in the west in the spring of 1940. The astonishing, unthinkable, and rapid conquest of Norway, Denmark, the Netherlands, Belgium, and France between April and July 1940 resulted in a completely new situation for unprepared Great Britain. After the defeat and signature of armistice with France in June 1940, Hitler was convinced the British would sue for peace. When peace initiative was dragging, he ordered preparations for the seaborne invasion of the British Isles, code named Operation *Seelöwe* (sea lion). The Germans were confident and regarded crossing the Channel as no more than an extended river crossing. The German navy and army gathered all available resources along the Channel in preparation for the invasion. They assembled barges and landing crafts in the French and Belgium Channel ports and set aside a force totaling 20 divisions for the operation. The 16th German Army would embark at Rotterdam, Ostend, Dunkirk, Calais, and Boulogne and would land in southern England between Bexhill and Folkestone. The 9th Army would embark at Boulogne and Le Havre and land between Portsmouth and Brighton. The 6th Army would embark at Cherbourg and land at Lyme Regis. In addition, paratrooper droppings were planned in southern England, notably at Folkestone and Brighton.

However, the safe movement of the landing forces depended on Hermann Göring's air force rather than the small German navy. Indeed, the British Home Fleet was a force that dwarfed anything the Kriegsmarine could put to sea. For the Germans, achieving air superiority was an essential prerequisite to any invasion. If the German Luftwaffe could destroy or paralyze the Royal Air Force and drive the Royal Navy out of the English Channel even temporarily, the invaders had a good chance for a successful landing with acceptable casualties. Whether Hitler would have actually attempted the invasion of Great Britain is still a debated question. In retrospect, it was probably only a bluff intended to bring Britain to heel. Hitler offered Britain the prospect of an "honorable peace," but at the same time, as evidence of intent, he ordered in July 1940 a massive air offensive, which came down in history as the Battle of Britain. Hitler, however, showed a remarkable lack of interest in Operation Sea Lion, taking his first and only vacation of the war during its preparation.

By the middle of September 1940, air war still raged in the British skies, and the date of the invasion had already been postponed three times. Finally, the Luftwaffe was never able to whittle the Royal Air Force strength down to an acceptable level for risking Operation Sea Lion. In October 1940, in spite of numerous and deadly air attacks, it was obvious that the Luftwaffe had failed to destroy the Royal Air Force and break the British people's resistance and morale. As a result, Hitler lost interest in the invasion plan altogether. The Battle of Britain had been a failure for Nazi Germany, and in October 1940, unknown to the British, Hitler canceled Sea Lion but maintained air attacks (the so-called blitz from September 1940 to May 1941) merely as a means of bringing military and diplomatic pressures.

Meanwhile in Britain, anti-invasion defenses of all types were planned and hastily carried out.

By the spring of 1941, the state of Britain's anti-invasion defenses had much improved, with many more trained and equipped men becoming available and field fortifications reaching a high state of readiness. When Germany invaded the Soviet Union on June 22, 1941, it became unlikely that there would be any attempt for a landing as long as that conflict was undecided—from the British point of view, at the time the matter hung in the balance. In July 1941, the preparation and construction of field fortifications in Britain was greatly reduced as the threat of a full-scale invasion had become improbable. Besides, on December 7, 1941, the Japanese surprise air attack on the American fleet at Pearl Harbor (Hawaii) resulted in the United States entering the war on Britain's side. With America now as an active ally, hope was revitalized, and soon resources and supplies flooded into the United Kingdom, effectively ending the danger of invasion after two years of grim uncertainty.

Sea Lion, the plan to invade Britain, was never revived. There were other ways of isolating, weakening, and defeating Britain. Hitler decided to slowly starve the British by launching a submarine blockade campaign in all supply waterways between the United States and Britain. That enormous naval contest became known as the protracted Battle of the Atlantic.

Anti–Invasion Preparations

Air Raid Shelters

The air raids carried out by zeppelins and Gotha heavy bombers during World War I had clearly demonstrated that given the tremendous advances in military aviation in the 1920s and 1930s, the main threat to British national security was the bomber airplane. Motivated by a public fearful of terror attacks from the sky, the British government prepared for the worst. In 1938, because of Nazi Germany's aggressive foreign policy threat, fear of war grew alarmingly and the British government ordered the mobilization of the volunteer members of Air Raid Precautions (ARP, created in 1935 and headed by Sir John Anderson). The ARP wardens operated wailing air raid sirens announcing the approach of enemy bombers and ensured that most people had time to take cover before the raid actually started. After the raid, the wardens sounded the "all clear" so that rescue teams could save and assist victims.

In the 20th century, helpless civilians were often vital workers producing the weapons of war and therefore had to be protected in purpose-built shelters. However, little of any significance or coherence happened in Britain until war actually broke out. Then the engineer William Patterson was commissioned to design a small and cheap shelter that could be erected in civilian people's gardens. In 1940, when German air attacks started, nearly 1.5 million of these so-called Anderson shelters were distributed to people living in areas expected to be bombed. Made from six curved metal sheets bolted together at the top with steel plates at either end and measuring 6 feet 6 inches by 4 feet 6 inches (1.95 meters by 1.35 meters), the individual shelter could accommodate six people. These shelters were half-buried in the ground and covered with earth heaped on top. The entrance was protected by a steel shield and an earthen blast wall. Other prefab air raid shelters were the so-called Stanton shelters designed and manufactured by the Stanton Ironworks Company in Derbyshire. Stanton shelters consisted of sections of precast concrete arched-shaped sides, bolted together to form a standard shelter typically for 50 men. The entrance could be brick lined and fitted with concrete steps (if required). The unit often included an emergency escape hatch at the rear. The shelters were built above ground or semi-sunk, but for added protection and concealment purposes, blast walls were added and they could be covered with dirt and turf.

In March 1940, the government began to build communal shelters designed to protect about 50 people living in the same area. Made of brick and concrete, they provided more protection than garden metal shelters. However, within a couple of months, there was a severe shortage of cement and their construction slowed down. Besides, they were noisy, dark and damp, impossible to keep waterproof, and very uncomfortable.

In a chaotic situation, improvised shelters and trenches of various designs were hastily dug in the parks of large towns. Indeed, many improvisations were done, for example, using the basements of the London Fruit & Wool Exchange or the Tilbury railway arches in Stepney, which were turned into large public shelters for thousands of people. Londoners also used underground tube stations as shelters during the Blitz. Subway stations were popular because they were dry, warm, and rather safe but extremely crowded and deprived of sanitation.

The British government also started issuing the so-called Morrison family shelters. Named after the Home Secretary Herbert Morrison, the shelters were made of very heavy steel and could be put in the living room and used as a table. One wire side lifted up for people to crawl underneath and get inside. Morrison shelters were fairly large and provided sleeping space for two or three people.

Eventually, the government decided to build large public concrete shelters far below ground in Central London. Each one of these shelters could have accommodated about 8,000 people. They never fulfilled their purpose, though, because they were not completed when the Blitz offensive was over.

It should be noted that England, and particularly London, was the target of a second German air offensive between June 1944 and March 1945. This time, the Germans used the so-called Vergeltungswaffen ("vengeance weapons") V1 flying bomb and the V2 rocket. Hitler's vengeance weapons had no real strategic purpose; their inability to be aimed at any precise target made them the ultimate World War II terror weapons. Vergeltungswaffen were also launched on Antwerp and Liège in Belgium, and the total terror bombing resulted in an estimated 18,000 victims, mostly civilians.

Home Guard

While under the threat of an invasion, the British met the crisis in a considerable hurry. The government ordered the placement of antiaircraft guns and barrage of tethered balloons over London and all other important British cities. Plans were made and carried out for the evacuation of children and the elderly from Britain's large cities. Important preparations were made in the summer of 1940, and the construction of defense works continued in 1941. More than merely passive resistance was expected from the population. On May 14, 1940, Secretary of State for War Anthony Eden announced the creation of the Local Defense Volunteers—later to become the Home Guard—which by the end of June 1940 totaled nearly half a million volunteers. Although the Home Guard has been gently and affectionately mocked in the BBC sitcom *Dad's Army* by Jimmy Perry and David Croft, it was by no means a ragtag force. Many volunteers were World War I veterans who were still fit, determined, and had more combat experience than many regular army men. Other members were young men without experience but were enthusiastic (like the character Frank Pike played by Ian Lavender), eagerly waiting to go into the forces. The high number of volunteers revealed the spontaneous mobilization and the seriousness with which ordinary people took the threat of invasion in the dramatic summer of 1940.

Basic defensive measures included blackouts and the removal of signposts, milestones, and railway station signs making it more likely that the enemy would become confused. As it was well known that German tanks worked on regular car petrol, gasoline pumps were removed from service stations near the coast and there were careful preparations for the destruction of those that were left. Detailed plans were made for the blowing up of anything that might prove useful to the invader such as port facilities, key roads, bridges, and rolling stock.

In certain vulnerable areas (e.g., along beaches), nonessential citizens were evacuated. The ringing of church bells was banned, and henceforth bells would be rung only by the military or the police to warn that an invasion or an attack by Fallschirmjäger (parachutists) was in progress. In 1941, in towns and villages, invasion committees were formed to cooperate with the military, and plans for the worst were made should their communities be attacked, isolated, or occupied. Three reserve groups of policemen were mobilized and armed in order to assist the British armed forces in the event of an invasion.

Coastal Defenses

Between the two world wars, the various British governments had concentrated on the establishment of air defenses and the maintenance of naval superiority, while coastal fortifications were neglected or even ignored. The fall of France in May 1940, therefore, found the defenses of Britain completely unprepared to resist a German attack from the sea. A German invasion of Britain seemed imminent now that they controlled all main ports in the North Sea including Antwerp in Belgium and Rotterdam in the Netherlands, as well as the French Channel ports like Calais, Boulogne, Le Havre, Cherbourg, and Brest. A German invasion would involve the landing of troops and equipment somewhere on the coast, and the most exposed areas were obviously the south and east coasts of England. The priority was the coastline of Dorset, Hampshire, Sussex, Kent, Essex, and Suffolk. Here, ancient Henrician castles, Napoleonic Martello towers, and old 19th-century Palmerston forts found themselves back in use as observation posts, artillery emplacements, and infantry machine gun positions, while emergency coastal batteries were hastily established to defend the major ports and likely landing places. Beaches were naturally the most vulnerable spots where a landing could take place, and in the summer of 1940, Britain's defenses focused on holding beaches as long as possible. Beaches as probable landing places were blocked with entanglements of barbed wire, extensive minefields, with both antitank and antipersonnel mines on and behind them. On many of the more remote beaches, this combination of wire and mines represented the full extent of the passive defenses. The southern beaches were overlooked by armed pillboxes of various types described below. These were sometimes placed low down to get maximum advantage from enfilading fire, whereas others were placed high up making them much harder to capture. Searchlights were installed at the coast to illuminate the sea surface and the beaches for artillery fire.

Observation bunker. This concrete bunker can still be seen at Coalhouse Fort, built in the 1860s, near Tilbury (Essex).

Parts of the wetland Romney Marsh in Kent were flooded and there were plans to flood more of the marsh if the invasion were to materialize. For this purpose, the Royal Military Canal, built between 1804 and 1809 against Napoleon I, was refurbished and reinforced with pillboxes. Piers situated in large numbers along the south coast of England in the Channel's summer resorts that could facilitate the landing of enemy troops were disassembled, blocked, or otherwise rendered useless or even demolished. Many small islands and peninsulas were fortified to protect inlets and other strategic targets.

Preparation went further than just south England. In the Firth of Forth in east central Scotland, Inchgarvie Island was heavily fortified with several gun emplacements. This provided invaluable defense from seaborne attacks on the Forth Bridge and Rosyth dockyard, approximately a mile upstream from the bridge. Farther out to sea, Inchmickery Island, 1.6 miles (2.6 kilometers) north of Edinburgh, was similarly fortified with gun emplacements on the coast to the north in North Queensferry and south in Dalmeny and Inchmickery Island.

The British engineering corps also experimented with petroleum floating on the surface of the sea just off some beaches. When enemy landing craft would have approached, the fuel would have been ignited, setting them ablaze. Although many projects of using fuel proved fruitless, a number of practical weapons were developed. Fire traps and flame fougasse barrels were installed at 7,000 sites, mostly in southern England and at 2,000 more sites in Scotland. Early experiments with floating petroleum on the sea and igniting it were not entirely successful, but by early 1941, a flame barrage technique was developed. Nozzles were placed above high-water mark with pumps producing sufficient pressure to spray fuel which produced a roaring wall of flame over. Such installations consumed considerable resources, and although this weapon was impressive, its network of pipes was vulnerable to prelanding bombardment. Considered ineffective, initially ambitious plans were cut back to cover just a few miles of beaches.

Fire control tower at Bawdsey. Located south of Norwich in Suffolk, this two-story reinforced concrete building with cantilevered roof housed high-precision range finders.

It seems likely the British would have used poison gas against German troops landing on their beaches. General Alan Brooke, in an annotation to his published war diaries, stated that he *"had every intention of using sprayed mustard gas on the beaches."* Mustard gas was manufactured as well as chlorine, phosgene, and Paris green. Poison gases were stored at key points for use by bomber command and in smaller quantities at many more airfields for use against invaders. Bombers and crop sprayers would shower landing craft and beaches with mustard and other poison gases.

All fortifications ready for use were quickly armed with whatever guns were

Artillery observation tower. King's Lynn Battery at Ongar Hill in Norfolk.

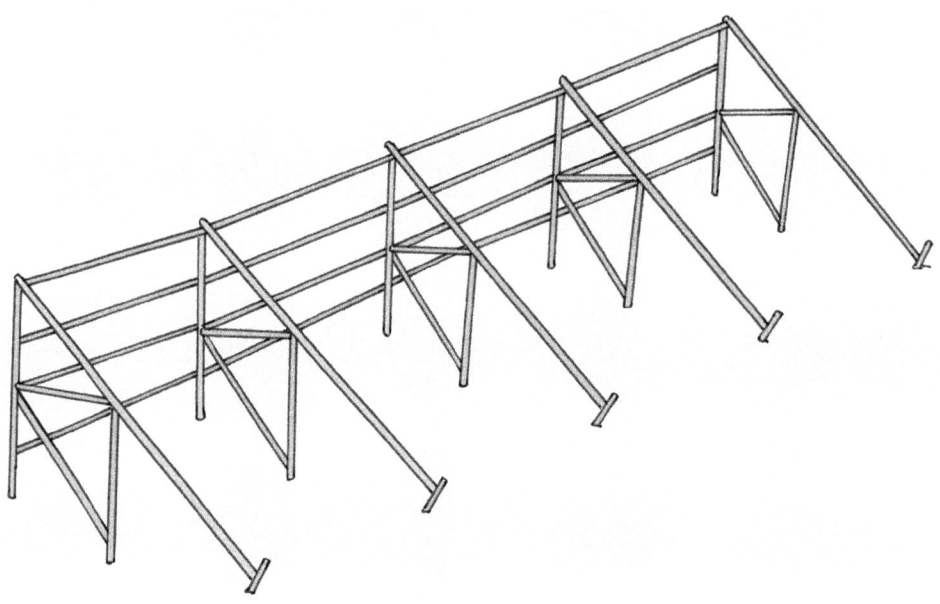

Admiralty scaffolding. Admiralty scaffolding (also known as beach scaffolding or Obstacle Z.1) was an anti-tank fence made of scaffolding tubes nine feet (2.7 m) high.

available, which mainly came from heavy railway artillery or from naval vessels scrapped since the end of World War I. These included 6-inch (152-millimeter), 5.5-inch (140-millimeter), 4.7-inch (120-millimeter), and 4-inch (102-millimeter) guns. At Dover, two 14-inch (356-millimeter) guns known as Winnie and Pooh were deployed.

There were also a small number of land-based torpedo launching sites. The shortage

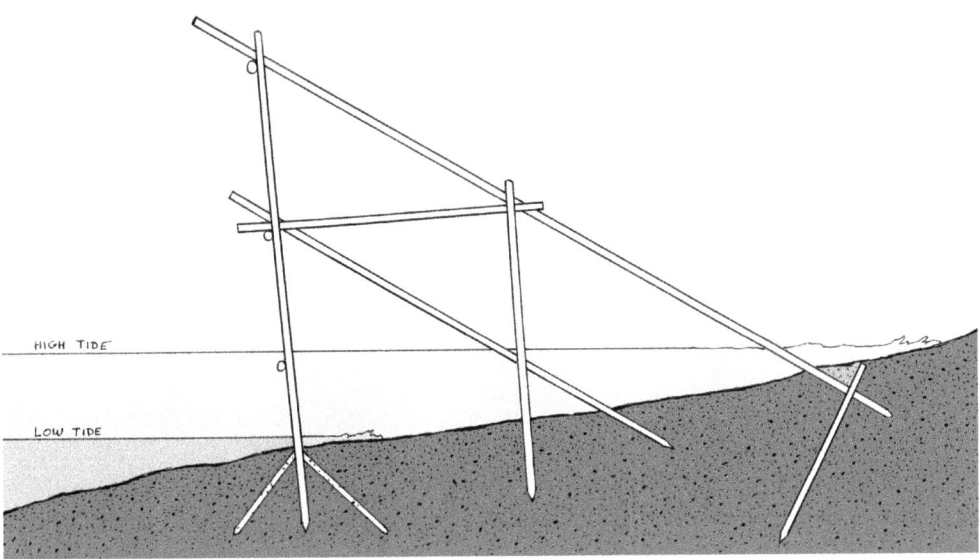

Admiralty scaffolding (cross-section).

of weapons of all kinds led to the desperate expedient of fabricating makeshift grenades and Molotov cocktails, made from bottles filled with tar, petrol, and paraffin ignited by primitive fuses.

Inland Fortifications

Stop Lines and "Islands"

The British also engaged in an extensive emergency program of field fortification farther back in the hinterland, behind the beaches in the countryside.

In May 1940, a Home Defense Executive was formed under General Sir Edmund Ironside, commander in chief of the Home Forces, to organize the defense of Britain. At first, defense arrangements were largely static and focused on the coastline (the so-called coastal crust) and, in a classic example of defense in depth, on a series of inland antitank "stop" lines. The stop lines were designated command, corp, and divisional according to their status. The longest and most heavily fortified was the General Headquarters antitank line, GHQ Line, which ran across southern England, wrapped around London, and then ran north to Yorkshire. It was intended to protect the capital and the industrial heartland of England. Another major line was the Taunton Stop Line, which defended against an advance from England's southwest peninsula. London and other major cities were ringed with inner and outer stop lines. Some 50 stop lines were constructed, some of the less important lines were just demolition belts, and not all lines were strong and completed.

Military thinking shifted rapidly. Given the lack of equipment and properly trained men, General Edmund Ironside had little option but to adopt a strategy of static warfare, but it was soon perceived that this would not be sufficient. Ironside was criticized

Trench. (1) Breastwork made of sandbags, (2) revetted side, (3) firestop.

for having a siege mentality, and on July 19, 1940, he was replaced by General Brooke. Under Brooke, new strategies and in-depth tactics were devised. More concentration was placed on defending the coastal crust, while inland, a hedgehog defense strategy of defended localities, centers of resistance, and antitank "islands" was established, each having all-round defense and each mutually supporting. Many of these antitank islands were established along the already constructed stop lines where existing defenses could be integrated into the new strategy and especially at towns and villages where there was a Home Guard to provide personnel. These were fortified with trench networks, gun and mortar emplacements, and pillboxes, barriers of scaffolding, sandbagged positions and loopholes pierced in existing buildings, removable roadblocks, barbed-wire entanglements, and land mines. Resistance islands included the involvement of civilians for a spirited defense and were generally placed in major road junctions, railway bridges, tunnels, and other weak spots, as well as river crossing points, villages, towns, and cities, which were seen as time consuming and costly to take. Home Guard troops were largely responsible for the defense of hubs, nodal points, and other centers of resistance.

Defense of Airfields and Open Areas

Open areas were considered vulnerable to invasion from the air by paratroops, glider-borne troops, or even with powered aircraft, which could land and take off again.

Open areas with a straight length of 500 yards (460 meters) or more within five miles (eight kilometers) of the coast or an airfield were considered vulnerable. These were blocked by trenches or, more usually, by wooden or concrete obstacles as well as old cars, vehicles, or whatever could work as an obstacle.

Securing an airstrip would be an important objective for the invaders. Airfields, considered extremely vulnerable, were protected by trench works and pillboxes, which faced inward toward the runway, rather than outward. Many of these fortification works and obstacles were specified by the Air Ministry. Some defensive designs were unique to airfields; these would not be expected to face heavy weapons, so the degree of protection was less and there was more emphasis on all-round visibility and sweeping fields of fire. It was difficult to defend large open areas without creating impediments to the movement of friendly aircraft.

Pickett-Hamilton Pillbox

Pickett-Hamilton pillbox.

Solutions to this problem included the pop-up Pickett-Hamilton pillbox. The Pickett-Hamilton pillbox was particularly design to defend British airfields from German landing and airborne troops. As any structure built above the ground of an airfield presented a hazard to aircraft, the Pickett-Hamilton was a retractable (or "disappearing") gun turret intended to pop up out of the ground only when a German paratrooper attack would occur. The cylindrical pillbox was made of concrete, and was 2.7 meters (9 feet) in diameter with a thickness of 22–25 centimeters (9–10 inches). It was buried in the ground, manned by two or three gunners standing and serving a light machine gun, and access was via an armored hatch on the top. The pillbox consisted of two precast concrete cylinders: the outer attached to the base section and the inner, which formed

Pickett-Hamilton pillbox (cross-section).

the moving "pop up" element. The inner section was moved up and down by a compressed air-driven jack. Because it was sometimes unreliable, the jack was backed up by a hydraulic hand pump worked manually by one of the crew. This was accomplished by a double-acting pump mounted on top of a cylindrical oil reservoir, which enabled the pillbox to be raised in around 8 seconds and lowered in about 20 seconds. When not in use, the pillbox was totally concealed, the concrete lid of the sliding inner cylinder lay flush with the ground surface; when brought up into action (as the illustration shows), it rose some two feet six inches above the surface to allow fire from either of its three embrasures. The main idea behind the "pop up" pillbox was to maximize the element of surprise. At the start of any enemy attack on the airfield, they would rise out of the ground with an all-round field of fire. Although in theory the idea was good, it is doubtful whether these pillboxes would have been useful in practice. They were prone to flooding by rain or infiltration, the raising mechanism was rather vulnerable, they were not large enough to accommodate heavy weapons and cramped for the crew, and no upward view was possible (apart from cursory looks through the entrance hatch). They were, however, a very good indication of the ingenuity of the wartime brain when it came to inventions. With each pillbox costing approximately £240 to build, tenders were sent out to various companies for their construction. Most of the concrete work was carried out by DWG of Barnstable in Devon, while the original castings for the raising mechanism were produced by the Cornish firm Willey and Co. Some 335 "pop up" pillboxes were constructed at almost every airfield in Britain.

Bison

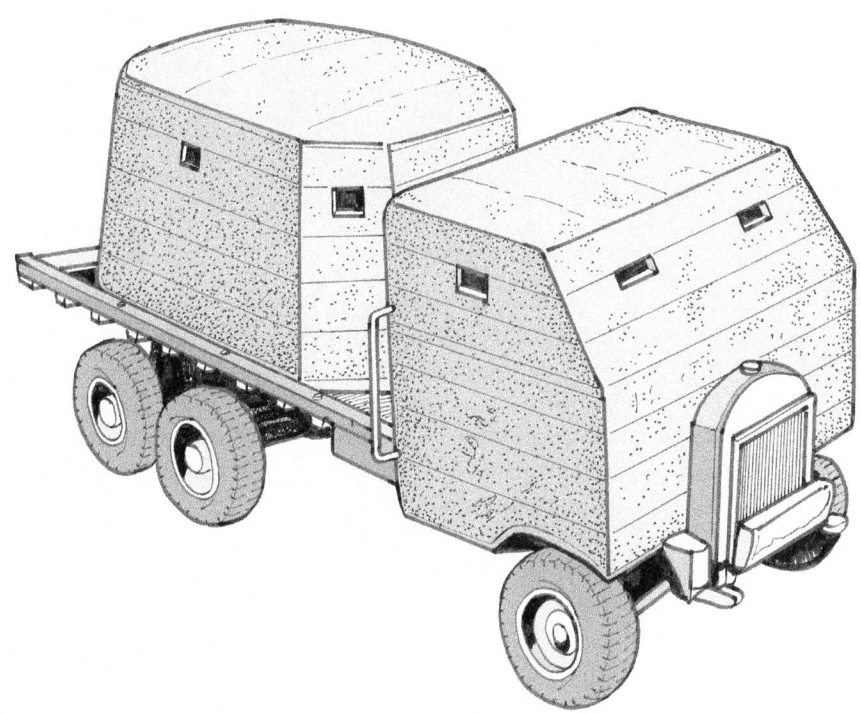

Bison (Thornycroft Tartar three-ton, six-by-four truck).

Another arrangement was a mobile pillbox (called Bison) that could be driven across runways in order to block the airfield. Based on a number of different truck chassis, it featured a fighting compartment protected by a layer of concrete. Various kinds of Bisons were used by the Home Guard and the Royal Air Force to protect airfields.

Bisons would certainly have been quite adequate to defend against the lightly armed German paratroopers who were the anticipated enemy at airfields. It is not clear how many Bisons were produced. Estimates vary between 200 and 300.

Mushroom Pillbox

Mushroom pillbox.

The so-called cantilever or "mushroom" pillbox was specifically designed for airfield defense by the F.C. Construction Company from Peterborough. As the name implies, this pillbox was circular with various diameters and was covered with an overhanging domed concrete slab, which was supported by a strong central X-shaped masonry pillar. It offered limited protection to defenders because of its large open embrasure but allowed for an excellent all-round defense owing to its 360° traverse. The circular embrasure featured a continuous tubular rail attached to the masonry allowing machine guns to be quickly moved in order to fire in all directions. Mushroom pillboxes, often partially sunk into the ground, were common features in many airfields in south England.

Allan Williams Turret

The Allan Williams turret was a small metal turret placed on top of a concrete or brick-lined cylindrical pit 1.2 meters (4 feet) in diameter sunk into the ground. The turret, which rose only 13 inches from the ground surface, was mounted on a ball race that enabled a complete 360° traverse. It was manned by a single soldier armed with either a rifle or a light Bren machine gun or a Boys antitank rifle. The weapon could be fired either through a front slot (which was further protected by a shutter) or through a circular top hatch for antiaircraft fire. This opening was also the only way in and out for the occupant who had to be a nonclaustrophobic man. The pit was prone to flooding, the

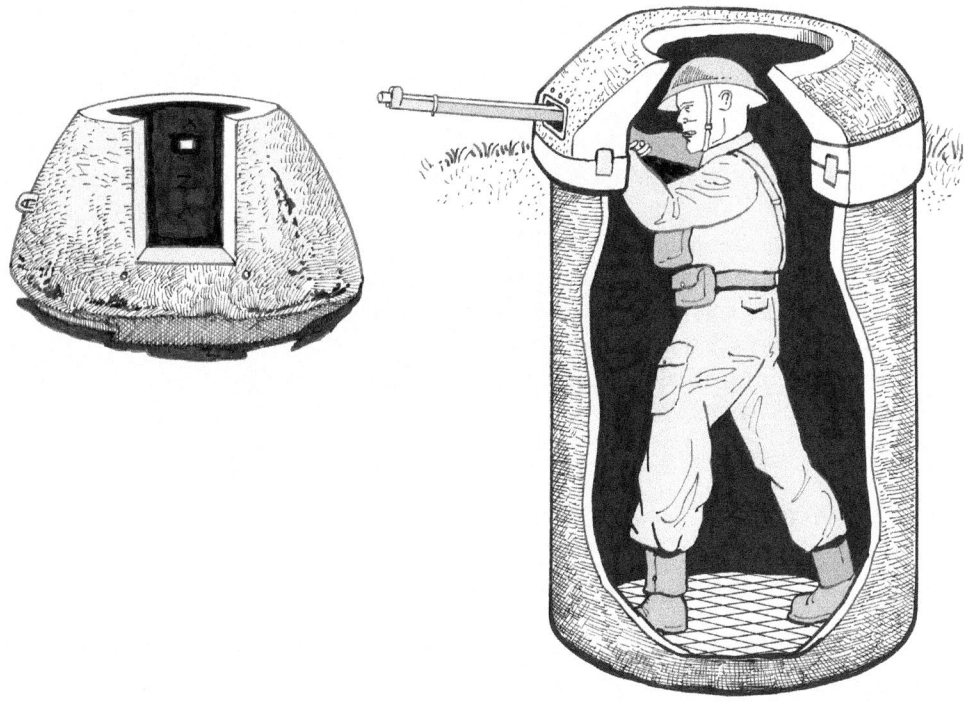

Tett turret.

view was limited, and it was also difficult to rotate the cupola while tracking targets, firing, and reloading at the same time. Because of all its disadvantages, the British army did not like this design, so most were placed at Royal Air Force airfields. The Allan Williams turret had a brief life as steel became scarce. It is estimated that about 200 of these turreted pits were installed around airfields in Britain. After the war, steel turrets were reclaimed for their metal, so extant examples are extremely rare.

A similar pit was the Tett turret, named after its inventor and designer H.L Tett. This was a private commercial venture manufactured by Burbridge Builders of Surrey. It was in fact the same design as the Allan Williams turreted pit only the cupola was not made of metal but of concrete. It was thus cheaper but also much heavier and presented the same general features and the same shortcomings.

Other Methods

Another device was the runway plough, designed to rip up aerodrome runways and railway lines if an invasion took place as a means of causing disruption making it useless for the invaders. The plow was hauled by a powerful Foden Trucks tractor, possibly via a pulley-and-cable system.

The defenses against air attacks also included many antiaircraft batteries, the construction of many airfield and airstrips for Royal Air Force interceptor fighters, and the establishment of a chain of warning observation posts as well as radar detection stations. In the last years of World War I and during the interwar period (1918–1939), the introduction of bombers led to the development of detection and warning devices.

A forerunner of electronic radar, spectacular acoustic mirrors were built on the south and northeast coasts in the 1930s, notably at Denge near Dungeness in Kent. They were made of large parabolic concrete structures intended to detect noise made by incoming enemy aircraft, thus providing an early warning. For a while, they did work, but the development of faster aircraft rendered them rapidly useless, as an approaching aircraft would be within sight by the time it had been detected. Besides, ambient noise made the mirrors difficult to use successfully, and the development of rather accurate radar electronic techniques rendered acoustic detection redundant.

Another way of preserving likely targets from air attack was to try to fool enemy airmen to drop their bombs elsewhere. For this purpose, decoy sites (coded Starfish) were installed consisting of braziers, fireworks, and explosives. Giving the impression at night of exploding bombs, fires, antiaircraft gunfire, and fighter tracers, they were meant to distract German bombers from their targets.

FW3 Pillboxes

The Directorate of Fortifications and Works

The field fortifications constructed throughout Britain included large numbers of trenches, outposts, and concrete armed pillboxes. The 1940 British pillboxes were visibly designed and built in a great rush.

In May 1940, the Directorate of Fortifications and Works (FW3) was set up at the War Office. Placed under the leadership of Major General G.B. Taylor, FW3's task was to provide a number of basic pillbox models designed on practice in World War I and based on experience gained in northern France in 1939. An important request stipulated that they should be cheap and could be built by unskilled soldiers and local labor at appropriate defensive locations. In June and July 1940, FW3 issued six basic designs for rifle and light Bren machine gun pillboxes, designated Type 22 to Type 27. In addition, there were designs for artillery emplacements suitable for either the Ordnance QF 2-pounder or the Hotchkiss 6-pounder gun (designated Type 28) and a design for a concrete medium machine gun emplacement. However, in typical British fashion, the whole scheme was rather loosely applied and constructed in a very hasty manner. As a result, there were many nonstandardized designs for pillbox-like structures for various purposes including light antiaircraft positions, observation posts, searchlight positions to illuminate the shoreline or the sky, and a number of pillboxes particularly designed to protect airfields from airborne enemy attacks. A small number of pillboxes had been constructed in World War I and, wherever possible, these were integrated into the 1940 defense plans. Some pillboxes predated the publication of the FW3 designs, and in many cases, some local commanders introduced modifications to the standard FW3 designs or introduced variants of their own. These nonstandard-design pillboxes could be produced in some numbers or completely ad hoc designs suited to local conditions. Other designs were produced as commercial ventures by private companies or totally improvised by the army, Royal Air Force, navy, and the Home Guard.

The degree of protection offered by these different types of pillboxes varied. Compared to French Maginot Line forts and German Westwall (Siegfried Line) bunkers, which had been carefully planned, designed, and built in the prewar years, British pillboxes were extremely primitive and vulnerable. It must be remembered that they were hastily designed and built in an emergency in a period of crisis with little time and little means available. Their primary purpose was to repulse the first wave of attackers—the German assault engineers or paratroopers. Once they had revealed themselves, the life expectancy of their occupants would have been very short indeed. Roofs and walls were usually between 30 centimeters (12 inches) and 1.1 meters (3 feet 6 inches) thick, which made them only bulletproof. Embrasures were available precast and factory produced to standard designs, but as these were in short supply, some embrasures were improvised from brick or concrete paving. Embrasures were frequently fitted with a steel or concrete-asbestos shutter. From March 1941, some pillbox embrasures were fitted with a Turnbull mount; this was a metal frame that could support various automatic weapons like the Bren light machine gun or Vickers machine gun. Internally, pillboxes and shelters were cramped and spartan with no accommodation for the crews, no ventilation, and no combat gas protection systems. Some internal concrete shelves and tables were provided to support weapons and some were whitewashed inside. Here again, it must be pointed out that they were not intended for prolonged comfortable sojourn but merely and simply intended as combat emplacements for a short time. The makeshift character of British pillboxes of 1940–1941 was accentuated by private commercial ventures, local improvisations, and the fact that the loose basic FW3 regular designs were adapted to local circumstances and depended on available building materials. As already said, local commanders introduced modifications to the standard FW3 designs or introduced features of their own, which resulted in completely ad hoc structures suited to local conditions. So, outwardly, two pillboxes of the same basic design could look a bit different. The height of a pillbox could vary significantly according to local needs. Some were half-sunk so that the embrasures might be as low as ground level for optimal grazing fire while others were raised to give a better view. Some included two stories formed of two standardized models built on top of each other. Those built into hillsides might lack embrasures on some walls. The entrance was obviously placed on the side away from the expected direction of attack. Its size varied as might be convenient, and there may be additional walls to protect the entrance, a freestanding blast wall or a steel door. Very often, shelters' and pillboxes' walls had a small area of unmortared brickwork at the rear of the structure at ground level constituting a kind of emergency exit. If the main access was blocked or severely damaged, occupiers could smash their way out at that weak spot. Appearance also varied because of the building materials used, which due to wartime conditions was sometimes of inferior quality. Although all FW3 designs were to be made of concrete, the reinforced concrete used in construction was generally conventional making use of thin steel rebars with floor, walls, and roof all mutually bonded. However, several instances are known where scrap or salvage metal bars were used such as parts of an old bed, park railings, or even discarded wire mesh. Where brick was used as a shuttering, the bricks essentially formed a mold into which concrete was poured, the bricks being left in place. Otherwise, the pillbox was formed using shuttering of wood (usually planks but sometimes plywood) and/or corrugated iron. Wood shuttering was removed, whereas corrugated iron was sometimes left in place. Construction often took advantage of whatever materials were available locally (e.g., at the coast, beach sand and pebbles would be used), and this expedient use of local materials had the added advantage of aiding camouflage. The rate of construction was frenetic: by the end of September 1940, 18,000 pillboxes and countless other preparations had been completed. About 28,000 pillboxes and other hardened field fortifications were constructed in the whole United Kingdom of which about 6,500 still survive.

Pillbox Type FW3/22

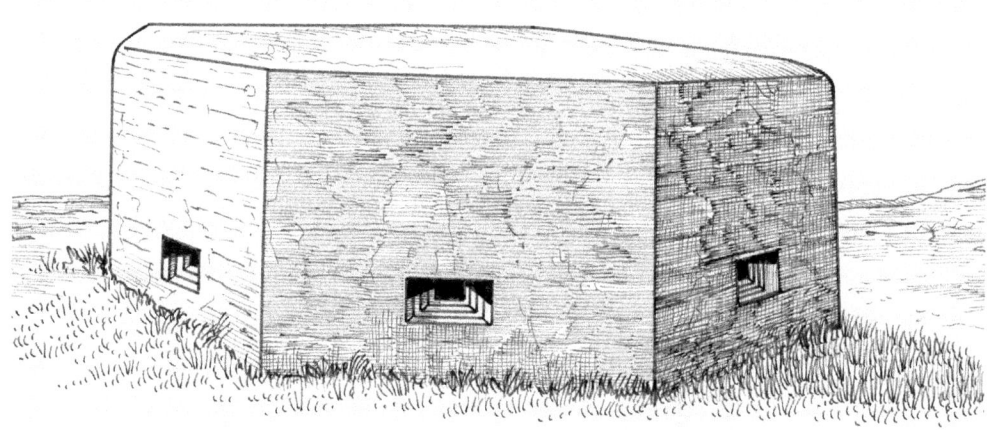

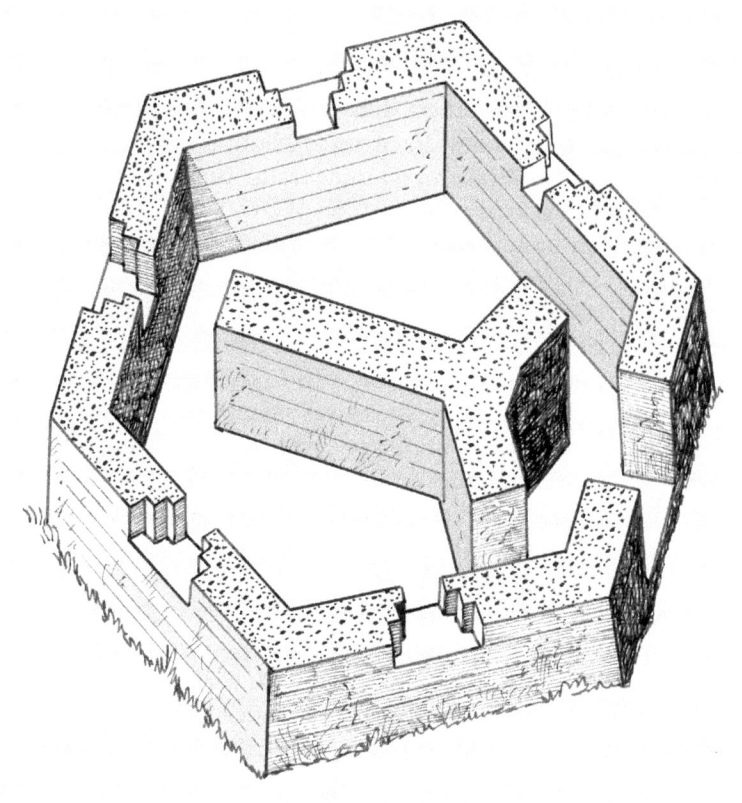

Type 22 was the second most common design with 1,209 recorded as being extant. In plan, it appeared as a regular hexagon and can easily be confused with Type 24. The walls of Type 22 were approximatively 30 centimeters (12 inches) thick by 1.8 meters (6 feet) long. The standard 12-inch thickness was bulletproof, and there existed a stronger version with walls about 40 inches (1 meter) thick. Most had the entrance at the back, and a loophole (embrasures) on each side of the hexagon. The loopholes were suitable for rifles or Bren light machine guns. Internally, Type 22 included a Y- or T-shaped anti-ricochet wall, which served two purposes. It helped support the weight of the roof, and in case of an enemy projectile being fired or thrown through one of the two embrasures, damage within the pillbox was limited.

Pillbox Type FW3/23

Type 23 was rectangular in plan, consisting of two parts, one open and the other covered with a roof. The open section was intended for a light antiaircraft weapon (Bren or Lewis). The roofed section was provided with four embrasures suitable for rifles and light machine guns and compartmented by an internal anti-ricochet wall. The pillbox had a width of 2.4 meters (8 feet) and a length of 4.9 meters (16 feet), and its walls were of the bulletproof standard 30 centimeters (12 inches). Access to the pillbox was either an entrance at ground level or usually over the wall of the open section by metal rungs set into the wall. Type FW3/23 was rather uncommon, and only 156 are recorded as still extant.

Opposite: Pillbox type FW3/22. The bottom illustration shows the horizontal section at the level of the embrasures.

Pillbox type FW3/23.

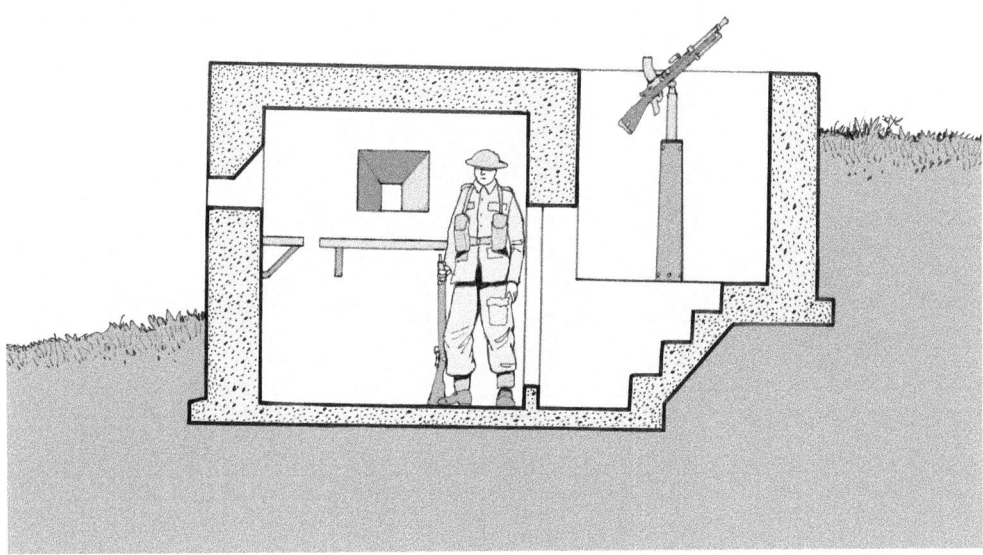

Cross-section pillbox type FW3/23.

Pillbox Type FW3/24

Commonly and easily mistaken for Type 22, Type 24 was a slightly irregular hexagon with five sides each fitted with an embrasure suitable for rifles and light Bren machine guns, and the rear wall was longer than the others with the entrance and two loopholes. The rear wall was c. 4.3 meters (14 feet) long, with the other walls being 2.2–2.5 meters (7–8 feet) in length. The walls were generally of bulletproof standard 30-centimeter (12-inch) thickness, but there was a larger thick-walled version with walls 90–127-centimeters (36–50-inch) thick, sometimes referred to as Type 29. Type 24 was provided with an internal anti-ricochet Y- or T-shaped wall. It was the most common pillbox with about 1,787 recorded as still being extant.

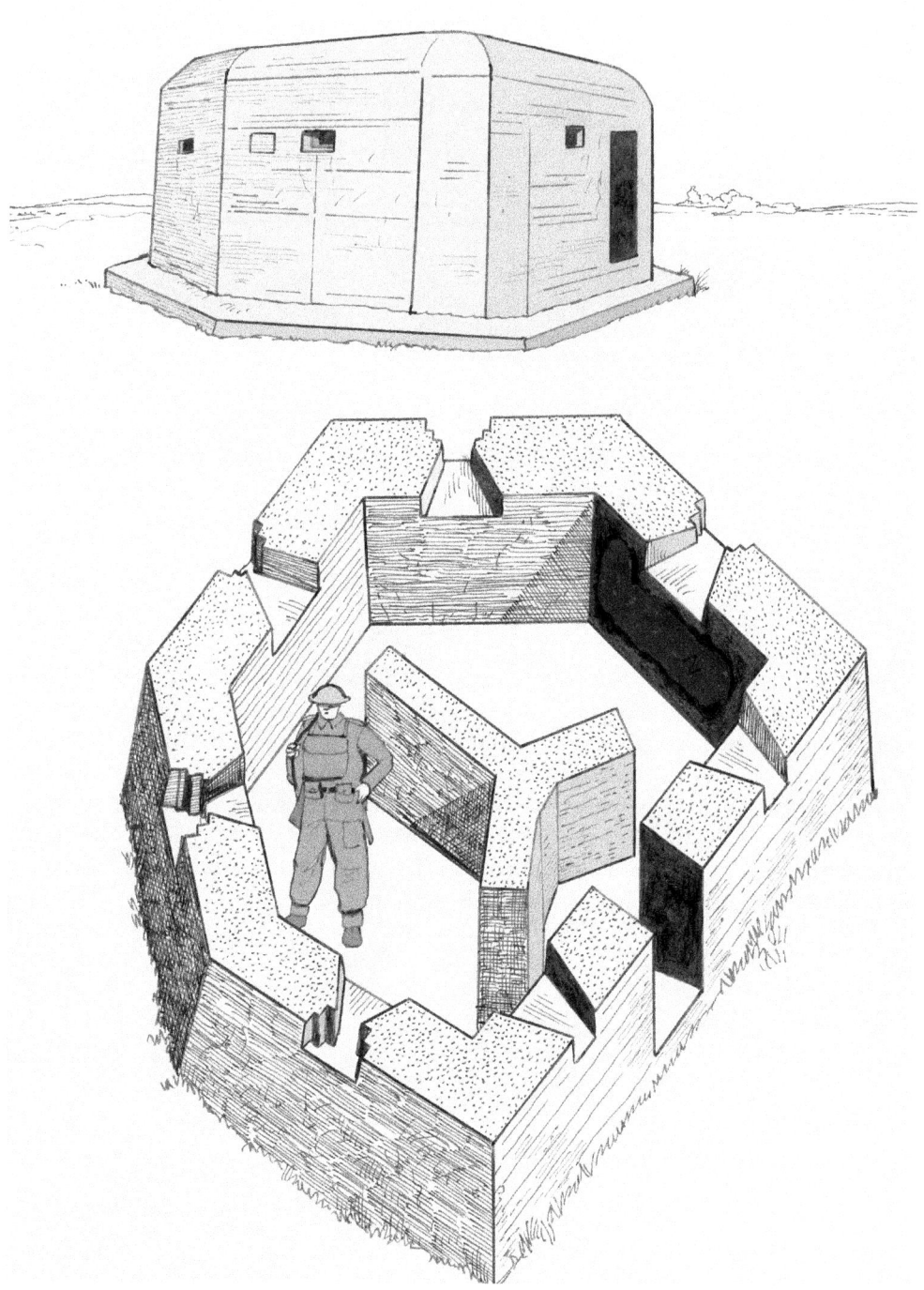

Pillbox type FW3/24.

Pillbox Type FW3/25

Type 25 was circular and small, being 2.4 meters (8 feet) in diameter. It normally had three loopholes suitable for infantry rifles or light machine guns, with the entrance at the back. The circular wall had a thickness of 30 centimeters (12 inches). Type 25 was made of reinforced concrete shuttered by corrugated iron plates produced by the Armco Company and was hence sometimes referred to as the Armco pillbox. It was not a common type, and only 46 are still extant.

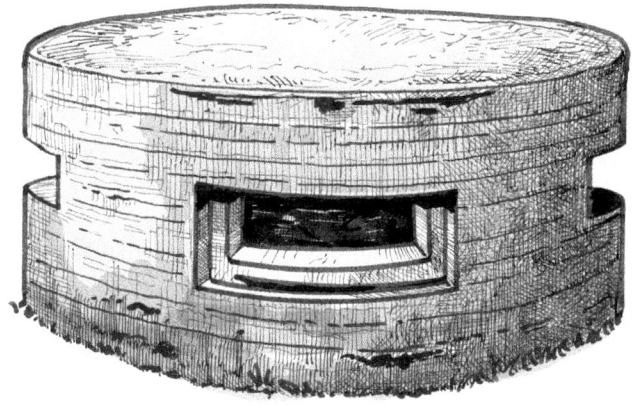

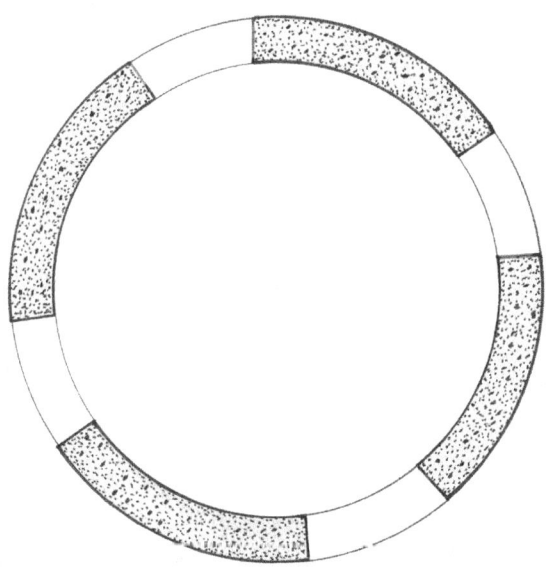

Pillbox type FW3/25.

Pillbox Type FW3/26

In plan, Type 26 was a simple square with walls 3 meters (10 feet) long. These were 46 centimeters (18 inches) thick with loopholes on each side and one entrance at ground level. This pillbox was not common, and about 200 are recorded as extant. There was a prefab variant of Type 26, which came as a kit and allegedly could be built in a day owing to inner and outer shuttering provided by precast concrete slabs slotted into reinforced concrete posts. This was sometimes referred to as Stent pillbox, after the company that produced its components, Stent Precast Concrete.

Part 7. British Fortifications in World War II

Pillbox type FW3/26.

Pillbox Type FW3/27

Of all FW3 designs, Type 27 was the most elaborate. It was hexagonal or octagonal in plan with walls between 3 and 3.5 meters (9 feet 9 inches and 11 feet 6 inches) long, and each side was provided with embrasures suitable for rifles or light infantry Bren machine guns. The pillbox was built to shellproof standard with walls 91 centimeters (36

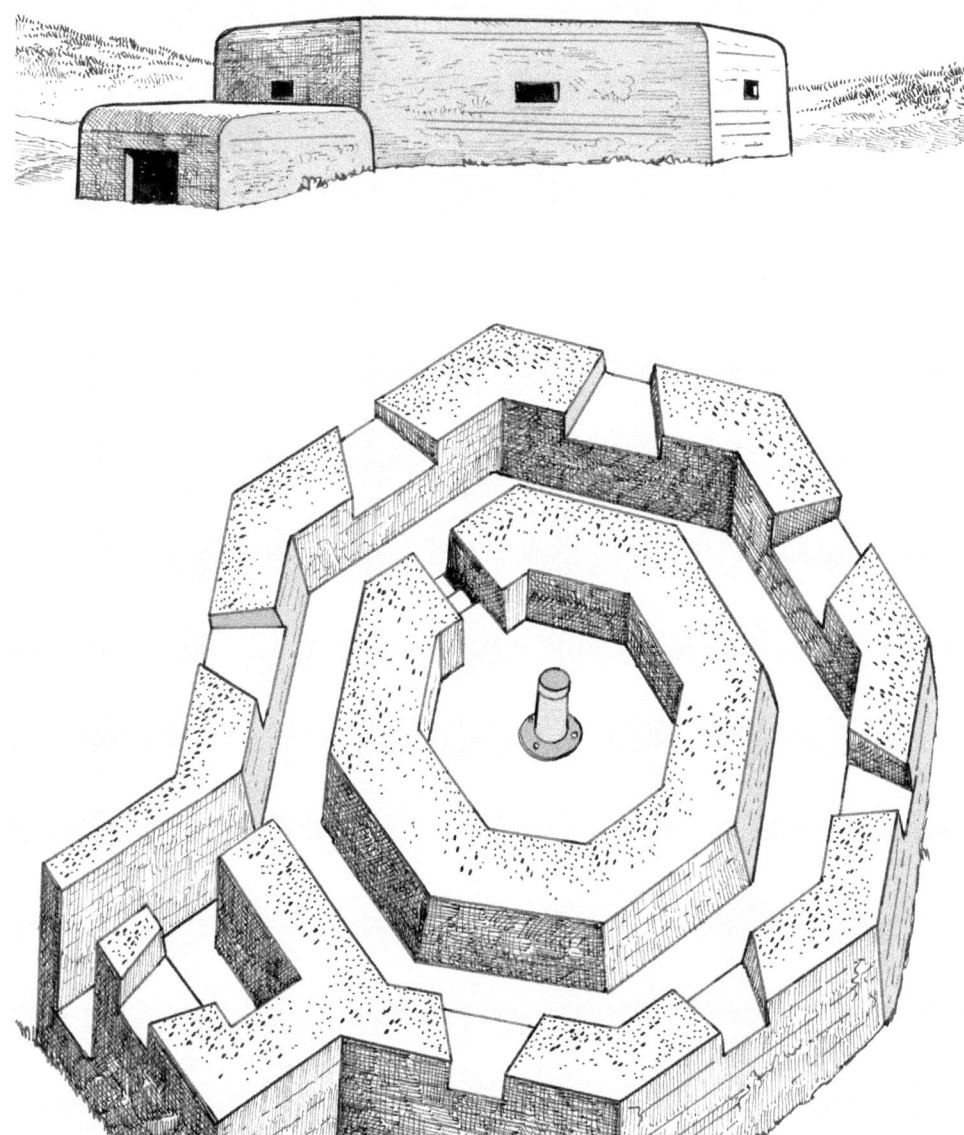

Pillbox type FW3/27.

inches) thick. The most notable features were a slightly raised central open pit placed on top for a light antiaircraft weapon and a defended entrance. Often located on or near airfields, this pillbox is rather uncommon with only 127 still extant.

Pillbox Type FW3/28a

Type 28 was the largest of all FW3 designs. It was nearly square, namely 6.1 meters by 5.8 meters (20 by 19 feet), and its walls were of shellproof standard with a thickness of 107 centimeters (42 inches). It was the only pillbox that could take artillery, as it was designed to house a QF 2-pounder antitank gun or other light artillery pieces like a Hotchkiss 6-pounder. It was fitted with a rather large frontal embrasure for the gun (with a traverse of 60°) and two smaller loopholes suitable for observation and use of rifles or light machine guns on the sidewalls. The entrance, placed at the rear, was large enough for moving the gun in and out. There was a wider, improved, and common variant, known as Type 28a (depicted below), with a frontal firing emplacement for infantry weapons separated from the artillery chamber by a concrete wall, allowing defense against an enemy head-on infantry attack. Type FW3/28 and its variants (of which some 350 are recorded as extant) were usually placed along a fixed line of defense, covering a bridge, or enfilading an antitank ditch. There were also numerous nonregular field gun emplacements based on Type 28 designed and built by local troops.

Pillbox type FW3/28.

FW3/28A Twin

Another (rare) variant on version FW3/28 was Type 28A twin, which had two large gun embrasures in two adjoining walls allowing firing in two different directions with two additional infantry firing compartments.

FW3/28A Twin.

Pillbox for Vickers Machine Gun

The Vickers machine gun pillbox, also known in Norfolk as Type 20, somewhat resembled Type 28, but it was smaller. Its plan was square with walls 4.3 meters (14 feet) long and 91 centimeters (36 inches) thick. Its front was fitted with a large embrasure

Pillbox for Vickers machine gun.

for a heavy solid and reliable water-cooled .303-inch (7.7-millimeter) belt-fed Vickers machine gun, whose tripod mount rested internally on a trapezoidal concrete table. The other walls were provided with loopholes suitable for infantry rifles and Bren light machine guns. The entrance, located at ground level either on the left or right side, was often provided with a blast wall for extra protection. Vickers machine gun emplacements were often placed in pairs for enfilading fire on beaches and coastlines and frequently dug in and covered with earth for additional protection and camouflage purpose.

Lozenge Pillbox

The so-called lozenge pillbox, found only in the North East of England and in the coastal crust beach defenses, was named after its stretched hexagonal shape. The front and rear walls were longer than the others enabling space for four frontal embrasures suitable for infantry rifles and/or light machine guns. The entrance (often protected by

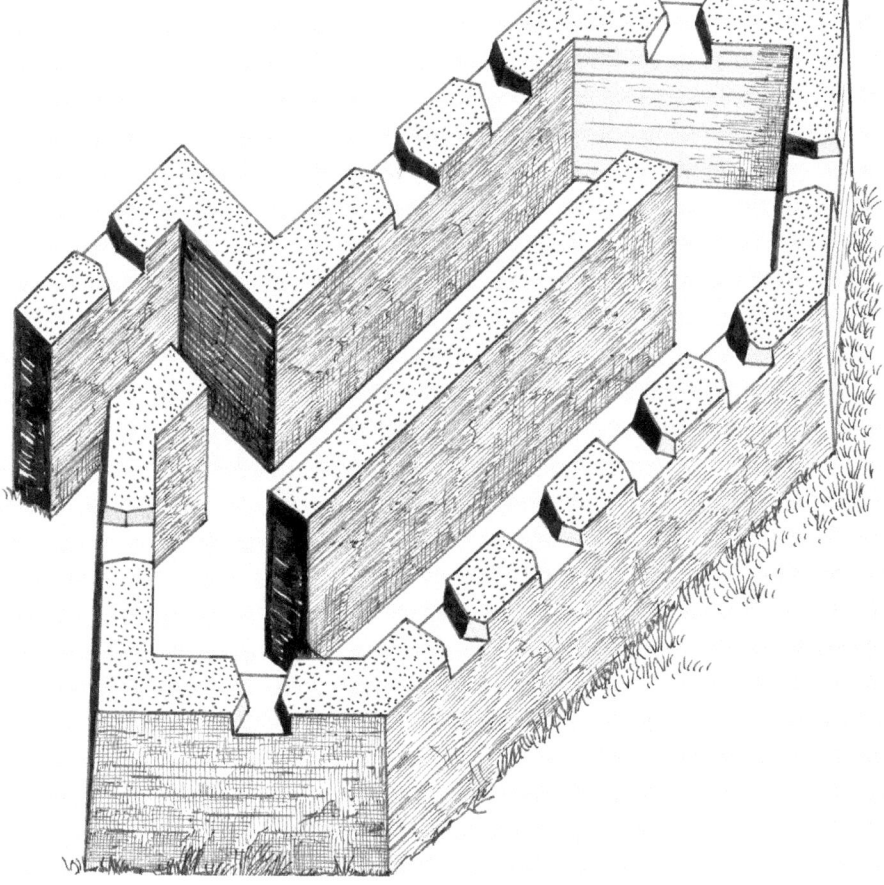

Lozenge pillbox (front view and horizontal section at the level of the embrasures).

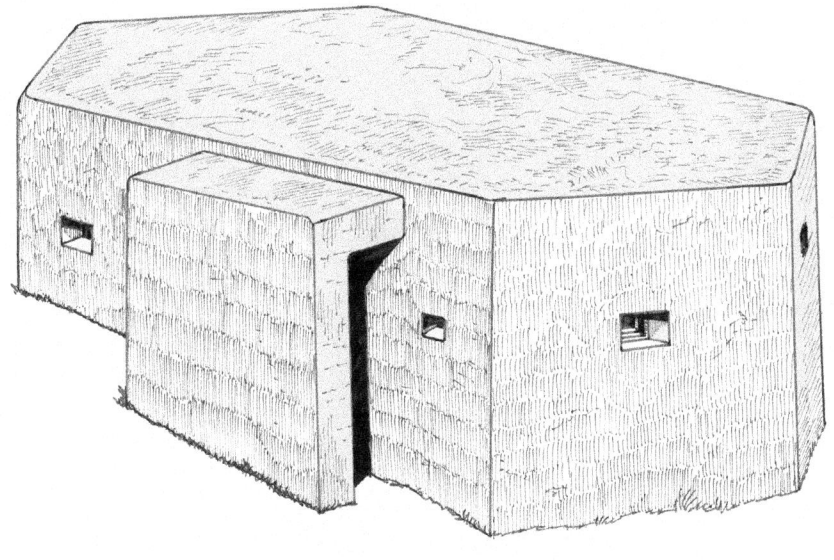

Lozenge pillbox (seen here from the back).

a blast wall) was placed in the rear wall with two additional loopholes. Internally, there was an anti-ricochet wall separating the inner space into two distinct firing chambers.

Eared Pillbox

The "eared" pillbox had the plan of an irregular hexagon. It was designed to house two Vickers machine guns firing through two wide and narrow loopholes facing forward. The two embrasures were at 90° to each other allowing for an arc of fire of 180°. The wall bulged out beneath the loopholes to allow space for accommodating water-cooling cans for the Vickers machine guns. The pillbox was particularly badly designed as its walls were rather thin, it was not fitted with rifle loopholes for close-range defense, and the two forward-facing entrances were exposed to enemy fire, thus leaving little possibility of exit once under attack. Eared pillboxes would certainly have been very unpopular positions. They were only used as part of the coastal crust defenses and, like the lozenge pillboxes, were only found on positions overlooking beaches in the North East of England.

Eared pillbox.

Lincolnshire Pillbox

This pillbox earned its name from the fact that it was built only in Lincolnshire. Basically, it was a variant of the FW3/23 including two roofed firing chambers (each fitted with three loopholes) and a central slightly raised emplacement for an antiaircraft machine gun.

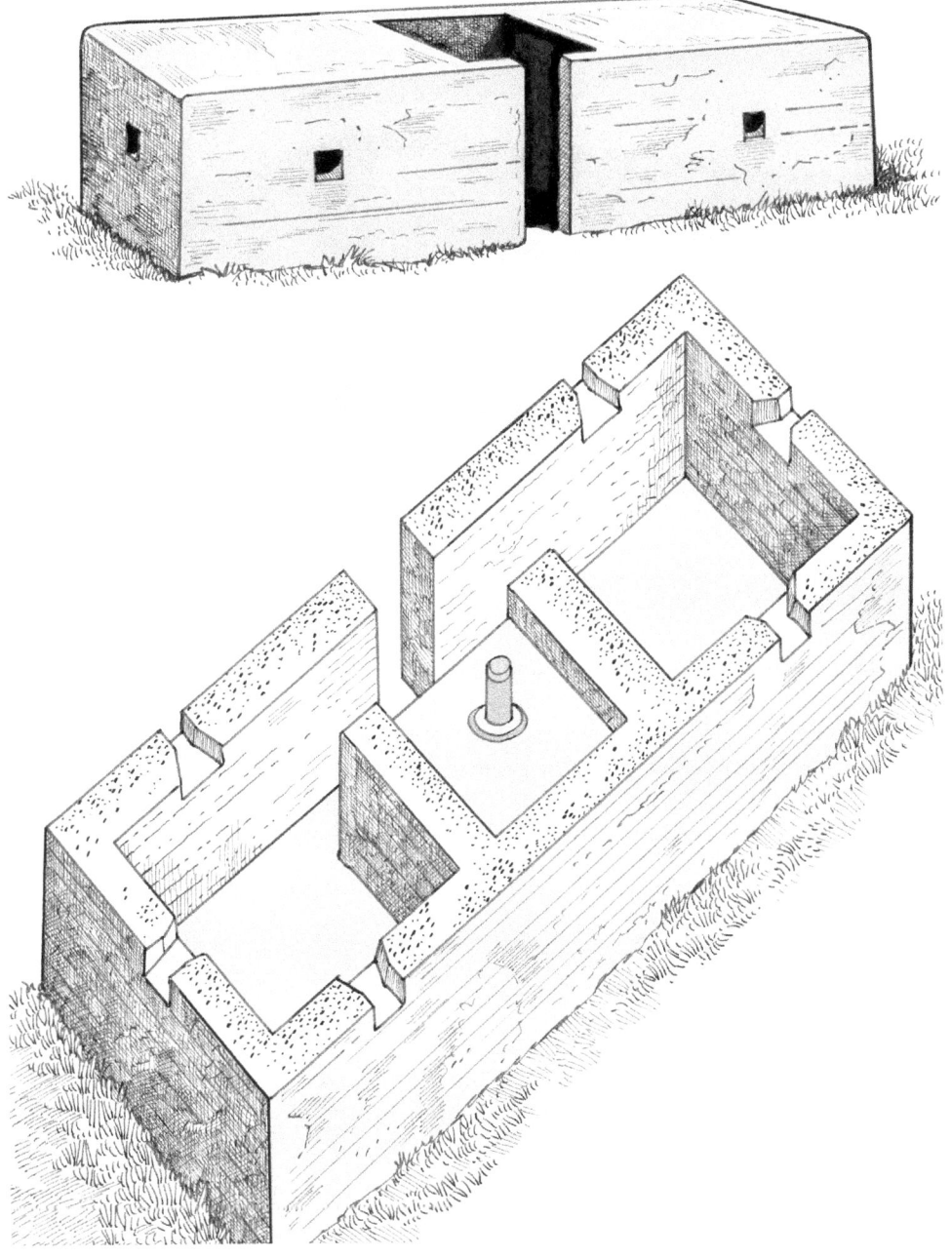

Lincolnshire pillbox.

Dover Quad Pillbox

The so-called Dover Quad pillbox (built only around high grounds commanding the port of Dover) was a square structure measuring 4 meters by 4 meters (13 feet by 13 feet). Each wall had two large embrasures suitable for rifles and light machine guns. The pillbox was covered with a thick overhanging concrete slab.

Dover Quad pillbox.

Norcon Pillbox

The Norcon pillbox, named after the company that designed and produced it as a private commercial venture, consisted of a nonreinforced concrete pipe usually 1.8 meters (6 feet) in diameter, 1.2 meters (4 feet) in height, 10 centimeters (4 inches) in thickness, and fitted with several cut loopholes. The original design had no roof, but later models had a roof made of timber and corrugated iron covered with earth. The vulnerable pillbox thus offered very poor protection, but it was an extremely convenient makeshift, quick to produce and cheap to build. Although extra protection could be provided by the use of piled earth and sandbags placed around and on top of it, very few were actually built.

Part 7. British Fortifications in World War II

Norcon pillbox.

Cross-section Norcon pillbox [1]. Loophole, (2) earth cover, (3) corrugated iron plate, (4) concrete wall, (5) entrance trench, (6) sandbags.

Ruck Pillbox

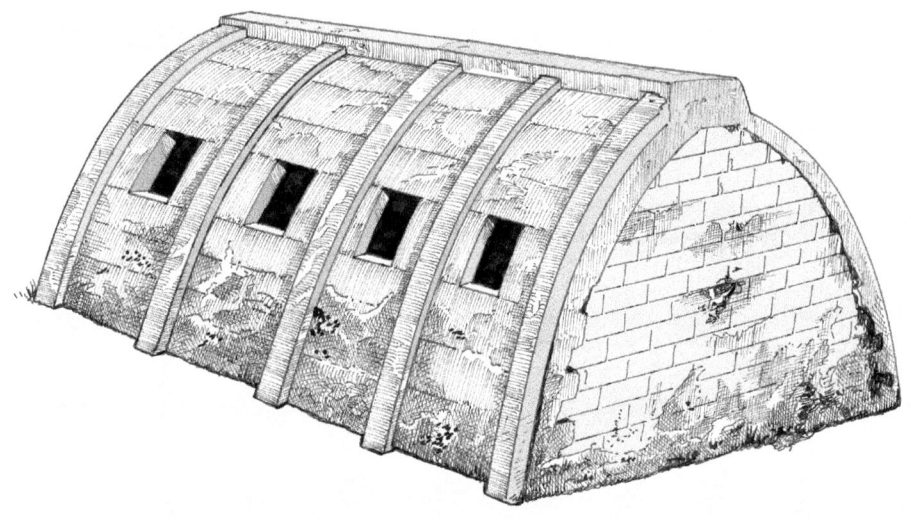

Ruck pillbox.

The Ruck pillbox (designed by James Ruck) was an infantry post made from prefabricated Stanton air raid shelter sections. The segments were 20 inches wide, and a pair of them formed an arch 7 feet high to which transverse struts were provided to ensure rigidity. These fitted into longitudinal bearers, which were grooved to receive the foot of each segment. Each pair of segments were bolted together at the apex of the arch, and each segment was also bolted to its neighbor, the joints being sealed with bituminous compound. The convenient handling of these segments enabled them to be transported with ease. The pillbox was designed to be partially buried or reinforced with sandbags and piled earth. It was intended for eight infantrymen. Although the Ruck pillbox was said to have poor application because of its limited field of fire, Northern Command placed orders for 6,000 in September 1940 and planned to site them at all antiaircraft batteries, searchlight positions, airfields, and other defended localities. Although 4,000 have been erected in Lincolnshire and along the coast of south England and although there was an order for 2,000 more, only a few are extant.

Spigot Mortar Emplacement

The roofless spigot mortar emplacement was made of concrete or brick, or simply revetted with earth and reinforced with sandbags. The weapon (also known as the 29-millimeter Blacker Bombard from its designer, Lieutenant Colonel Stewart Blacker) was mounted on a stainless-steel pin, which was set in a central concrete base. The weapon fired 20-pound high-explosive antitank bombs, which were propelled by black

Spigot mortar emplacement.

powder. The mortar had an effective range of 90 meters (100 yards) in its antitank role and up to 460 meters (500 yards) when firing a lighter antipersonnel 14-pound projectile. The emplacement's indented sides were recesses where ammunitions were kept and where soldiers who made up the gun crew (usually three) could crouch.

Obstacles

Antitank Obstacles

The primary purpose of the stop lines and the antitank islands that followed was to hold up the enemy, slowing down their progress, restricting access, and delaying attacks. The need to prevent tanks from breaking through was of key importance, and much ingenuity was devoted to the development of antitank devices, both weaponry and static hindrances. Consequently, the defenses generally ran along preexisting barriers such as rivers and canals, railway embankments and cuttings, steep hills, thick woods, and other natural obstacles. Where possible, usually well-drained land was allowed to flood making the ground too soft to support even tracked vehicles.

Thousands of miles of antitank ditches were dug, usually by mechanical excavators but occasionally by hand. They were typically 18 feet (5.5 meters) wide and 11 feet (3.4 meters) deep and could be either trapezoidal or triangular in section with the defended side being especially steep and revetted with whatever material was available.

Elsewhere, antitank barriers were made of massive reinforced concrete obstacles either cubic or pyramidal/cylindrical. The cubes generally came in two sizes: 5 feet (1.5 meters) or 3.5 feet (1.1 meters) high. In a few places, antitank walls were constructed—essentially continuously abutted cubes. An even more robust barrier to tanks was provided by long lines of antitank cubes. The cubes were made of reinforced concrete 5 feet (1.5 meters) to a side. Thousands were cast in situ in rows sometimes two or three deep.

Large cylinders were made from a section of sewer pipe 3–4 feet (91–120 centimeters) in diameter filled with concrete typically to a height of 4–5 feet (1.2–1.5 meters), frequently with a dome at the top. Smaller cylinders cast from concrete were also frequently found.

Anti-tank obstacles: (1) concrete cylinder; (2) hair spin; (3) rail.

Antitank pimples.

Part 7. British Fortifications in World War II

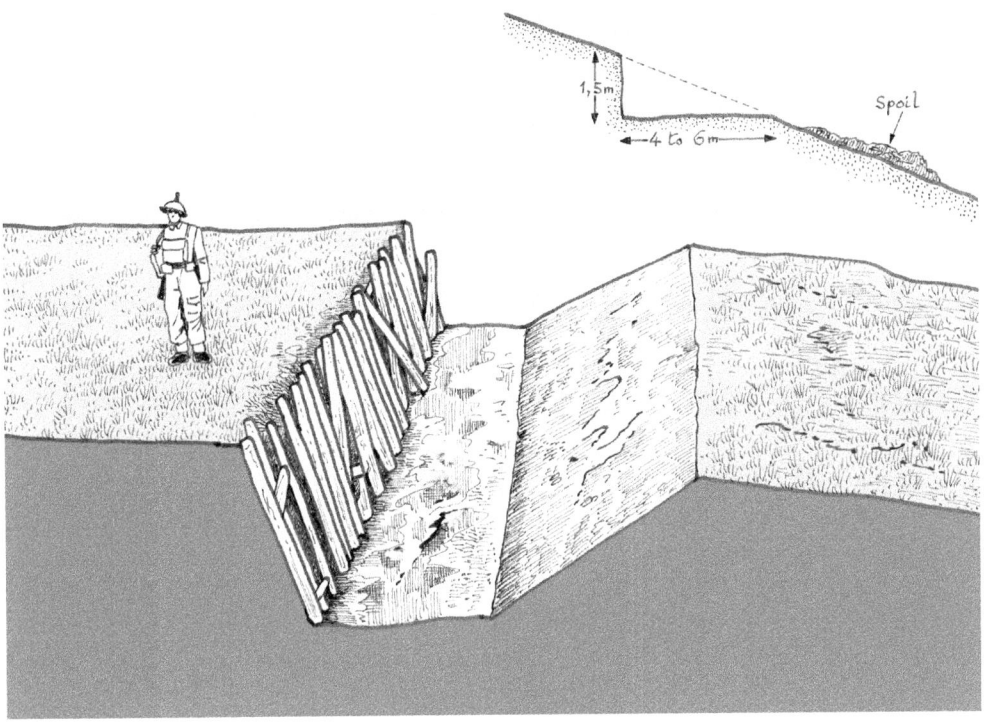

Antitank ditch (cross-section).

Pimples, popularly known as dragon's teeth, were pyramid-shaped concrete blocks designed specifically to counter tanks which, attempting to pass them, would climb up exposing vulnerable parts of the vehicle and possibly slip down with the tracks between the points. They ranged in size somewhat but were typically two feet (61 centimeters) high and about three feet (91 centimeters) square at the base. There was also a conical form.

Cubes, cylinders, and pimples as well as poles, beams, tetrahedral, or caltrop-shaped obstacles were deployed in long rows to form antitank barriers on the beaches and inland. They were also used in smaller numbers to block roads and highways the invaders were likely to come along. They frequently sported loops at the top for transport by a crane and when positioned for the attachment of barbed wire.

Roadblocks

Where natural antitank barriers needed only to be increased, concrete or wooden posts sufficed.

Roads offered the enemy fast routes to their objectives and consequently, they were blocked at strategic points. The simplest of the movable roadblocks consisted of concrete cylinders of various sizes but typically about three feet (0.91 meter) high and two feet (61 centimeters) in diameter; these could be manhandled into position as required. However, these would be insufficient to stop armored vehicles. One common type of mobile antitank roadblock consisted of massive concrete posts permanently installed at the roadside; these posts had holes and/or slots to accept horizontal railway lines or

Roadblock.

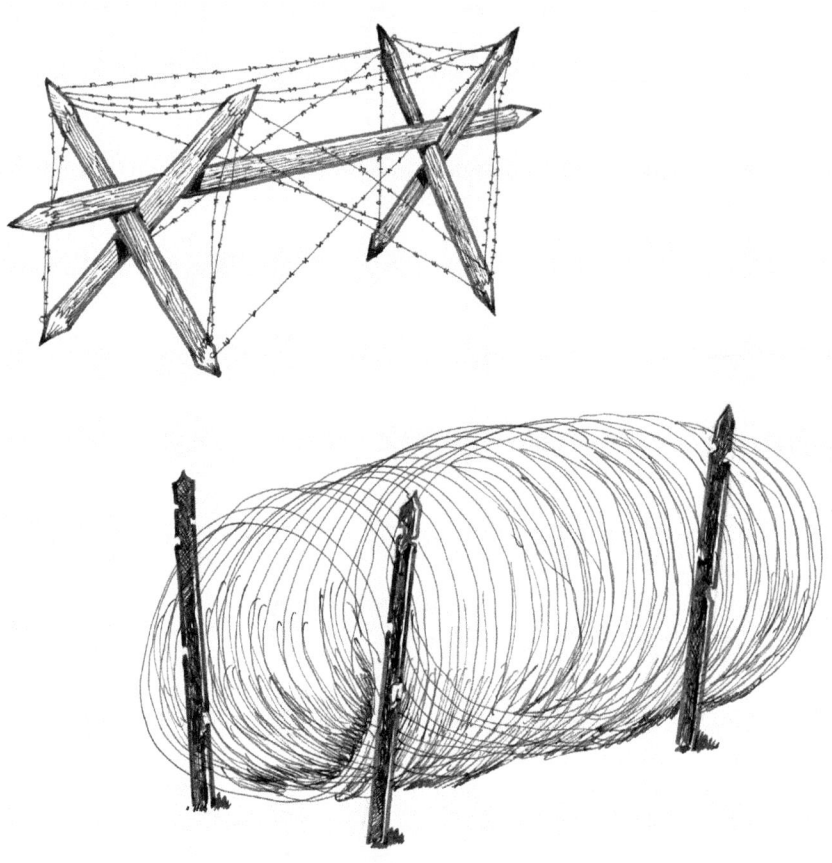

Knife rest and concertina wires.

Ankle-height barbed wires.

rolled steel joists. Similar blocks were placed across railway tracks because tanks could move along tracks almost as easily as they could along roads. These blocks would be placed strategically where it was difficult for a vehicle to go around—antitank obstacles and mines being positioned as required—and they could be opened or closed within a matter of minutes.

There were two types of socket roadblocks. The first comprised vertical lengths of railway line placed in sockets in the road was known as the hedgehog. The second type comprised railway lines or rolled steel joists bent or welded at around a 60° angle, known as hairpins. In both cases, prepared sockets about six square inches (152.40 square millimeters) were placed in the road, closed by covers when not in use, allowing civilian traffic to pass normally.

Another movable roadblocking system used mines. The extant remains of such systems superficially resemble those of the hedgehog or hairpin, but the pits were shallow, just deep enough to take an antitank mine. When not in use, the sockets were filled with a wooden plug allowing traffic to pass normally.

All obstacles and explosive devices were intended to delay or thwart enemy advance and funnel invaders into a killing ground where they would fall victim to machine gun or antitank weapons. Obstacles could not be ignored or carelessly bypassed.

Prepared Demolition

Bridges and other key points were prepared for demolition at short notice by preparing chambers filled with explosives. A depth charge crater was a site in a road (usually at a junction) prepared with buried explosives that could be detonated to instantly form a deep crater as an antitank obstacle. The Canadian pipe mine (later known as the McNaughton tube after General Andrew McNaughton) was a horizontally bored pipe packed with explosives—once in place this, could be used to instantly ruin a road or runway. Prepared demolitions had the advantage of being undetectable from the air—the enemy could not take any precautions against them or plot a route of attack around them.

Camouflage

With respect to camouflage, the British were indeed quite resourceful. Detailed instructions were given for the careful concealment of pillboxes, obstacles, and other field defenses. From a distance across the fields or from the air in an aircraft, all defenses were as much as possible concealed and blended with their local environment. Many bunkers were dug into the ground or inserted into a hedgerow or a hillside to provide the lowest possible profile, while others on flat terrain had soil piled up on the roof and sides. Deceiving paint schemes, camouflage netting, fresh-cut vegetation, and other expedients were widely used to help break up the outline and shadow of the pillboxes. Local materials were often employed—concrete made with beach sand, a covering of pebbles, gravel or stone from a nearby beach. This was not only a time-saving measure, but it also aided camouflage by merging the defenses into the background. Artists such as Roland Penrose (author of the *Home Guard Manual of Camouflage*), Stanley William Hayter, Julian Trevelyan, and many others were employed to design camouflage patterns. In built-up areas, pillboxes were disguised to look like a part of an adjacent building, carefully matching and provided with a roof to look as if they had always been there. In extreme cases, they were built inside or leaning against existing buildings. Some pillboxes were carefully painted to resemble a quite different civilian and innocent structure: a haystack, a pile of logs, a disused cottage, a barn, a closed seaside kiosk, a beach snack bar or a restaurant, a summerhouse, a bus stop shelter, or a railway signal box. It was not uncommon for pillboxes to be fitted with a dummy pitched roof to increase the deception. In some cases, the reinforced concrete roof was sculpted to make the distinctive form of a pillbox less obvious from the air. Some clearly visible bunkers, conversely, were only mock-ups of timber and canvas with heavy guns being painted telegraph poles. Dummy installations were thus built in order to present a strength that actually did not exist. Those illusive positions were meant to attract enemy fire. Some of these disguises bordered on the fanciful. In Somerset, along part of the Taunton Stop Line due to the shortage of material available, six pillboxes were coated with a mixture of cow manure and mud and topped with straw forming a kind of natural stack. Close to Axminster (East Devon), a square pillbox was disguised as a Romany trailer. During the summer months, a scarecrow "family" and a horse made of straw were dressed and suitably arranged around the caravan to visually hide and deceive what actually was an armed pillbox.

Volunteers of the Home Guard and regular soldiers were encouraged to use anything that would delay the enemy. In addition to hiding real weapons and fortifications, steps were taken to create the impression of the existence of defenses that were not real. Drainpipes stood in place of real guns, dummy pillboxes were constructed, and uniformed mannequins kept an unblinking vigil.

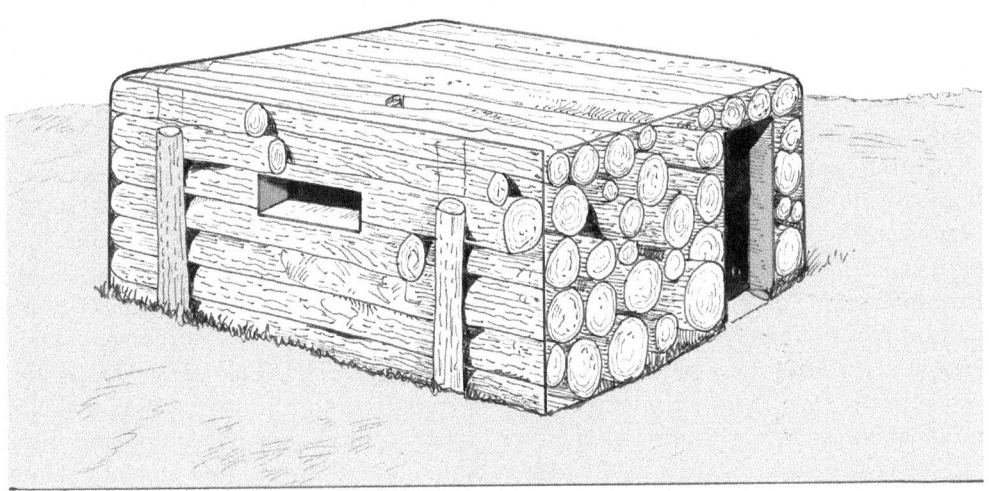

Camouflaged pillboxes. Top: Deceivingly painted like a pile of logs. Bottom: Looking like an innocent barn with the addition of a fake tiled roof.

Frantic use was made of any existing walls or buildings and loopholes for firing or passing heavy chains and cables through to form barriers strong enough to slow down or stop soft-skinned vehicles. Suspicious-looking parcels could be attached to strengthen the illusion of a mine, a time bomb, or a booby trap.

Sea Forts

Maunsell Forts

When FW3 pillboxes and other improvised bunkers were on the whole cheap, unsophisticated, and generally of poor quality, there were other impressive defense structures constructed in Britain during World War II. Among these were sea forts, known as the Maunsell Forts, named after their designer, the civilian architect Guy A. Maunsell. The idea originated from a steel-and-concrete cylindrical tower, which had been constructed in 1918 by the end of World War I in the English Channel off Shoreham in Sussex to obstruct the Nab rocks east of the Isle of Wight. Maunsell Forts were built in 1942 and 1943 as advanced positions off the Kent shores in the Thames estuaries and in the Mersey estuary in order to defend the approach to London and Liverpool, respectively. Their main purpose was to act as antiaircraft units for stopping German mine-laying aircraft, whose mines could disrupt a significant amount of shipping. They also served as warning posts for German bomber formations heading toward London and Liverpool using the rivers as landmarks. They also had the mission of deterring E-boat (rapid torpedo boats) raids. The Maunsell Forts were of two basic types: army and navy.

Army Fort

The army fort (built at Nore, Red Sands, and Shivering Sands) consisted of several armored boxlike towers comprising two levels and a roof on which their equipment was located. Each tower was mounted on four long concrete legs that rested on a concrete base anchored in the seabed. The lower level housed the officers' quarters, stores, and latrine facilities. The upper level was the main barracks area for the troops. The fort's equipment and weapons were on the open top-deck terrasse. Seven of these structures were placed close together and linked by steel footbridges. The central tower was the fire control post, and four of the outer positions mounted heavy 94-millimeter AA guns and 3.7-inch guns, with one other position mounting Bofors 40-millimeter guns. The towers with weapons formed a circle around the central control position. One of the towers was equipped with searchlights. The garrison for this type of fort was about 120 men.

Three army forts also occupied positions off the Mersey estuary leading to Liverpool on the west coast code named Queens, Formby, and Burbo. Originally, there were going to be three more offshore forts in the Liverpool Bay area, but they never materialized.

There was talk of constructing more of this type of army Maunsell Forts off the Humber estuary as well as in the seaports of Portsmouth, Rosyth in Scotland, and Belfast and Londonderry in Northern Ireland, but this never happened. The Humber estuary was protected by the already existing Bull Sand Fort (built in 1919) and Haile Sand Fort (1918) constructed on sand spits. Bull Sand, armed with four 6-inch Mark VII guns, returned to full service in 1939 with two of its guns replaced by twin 6-pound weapons for anti-boat service. The military removed Haile Sand's old armament of two 4-inch guns in 1928 and, in 1939, set up two 12-pound guns, which in 1940 were replaced with twin 6-pound guns. Older Victorian sea forts received new armament for similar purposes on other inlets and British harbors.

Navy Forts

The navy forts (established at Rough Sands, Sunk Head, Knock John, and Tongue Sands) were mounted on two large cylindrical concrete pillars, which were hollow, and placed on a barge-shaped concrete caisson. Once located to its strategic position, seawater was let to pour inside the hollow caisson, which sank and came to rest in the seabed. The large pillars contained seven levels with facilities for the garrison of about 120 men: magazines, storage rooms, infirmary, officers' and crew quarters, power generators, kitchen, latrines, etc. Resting above the big pillars, there was a large flat steel platform with the control room, eventually radar installation on top of a tall central tower, and armament consisting of 3.7-inch and 40-millimeter Bofors antiaircraft guns.

Assessment

Regarding the effectiveness of the Maunsell sea forts, it was reckoned that during World War II, the Thames estuary forts shot down 22 German aircraft and around 30 Fieseler Fi 103 flying bombs, better known as V1 or buzz bomb.

The offshore Maunsell Forts also achieved one naval victory: the sinking of a Kriegsmarine E-boat (a fast torpedo-armed light attack craft).

With regard to the Liverpool Bay series of forts, there is no record of them inflicting any casualties on the enemy. This may all seem ineffective considering how much time, effort, and resources were spent on the Maunsell project. But perhaps their true value was more as a deterrent as well as serving as useful intelligence gathering and warning posts.

Maunsell army fort at Red Sands.

By the early 1950s, all Maunsell Forts were decommissioned by the Ministry of Defence, but there was no attempt to destroy or dismantle them. They were simply abandoned by the military. All three Liverpool Bay forts were demolished by 1955 as they were deemed to be a hazard to shipping, and other forts had a curious fate.

There was an unsuccessful attempt by a certain Paddy Roy Bates (a British pirate radio broadcaster) to set up an independent nation (with flag and constitution) on the deserted Maunsell Fort Roughs located about 12 kilometers from the coast of Suffolk. The nation was called the Principality of Sealand; it was certainly a kind of joke and was never officially recognized by any other country as a sovereign nation.

At various stages during the 1960s, Maunsell Forts were used as pirate radio stations, notably Radio Caroline created by Ronan O'Rahilly, which was outlawed by the Marine Broadcasting Act in 1967.

As for the army forts, five still exist today in various states of disrepair. There is an effort called the Redsands Project whose objective (as of 2018) is to renovate the Redsands Army Fort and turn it into a museum since it is considered to be in the best condition of the surviving forts.

Churchill's Command Bunker

In early 1939, the cellar of His Majesty's Treasury building located at Clive Steps, King Charles Street near St James's Park in southwest London was converted to function as a secret command center, known as the Cabinet War Rooms. The large cellar, intended to shelter Prime Minister Winston Churchill and his government during the Blitz, was equipped with a ventilation system from a ship, but it lacked comfort. In contrast with Hitler's strongly built concrete bunker located under the Chancellery in Berlin, Churchill's secret command center was spartan and badly protected, notably against air attack as the ceiling was made of the original wooden floor. It had certainly not resisted an aerial bomb direct hit. It was only by the end of 1940 that a concrete protection 1.5-meter-thick roof was installed. The bunker consisted of a main entrance, corridors, and about 40 rather small rooms including meeting rooms for the army headquarters and for the war cabinet, a map room, an office for Churchill, and several working rooms for his secretaries, administrative personnel, and ministries, associates, and collaborators. There was also sleeping, eating, and living accommodation for the prime minister, his wife, and his daughter, for the bodyguards, and for senior officers, guests, and personnel. Churchill, however, rarely slept in the underground shelter, preferring to sleep at 10 Downing Street or the No. 10 Annexe, a flat in the New Public Offices directly above the Cabinet War Rooms. The command complex also included a studio with broadcasting equipment, a communication switchboard exchange, and an office with a direct transatlantic telephone hotline with America, which allowed Churchill to securely speak with U.S. president Franklin D. Roosevelt. Until the surrender of Japan in August 1945, the British government constantly used the command bunker where important wartime decisions were taken.

After the war, the Cabinet War Rooms became redundant and was abandoned. In the 1980s, its historical importance was however recognized, and since April 1984, the bunker has become a museum (commonly called the Churchill War Rooms or Churchill Museum) open to the public and managed by the Imperial War Museum.

Evaluation

By 1942–43, the threat of invasion had greatly diminished, and the need for costly and elaborate anti-invasion defenses had become redundant.

The question of whether the 1940–41 British defenses would have been effective is disputed and remains an open and somewhat pointless debate, as suppositions are endless. We shall never know; we can only speculate. In mid-1940, the preparations relied heavily on field fortifications, and World War I had made it clear that assaulting prepared defenses with infantry was deadly and difficult. However, similar preparations in Belgium and the Netherlands had been overrun by well-equipped assault engineers and German Panzer divisions in the early weeks of 1940. The German army had also dealt with the immeasurably stronger French Maginot Line fortifications, so there is no doubt the poor FW3 pillboxes (only light ammunitions–proofed) would have probably been quickly overwhelmed. Besides, with so many British Expeditionary Force vehicles, heavy weapons, and armaments lost and left behind at Dunkirk, British forces were woefully ill-equipped to take on German armor. On the other hand, although British preparations for defense were ad hoc, so were the German invasion plans: a fleet of 2,000 converted barges and other vessels (notably catamaran Siebel ferryboats) had been hurriedly made available and their fitness was debatable; in any case, the Germans could not easily land troops with all their heavy equipment. Until the Germans would have captured a port like Dover, Ramsgate, Brighton, or Folkestone, for example, the invaders would have been short of support weapons like tanks and heavy guns.

The later experiences of the Canadian Army during the disastrous Dieppe raid in 1942, the American forces at Omaha Beach, Normandy on D-Day June 6, 1944, and U.S. Marines taking on Japanese defenders on Pacific Islands showed that under the right conditions, a defender could inflict enormous casualties to an amphibious assaulting force, significantly depleting and delaying enemy forces until reinforcements could be deployed.

In the event of a German invasion, the Royal Navy would of course have rapidly sailed to the landing places. The German forces would perhaps have been able to land and gain a significant beachhead, but intervention by the Royal Navy would probably have been decisive.

Following the failure to gain even local air superiority in the Battle of Britain, Operation Sea Lion was postponed several times and canceled definitely. Hitler and his generals were aware of the problems of an invasion. Hitler was not ideologically committed to a long war with Britain, and many commentators suggest that German invasion plans were a feint never to be put into action. This, too, is and endless debate.

Although Britain may have been militarily secure in 1940, both sides were aware of the possibility of a political collapse. If the Germans had won the aerial Battle of Britain, the Luftwaffe would have been able to strike anywhere in southern England, and with the prospect of an invasion, maybe the British government would have been obliged to come to terms under pressure. The extensive anti-invasion preparations demonstrated to Germany and to the people of Britain that whatever happened in the air, the United Kingdom was both able and willing to defend itself. The British determination to resist and fight Nazi Germany was expressed by Prime Minister Winston Churchill in a famous galvanizing speech delivered in the House of Commons on June 4, 1940: "*We shall fight on the beaches, we shall fight on the landing grounds, we shall fight in the fields and in the streets, we shall fight in the hills; we shall never surrender."*

German Fortifications in the Channel Islands

The only British territories conquered and occupied by Nazi Germany during World War II were the Channel Islands (Guernsey, Jersey, Sark, and Alderney) located off the western coast of the French Cotentin peninsula in Normandy. The Channel Islands were incorporated into the German Atlantic Wall—a continental fortification line stretching from Norway in the north to the Franco-Spanish border in the south. Its aim was to repulse any Allied attempt to land and open another front in Western Europe. The British Channel Islands were extremely well fortified and represented the most formidable sections of the Atlantic Wall. For the Nazi regime, the British Channel Islands offered an immense propaganda value, as the occupation could be presented as an early sign of Nazification of the British people. Strategically, they formed a fore post at sea allowing surveillance and defense of the west Cotentin peninsula and northern Brittany. Because Hitler believed that the British were awaiting a suitable opportunity to regain possession of them, he became increasingly anxious to make the islands secure. For psychological and propaganda reasons, which went far beyond military considerations, Hitler was obsessed with the Channel Islands. Therefore, he decreed in October 1941 that the islands of Jersey, Guernsey, and Alderney had to be turned into impregnable sea fortresses. Designated AOK 319 Division, the islands were occupied by 21,000 soldiers belonging to the 319th Infantry Division, the German force eventually reached a strength of 40,000 soldiers including a tank regiment, artillerymen, Flak gunners, signal troops, and Luftwaffe and naval personnel. The defense of the Channel Islands was deemed so important that military engineers' creativity was given free rein. The Channel Islands present outstanding and original concrete structures in contrast to the other repetitive archetypal designs along the rest of the Atlantic Wall. There are still amazing and impressive multistory observation towers placed on top of spectacular and craggy cliffs. There were also several superheavy long-range coastal installations—notably, the formidable Mirus battery of 30.5-centimeter (12-inch) guns on Guernsey, numerous Flak (antiaircraft) batteries, fortified resistance points, groups of resistance points, and no less than eight kilometers of continuous concrete antitank walls. The German Organisation Todt (a group of militarized building companies) also built a large underground hospital at Meadowbank, Saint Lawrence. A monumental gate still gives access today to a maze of tunnels including operating theaters, quarters for doctors, orderlies, and nurses, wards, offices, and supply stores. Owing to Hitler's maniacal pride and obsessive fear, the defenses of the Channel Islands were an exaggerated and unnecessary deployment of resources. They only represented a pointless drain in men, equipment, and labor force.

If for the port of St. Helier and several German gun batteries, the Channel Islands were not attacked by the Allied air force, but for five dark years, the population was cold, hungry, and oppressed and left in total ignorance of world news, except for what was gleaned from secret wireless sets, used at serious risk; some islanders were fined, deported to Nazi concentration camps, or shot for offenses like listening to the forbidden BBC and spreading news and sabotage. After D-Day (June 6, 1944), Marshals Gerd von Rundstedt and Erwin Rommel appealed urgently to Hitler for the numerous idle troops in the Channel Islands to help stem the tide of the Allied advance in Normandy but to no avail. Hitler's most prized conquest was to remain fully staffed. The powerful concrete fortresses bristling with weapons never saw combat; they were simply ignored by the Allies. The German occupation troops in the British Channel Islands surrendered on May 9, 1945.

Other Theaters of Operation

The stunning victories by the Germans early in World War II showed that fixed fortifications like the Maginot Line were worthless if there was room to circumvent them. At the Battle of Sevastopol, Red Army forces successfully held trench systems on the narrow peninsula for several months against intense German bombardment. The Western Allies in 1944 successfully broke through the incomplete German Atlantic Wall through a combination of surprise, amphibious landings, naval gunfire, air attack, and airborne landings and assaults. Combined arms tactics where infantry, artillery, armor, and aviation cooperating closely seemed to render trench warfare inefficient. However, not all entrenchment were redundant. It was (and still is today) a valuable and rather cheap method for reinforcing natural obstacles to create a line of defense. For example, at the Battle of Stalingrad (August 1942–February 1943), soldiers on both sides dug trenches within the ruins. In addition, before the start of the Battle of Kursk (July–August 1943), the Soviets constructed a system of field defense more elaborate than any other they built during World War II. These defenses succeeded in stopping the German armored pincers meeting and enveloping the salient. Also, at the start of the Battle of Berlin (April–May 1945), the last major assault of World War II, the Russians attacked over the river Oder against German troops dug in on the Seelow Heights, about 50 kilometers (30 miles) east of Berlin. Entrenchment allowed the Germans, who were massively outnumbered, to survive a bombardment from the largest concentration of artillery in history; as the Red Army attempted to cross the marshy riverside terrain, they lost tens of thousands of men to the entrenched Germans before breaking through.

Home Britain was, of course, not the only place where the British army built fortifications. World War II was characterized by mobile warfare with huge armies crossing oceans and offensives moving on a continental scale. However, on every World War II static front, permanent, field, and semipermanent fortifications were established. The best example would be the previously discussed German Atlantic Wall, which stretched from northern Norway to southern France with the purpose of repulsing an attack on the coasts of the Atlantic Ocean along the western facade of Europe. Even the battles in North Africa (June 1940–May 1943), although dominated and characterized by rapid and vast movements of motorized and armored units, saw the use of field fortifications, e.g., during the siege of Tobruk, the Battle of Gazala, the Battle of El Alamein, and the campaign of Tunisia.

Tobruk

The siege of Tobruk was a battle between Axis and Allied forces in North Africa during the Western Desert Campaign. The siege started on April 10, 1941, when an

Italo-German force under the command of Lieutenant General Erwin Rommel attacked the port of Tobruk in Libya. It was vital for the Allies' defense of Egypt and the Suez Canal to hold the town with its harbor, as this forced the Axis forces to bring most of their supplies overland from the port of Tripoli, across 1,500 kilometers (930 miles) of desert, as well as diverting troops from their advance. The small city and harbor of Tobruk was subject to repeated ground assaults and almost constant artillery shelling and air bombing. The British and Commonwealth defenses ran in a semicircle from coast to coast. They consisted of concentric rings of barbed wire, tank traps, trenches, minefields, dugouts, and concrete pillboxes. The defenses were divided into two main lines. The outer red line was the main defense positions. It was constituted by the original Italian fortifications including concrete bunkers, antitank ditches, and minefields. The inner blue line was established behind it to protect the artillery positions, and the old Italian fort Pilastrino was used as headquarters. There was a third line, coded green, which was based on ancient Italian forts Perrone, Airenti, Solaro, and Marcusa. The green line was constructed to defend the city port itself and was to be manned and defended by service personnel in the event of a breakthrough and collapse of the red and blue lines. However, the blue and green lines were not completed, and work on them went on for many months. Yet in the face of this, Tobruk was not only fed and maintained, but it was turned into a dump. During dark moonless nights, the British Navy managed to ferry food, fuel, matériel, weapons, and ammunition. There was a pumping

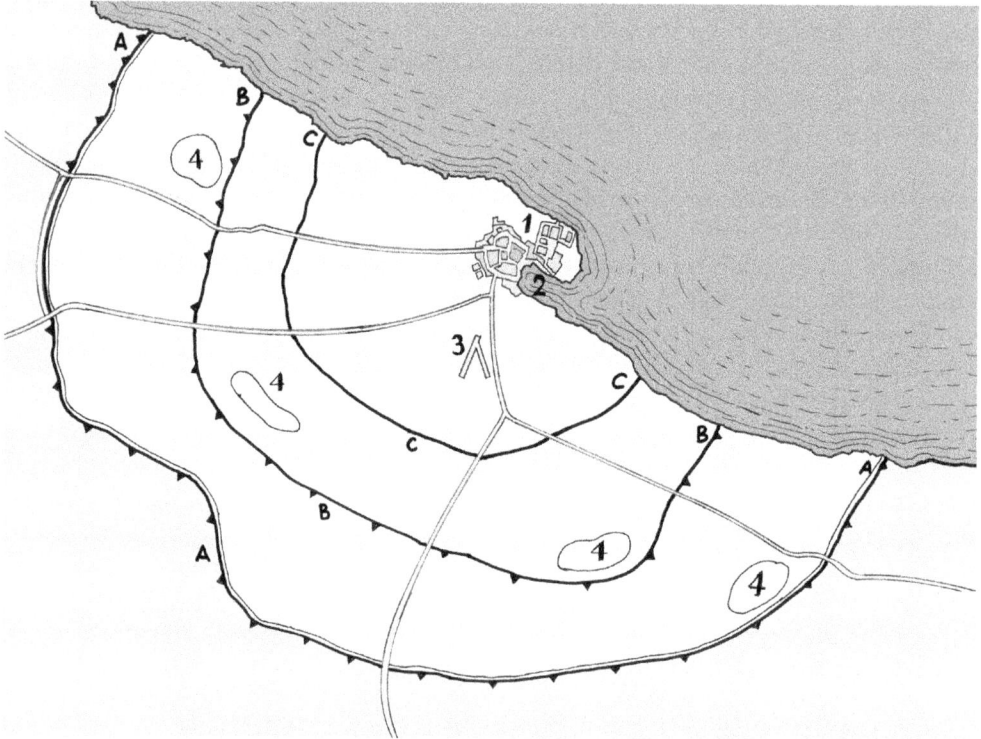

Map of the fortifications of Tobruk in Libya. AA: outer red line, BB: secondary blue line, CC: inner green line. (1) Tobruk city, (2) harbor, (3) El Gubbi airfield, (4) scattered antiaircraft guns positions.

station just northwest of the town capable of bringing water from artesian wells and two water distilleries built by the Italians. As the torrid summer wore on, the defense system was worked out with elaborate detail. The perimeter was divided into interdependent sectors, and all the defenses were designed to swivel around to the flank or the rear and form a new front inside the sector at a moment's notice. All signal wires and communications were duplicated by alternative routes, so if one line was cut out, another could instantly be switched on. There was a complex technique of mutual assistance among the closely interlocked subdivisions of the front line, and reserve forces for counterattacks were always ready in the rear. The siege of Tobruk ended in November 1941, when the port was relieved by the 8th Army (Operation Crusader).

Gazala Line

Another example of World War II British field fortifications is the so-called Gazala Line.

The Battle of Gazala was an important battle of the Western Desert Campaign. It was fought in Libya from May 26 to June 21, 1942. The combatants on the Axis side were the Panzer Army Afrika consisting of German and Italian units and commanded by "Desert Fox" Erwin Rommel. The opposing Allied forces (the famous 8th Army) was commanded by Lieutenant General Neil Ritchie under close supervision of the commander in chief of the Middle East, General Sir Claude Auchinleck.

At the start of February 1942, the British 8th Army dug itself into a fortified line 30 miles west of Tobruk. The line was not a continuous linear defense but a network of strongholds known as "boxes." It extended over some 50 miles from the Mediterranean Sea in the north to Bir Hakeim in the desert inland to the south. Across that line, the two armies faced each other for the next four months while they built up their strength. The boxes were fortified areas, and each of them was an irregular circle or oval, about three to four kilometers in width and five or six in length. Each box included a multitude of foxholes, combat trenches and communication saps, armed emplacements, underground kitchens, field infirmaries and dressing stations, dugouts, shelters, and supply dumps in field fortification style. Everything was leveled and camouflaged, the trenches were narrow, and unless a bomb or a shell fell on them directly, little damage was done. Combat positions were revetted with sandbags, rocks, sand-filled empty ammunitions boxes, and empty fuel drums filled with dirt. Tents and shelters as well as tarps and cargo truck covers were pitched over prone shelters and slit trenches so that the occupants slept below ground level and had protection from the hot sun during the day and from the freezing cold at night. Every foot of elevation provided an advantage for observation, fields of fire, and cover, but high ground and lack of vegetation in some areas made camouflage a challenge. Dust clouds and vehicle tracks made it even more difficult. The heavy armament was placed in open emplacements protected by entrenchments and sandbags, all of them covered by tarps and concealed under camouflage nets. Artillery included 75-millimeter guns, antitank guns (caliber 75, 47, and 25 millimeters), mortars, machine guns, Bofors antiaircraft guns, and heavy antiaircraft machine guns. Barbed wire and manmade obstacles in general were little used because of the lack of materials, the broad frontages, and the wide-empty horizons that enabled enemy ability to outflank positions. Instead, extensive use was made of large minefields with antipersonnel and antitank mines (exploding under a pressure of 200 kilograms), making

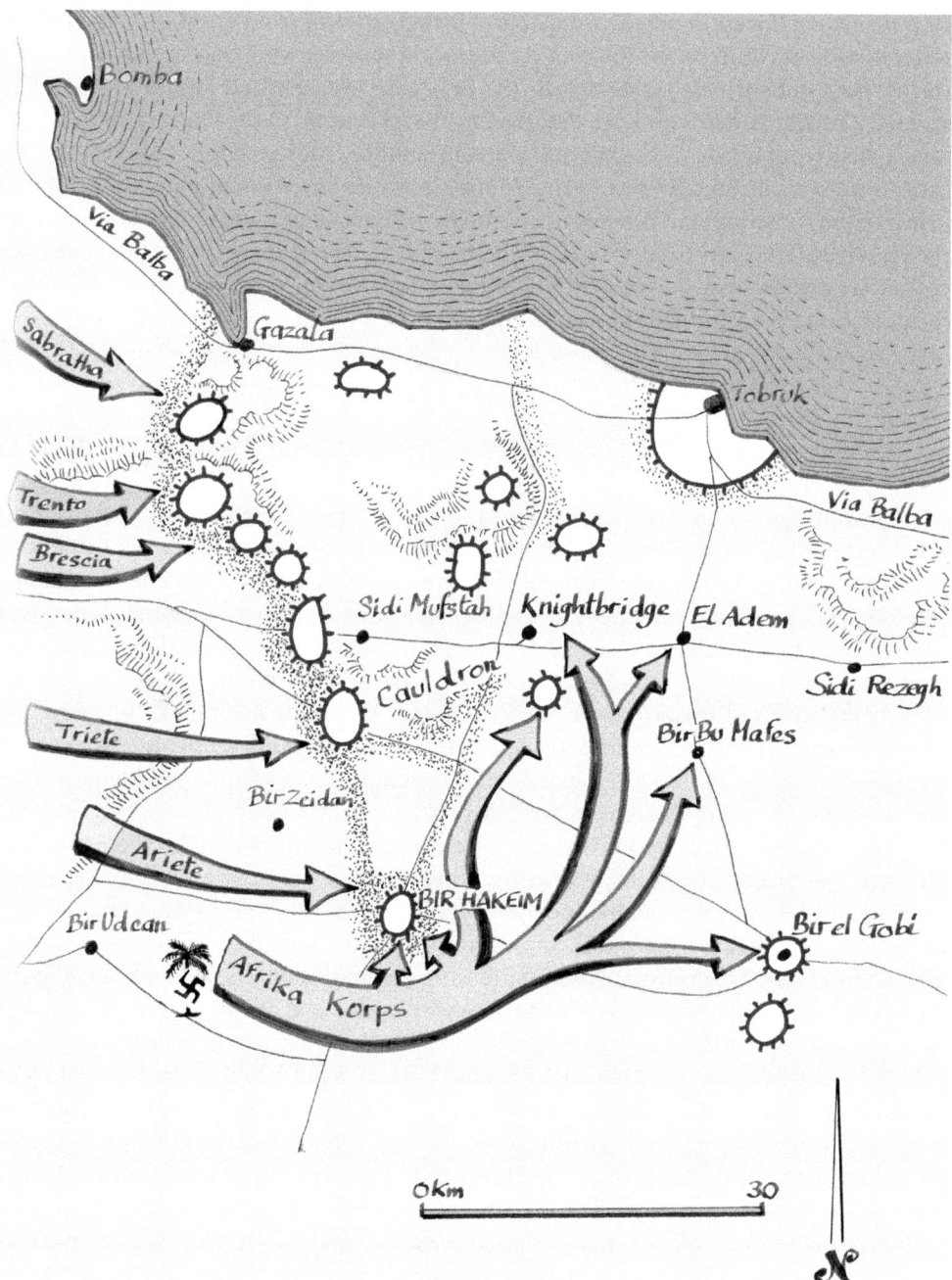

Map of the Gazala Line.

the terrain a huge death trap. There were gates and narrow patrolling lanes in the minefields in order to maintain contact with neighboring boxes and send mobile detachments whose tasks were to patrol and make reconnaissance operations, launch local raids and ambushes, collect information, lay mines, detect and destroy infiltrators, and obstruct enemy mine-clearing teams.

In late May 1942, Rommel drove his armored forces round the southern flank of the Gazala position to engage British armor from the rear. Despite successes in this engagement, Rommel found himself in a precarious position. Indeed, the continuing resistance of the Free French force in the Bir Hakeim box made his supply lines long and difficult. Rommel pulled back into a position abutting the British minefields, a defensive position termed "the Cauldron," creating a threatening presence in the midst of the British forces, which was impossible to ignore. When Ritchie attacked, his armored forces were decimated, and Rommel maintained the initiative until the British withdrew from the Gazala Line. The battle was considered the greatest victory of Rommel's career. However, Rommel's force was now nearing exhaustion, and Auchinleck was able to check his advance at the First Battle of El Alamein in July 1942. The Libyan Desert and North African War ended in May 1943 when the retreating Axis forces were defeated in Tunisia and surrendered.

Malta

The 20th century saw the rundown of the Mediterranean fleet as Britain withdrew its best ships to face the menace of the German navy in home waters. Throughout World War I, the Mediterranean stations of Malta and Gibraltar provided convoy facilities, and Malta succored the wounded from Gallipoli and the Middle East campaign. By the 1930s, the whole defense situation had again changed with the introduction of airpower against which Malta was inadequately protected. The coastal defenses of the two naval stations were very much as they had been at the beginning of the 20th century, mainly consisting of 9.2-inch and 6-inch quick-firing guns in their original battery positions, the marks of the guns from time to time upgraded, supplemented by Oerlikon cannon. During World War II, no major fleet action was attempted against either naval station, although planning for an Axis invasion of Malta was prepared to an advanced stage. The Italian and German bombers and fighters based in airfields in Sicily were a mere hundred miles from Malta and seemed poised to overpower any antiaircraft defenses that might be installed. When the decision was taken to hold on to Malta at all costs in the face of air bombardment, fighter and antiaircraft protection was built up and reinforcements, food, and fuel pushed into the island against stupendous odds. From June 1940 until November 1942, with high losses and large destruction, Malta was able to withstand a grim siege until the Allies defeated the Axis armies in North Africa.

Gibraltar

Between the world wars in the period 1918–1939, the fortifications of Gibraltar were modernized and enlarged with the construction of concrete pillboxes and troop shelters and the excavation of many additional tunnels and casemates into the rock. A 1.8-kilometer runway was built partly in the sea at the beginning of World War II in 1939 by the Fleet Air Arm of the Royal Navy as an emergency airfield for southern Europe and North Africa. The threat of attack on Gibraltar during World War II never materialized because the Spanish dictator General Francisco Franco remained neutral. Although Fascist Italy and Nazi Germany had greatly helped him establish his dictatorship in 1939, the prudent and cunning Franco stubbornly refused to join the Axis and to even allow the passage of a German strike force across Spanish territory.

New Zealand

Right before and during World War II, due to the fear of invasion by Japan, additional modernized coastal batteries with long-range guns, antiaircraft batteries, and concrete machine gun pillboxes were established particularly around the main harbors on the North Island (Auckland, Wellington, Tauranga, New Plymouth, and Gisborne). The fortifications either developed and improved using existing fortifications or new ones built from scratch included fire control and observation posts, underground bunkers sometimes with interconnected tunnels containing magazines, supply, and plotting rooms and protected engine rooms supplying power to the gun turrets and searchlights. There were also kitchens, barracks, and officers' and NCOs' quarters. Radar stations were installed allowing long-range detection and so improving the air defense.

Australia

Australia, too, was under threat of a Japanese invasion, and all available ordnance was pressed into service, including some obsolete guns and field guns adapted for coastal defense. Fortifications were reinforced particularly in Queensland (Brisbane, Cairns, Townsville, and Torres Strait), New South Wales (Newcastle, Port Stephens, Sydney, and Port Kembla), Papua New Guinea (Lae, Port Moresby, and Rabaul), and in Western Australia (Albany, Fremantle, and Geraldton). Presumably, there was a plan known as the Brisbane Line. This was a controversial defense proposal allegedly formulated in 1942 by the Menzies government to concede the northern portion of the Australian continent (roughly from Brisbane to Perth) in the event of a successful invasion by Japan. Instead,

Observation and fire control station at Changi Beach, Singapore.

defenses would have been concentrated in the vital industrial regions between Brisbane and Melbourne. In the end, this "defeatist" scheme seems to have been only an internal electoral dispute between Australian political rival parties.

Conclusion

Post–World War II Britain

The United Kingdom survived World War II, but its power, influence, and prestige were seriously damaged. The human cost was far less than in World War I, and casualties were 233,042 dead, 57,472 missing, and 275,975 wounded. Economic losses, on the other hand, were far heavier than in 1918, for the war effort had swallowed up most of Britain's reserves. So in 1945, victorious Britain emerged from war a debtor nation with an empire—still the largest in the world and stubbornly clinging to old pretentions of global power. When Churchill met Stalin and Roosevelt at Yalta, one observer had likened the trio to the Roman triumvirs who held power after Julius Caesar's assassination. Obviously, Stalin and Roosevelt were Octavius and Mark Anthony, while Churchill was the all-but-forgotten Lepidus. Despite being on the winning side, victorious, glorious but badly mauled Britain and its colonial empire would not recover from the geopolitical shifts caused by World War II. By then, Britain entered a period of terminal decline, and this pertains to France as well.

Decolonization

A general movement of decolonization took place after 1945. India was the first and largest area to be shed in 1948, and then the Middle East and then Africa followed. The British tried to repress a communist uprising in Malaya from 1948 to 1960 known as the Malayan insurgency. Various Caribbean and Pacific possessions also went their separate ways. By the 1950s, the British, impoverished by two wars and the loss of their huge empire, had reduced their commitments and lost interest in global power. The Royal Navy was withdrawn from Malta, and the island became independent in 1961. Gibraltar was, for the time being, retained out of sentiment and from an inability to resolve the conflict of interests between the local population and Spain, but more and more, the place assumed the aspect of a holiday resort rather than a great naval base. The last of the significant colonies to be lost was Hong Kong, which was returned to China in 1997.

Cold War

Soviet Russia and the United States had become, through their industrial and armed strength, superpowers, leaving the United Kingdom, France, as well as defeated Japan, Germany, and Italy to occupy much humbler positions.

The post–World War II period was marked by the so-called Cold War. With Communist countries under the dictatorial leadership of the USSR opposing the capitalist democracies led by the United States, the Cold War (the term was coined in 1945 by the English writer George Orwell) was a geopolitical contest of nerve, diplomatic maneuvers, hysterical arms races, propaganda and psychological warfare, technological competitions in space exploration, childish rivalry in sport events, intelligence gathering, and subversion in which each side was continually nervous about the other's hidden intentions and capability for making mischief. Both the United States (since 1945) and the USSR (since 1949) possessed atomic bombs, and as a result, the stakes of the Cold War were perilously high. Indeed, the period was marked by a number of serious crises and proxy wars, e.g., the Berlin blockade (1948–1949), the French Indochina War (1946–1954), the Korean War (1950–1953), the Hungarian Revolution (1956), the Berlin Wall (established in 1961), the Cuban missile crisis (1962), the Russian invasion of Czechoslovakia (1968), and the Vietnam War (1955–1975). During that period of tension, Anglo-American solidarity was as vital as it had ever been during World War II. However, Britain—although it was a member of the North Atlantic Treaty Organization and although it possessed nuclear weapons developed in the late 1950s—was only a second-rank power.

The Cold War ended with the dissolution of the Soviet Union in December 1991.

From April to June 1982, Great Britain was involved in a nondeclared war against Argentina, which had invaded the Falkland Islands located in the southern Atlantic Ocean.

Post–World War II Fortifications

The great majority of Britain's 1940 static defenses were dismantled, a process that started even before the end of the war. Ditches and trenches were filled, loopholes blocked up, wood and metal obstacles recycled, and many pillboxes destroyed.

After the war, farmers, across whose lands defensive structures had been built, were, in addition to receiving compensation, paid to fill in ditches and trenches and to demolish pillboxes. Today, hardly anything remains of the antitank ditches, but at the time, they must have been the most conspicuous of all the fortifications; a few remain, much humbled, as meadow drains or field boundaries. In the case of pillboxes, the challenge of demolishing such structures was considerable and it seems that most farmers preferred to use them as barns or sheds.

Today, it is very rare to find any part of Britain's defenses other than those made of concrete. Immediately after the war, there were more pressing matters to attend to than conserving the litter of a battle that never happened. For decades, weather, nature, erosion, and modern construction have destroyed many structures. Many coastal fortifications have tumbled into the sea or have sunk into the sands on which they were built. Other defenses have succumbed to road improvements or have been demolished to make way for modern developments. For many of those that remain, neglect and the attention of nature have achieved a degree of camouflage greater than that of during World War II. Years after the war, memories faded, and nobody could be sure how many pillboxes and related hardened field defenses had survived, or indeed how many had been constructed in the first place. In the late 1970s, the journalist Henry Wills began researching on the topic, eventually leading in 1985 to the publication of *Pillboxes: A Study of UK Defenses*. Interest was stimulated, both public and professional, and local surveys were carried out, culminating in the Defence of Britain Project, which took place from 1995 to 2002, attempting to record all known military defense sites. From this and other investigations, it is estimated that some 28,000 pillboxes and other hardened field fortifications were constructed in the United Kingdom of which about 6,500 still survive. For many pillboxes, a new use has been found. Type 28s, being internally spacious and having a large rear entrance, are probably the most amenable to reuse; on farms and in gardens, they serve as cattle sheds and storage lockers.

In the postwar context of uncertainty, the Cold War, threat, and global destruction by nuclear fire, purely military fortifications have become obsolete. It was clear that a new total war would not be a war of invasion but of obliteration by atomic bombs. In 1956, the British Coastal Artillery was disbanded, and many forts and batteries no longer had a national military purpose. All coastal fortifications were deactivated and all guns and installations removed from active service. Only a few were retained, though, as useful military matériel storages, accommodation, and training grounds. So after nearly 1,000 years of fortification, Great Britain's coasts were undefended.

However, some new fortifications emerged during the Cold War. Made proof against nuclear, radioactive, chemical, and bacteriological warfare, they had three major purposes.

The first category included active military installations and bases, notably underground silos (vertical tubes) surrounded by the necessary apparatus control centers, and stores for launching ballistic missiles and rockets armed with nuclear warheads, offering a chance of launching attacks of retaliation.

The second category included passive shelters for the physical protection of a fraction of the population, those happy few deemed indispensable to the future of humankind after an irreversible global, nuclear holocaust.

A third category included shelters for the rest of humanity with capacity and scale of protection varying from do-it-yourself shelters, caves, warrens, and basements to large and spacious well-provisioned government-built subterranean public structures. Admittedly, civilian shelters are basically refuges for frightened people in times of trouble, like the brochs of ancient times and the castles of the medieval period, but most of them would not withstand blast, fire, and radioactive fallout. It is well known that the only and best defense against the atomic bomb is not to be where it goes off. In a land bombed back into a totally poisoned Stone Age, the point of such short-term absurd protective structures is, of course, an open question.

The atomic bomb brought an end to World War II, but it did not bring an end to war itself. Traditional permanent fortifications have disappeared from the map, and there is no longer a place in coastal defense for long-range artillery firing at warships 30 kilometers away when rockets and missiles could sink the same battleship sailing on the other side of the ocean. However, cheap field fortifications are still widely used in conventional warfare. Even today, barbed wire is still a formidable obstacle, while a simple ditch, a foxhole or a trench dug in the ground, a simple parapet made of dirt, a Hesco barrier (a revival of the old gabion filled with earth), or a breastwork of sandbags are still extremely efficient for protecting against small-caliber projectiles and hand grenade splinters.

British Fortifications Today

Remainders of British Fortifications Today

Although many fortifications have disappeared, Britain has fortunately preserved a lot of them.

From the Tudor era, only a few fortifications have come down to us. The best preserved and most spectacular are without doubt Henry VIII's coastal forts in south England (Deal, Walmer, St. Mawes, and several others).

Many 17th- and 18th-century bastioned fortifications have been demolished, but there are a number of forts (Fort Tilbury near London and Fort George near Inverness, for example) and towns still displaying urban bastioned enceintes in various styles, notably Berwick-upon-Tweed and on Jersey in the Channel Islands.

From the Napoleonic Wars and from the 19th century, there are still many Martello towers, coastal and land forts, and redoubts. A number of Palmerston forts (at Portsmouth, Plymouth, Pembroke, Chatham, and Dover), sea forts (Spitbank Fort and Horse Sand Fort), and other strongholds are still extant.

Most World War II pillboxes have disappeared, but a number of 20th-century concrete bunkers and Maunsell sea forts still exist. Interesting and impressive concrete remnants of the World War II German Atlantic Wall can still be seen in the Channel Islands.

Numerous British fortifications are also preserved in overseas countries that were once parts of the empire: notably in Malta, Gibraltar, the United States, Canada, India, the Caribbean, Australia, New Zealand, and Africa. In Portugal, some of the Napoleonic fortlets of the Lines of Torres Vedras are preserved.

New Purposes

After decades of neglect, there is fortunately a renewal of interest in military architecture, and local, national, private, and governmental charities, trusts, membership organizations, and nonprofit associations have risen to preserve castles, strongholds, forts, and bunkers, one of the oldest being the National Trust founded in 1895. Many archaeological sites, historical buildings, monuments, and fortified places are now legally protected by the status of Scheduled Monuments. It is indeed important that these unique and sometimes rare historical artifacts be properly preserved for future generations.

The very nature of fortification, including high walls, deep ditches, difficult access, solid gates, and underground chambers and magazines, has made many ancient castles, discarded forts, obsolete citadels, and large empty subterranean bunkers suitable as safe repositories for all sorts of things and people, which have to be kept under lock and key. Fortifications were originally intended to prevent violent entrance, but they could also avoid escape. Throughout history, old, discarded fortifications have often been employed as prisons and prisoner of war camps. As treasure houses, they could be used to safely store various historical artifacts, precious pieces of art, security documents, equipment, archives, and other administrative records, which need a carefully controlled environment for their conservation. Some old forts are still used as training grounds by military forces, fire brigades, police forces, and Scout associations. Others are now under the control of town councils or natural preservation departments. Some dark and damp basements or pillboxes have been converted to make artificial caves for bats and other animals. Many strongholds have been turned into museums opened to the public. Many former military fortified places now serve a variety of civilian private purposes like concert or exhibitions halls and for hosting all forms of cultural and entertainment events. Many are used as conference centers, business events, care organizations, wedding venues, offices, restaurants, guesthouses, and hotels. The former fortifications help create jobs and boost the economy while forming at the same time pleasant recreation places and tourist attractions, as well as an important basis for cultural and historical conservation and a living asset for understanding Great Britain's rich, diverse, and fascinating past.

Appendices

Appendix 1

British Kings and Queens, 1485–2022

House of Tudor

Henry VII 1485–1509
Henry VIII 1509–1547
Edward VI 1547–1553

Lady Jane Grey 1553
Mary I 1553–1558
Elizabeth I 1558–1603

House of Stuart

James I 1603–1625
Charles I 1625–1649
Commonwealth (led by Oliver Cromwell) 1649–1660
Charles II 1649–1685

James II 1685–1688 (deposed)
William III and Mary II 1698–1694
William III (alone) 1694–1702
Anne 1702–1714

House of Hanover

George I 1714–1727
George II 1727–1760
George III 1760–1820

George IV 1820–1830
William IV 1830–1837
Victoria 1837–1901

House of Saxe-Coburg-Gotha
(called Windsor since 1917)

Edward VII 1901–1910
George V 1910–1936
Edward VIII 1936

George VI 1936–1952
Elizabeth II 1952–2022
Charles III 2022–today

Appendix 2

British Prime Ministers, 1721–1945

Prior to the creation of the United Kingdom in 1707, the Treasury of England was led by an officer called the Lord High Treasurer. From Tudor times, the Lord High Treasurer was regarded as one of the Great Officers of State and was often (though not always) the dominant figure in the government. There is no specific date when the office of prime minister first appeared, as the function was not created but evolved over a period of time. However, it is generally considered that the office appeared in 1721 with the tenure of the First Lord Robert Walpole. Since then and until today, the prime minister of the United Kingdom is the political leader of the country, the head of Her/His Majesty's Government, and still holds the title of First Lord of the Treasury. Traditionally, the reigning monarch appoints as prime minister the person most likely to command the confidence of the House of Commons—an elected body consisting of 650 members known as members of parliament (MPs) meeting in the Palace of Westminster in London. The chosen prime minister is typically the leader of the political party or coalition of parties that holds the largest number of seats in that chamber.

Robert Walpole (1721–42)
Spencer Compton (1742–43)
Henry Pelham (1743–54)
Thomas Pelham-Holles (1754–56; 1st tenure)
William Cavendish (1756–57)
Thomas Pelham-Holles (1757–62; 2nd tenure)
John Stuart (1762–63)
George Grenville (1763–65)
Charles Watson Wentworth (1765–66; 1st tenure)
William Pitt the Elder (1766–68)
Augustus Henry Fitzroy (1768–70)
Frederick North (1770–82)
Charles Watson Wentworth (1782; 2nd tenure)
William Petty-Fitzmaurice (1782–83)
William Henry Cavendish-Bentinck (1783; 1st tenure)
William Pitt the Younger (1783–1801; 1st tenure)
Henry Addington (1801–4)
William Pitt the Younger (1804–6; 2nd tenure)
William Wyndham Grenville (1806–7)
William Henry Cavendish-Bentinck (1807–9; 2nd tenure)
Spencer Perceval (1809–12)
Robert Banks Jenkinson (1812–27)
George Canning (1827)
Frederick John Robinson (1827–28)
Arthur Wellesley (1828–30; 1st tenure)
Charles Grey (1830–34)
William Lamb (1834; 1st tenure)
Arthur Wellesley (1834; 2nd tenure)
Robert Peel (1834–35; 1st tenure)
William Lamb (1835–41; 2nd tenure)
Robert Peel (1841–46; 2nd tenure)
John Russell (1846–52; 1st tenure)
Edward Geoffrey Stanley (1852; 1st tenure)
George Hamilton-Gordon (1852–55)
Henry John Temple (1855–58; 1st tenure)
Edward Geoffrey Stanley (1858–59; 2nd tenure)
Henry John Temple (1859–65; 2nd tenure)
John Russell (1865–66; 2nd tenure)
Edward Geoffrey Stanley (1866–68; 3rd tenure)
Benjamin Disraeli (1868; 1st tenure)
William Ewart Gladstone (1868–74; 1st tenure)

Benjamin Disraeli (1874–80; 2nd tenure)
William Ewart Gladstone (1880–85; 2nd tenure)
Robert Cecil (1885–86; 1st tenure)
William Ewart Gladstone (1886; 3rd tenure)
Robert Cecil (1886–92; 2nd tenure)
William Ewart Gladstone (1892–94; 4th tenure)
Archibald Philip Primrose (1894–95)
Robert Cecil (1895–1902; 3rd tenure)
Arthur James Balfour (1902–5)
Henry Campbell-Bannerman (1905–8)
H.H. Asquith (1908–16)
David Lloyd George (1916–22)
Bonar Law (1922–23)
Stanley Baldwin (1923–24; 1st tenure)
Ramsay Macdonald (1924; 1st tenure)
Stanley Baldwin (1924–29; 2nd tenure)
Ramsay Macdonald (1929–35; 2nd tenure)
Stanley Baldwin (1935–37; 3rd tenure)
Neville Chamberlain (1937–40)
Winston Churchill (1940–45; 1st tenure)

Bibliography

Beanse, Alec, and Gill Roger. *The London Mobilisation Centres.* Hampshire: David Moore Publishing, 2011.
Bird, C. *Silent Sentinels: A Study of the Fixed Defences Constructed in Norfolk During WWI and WWII.* Dereham: Larks Press, 1999.
Bowyer, Richard. *Dictionary of Military Terms.* London: A&C Black, 1999.
Bragard, Philippe, Johan Termote, and John Williams. *A la découverte des villes fortifiées (Kent, Côte d'Opale et Flandre Occidentale).* Dunkerque: Syndicat mixte de la Côte d'Opale, 1999.
Clayton, Hugh. *Royal Faces, 900 Years of British Monarchy.* London: Her Majesty's Stationery Office, 1977.
Cruickshank, Dan. *Invasion, Defending Britain from Attack.* Basingstoke: Boxtree, 2001.
Delderfield, Eric R. *Kings and Queens of England and Great Britain.* Newton Abbot: David & Charles, 1972.
Diderot, Denis, Jean Le Rond d'Alembert, and Louis De Jaucourt. *Encyclopédie, arts militaires.* Paris: Inter-livre Editions Paris, 1986.
Dupuy, Trevor N. *The Evolution of Weapons and Warfare.* New York: Da Capo, 1984.
Eyre, A.G. *An Outline History of England.* Harlow: Longman, 1971.
Fa, Daren, and Clive Finlayson. *The Fortifications of Gibraltar 1068–1945.* Oxford: Osprey, 2006.
Foot, William. *The Battlefields That Nearly Were: Defended England 1940.* Stroud: Tempus, 2006.
Harrington, Peter. *The Castles of Henry VIII.* Oxford: Osprey, 2007.
Harrington, Peter. *English Civil War Fortifications 1642–51.* Oxford: Osprey, 2003.
Her Majesty's Government. *Consolidated Instructions to Invasion Committees in England and Wales.* London: Her Majesty's Stationery Office, July 1942.
James, Lawrence. *The Rise and Fall of the British Empire.* London: Abacus Books, 1994.
Kamps P.J.M., ed. *Terminologie Verdedigingswerken, Inrichting, Aanval en Verdediging.* Utrecht: Stichting Menno van Coehoorn-De Walburg Pers, 1999.
Kauffmann, J.E., and Robert M. Jurga. *Fortress Europe: European Fortifications of World War II.* New York: Da Capo, 2002.
Keegan, John. *A History of Warfare.* London: Hutchinson, 1993.
Kightly, Charles. *Strongholds of the Realm.* London: Thames & Hudson, 1979.
Lampe, David. *The Last Ditch: Britain's Resistance Plans Against the Nazis.* London: Greenhill Books, 2007.
Le Hallé, Guy. *Précis de la fortification.* Paris: PVC Editions, 1983.
Lendy August, Frederick. *Treatise on Fortifications or Lectures Delivered to Officers Reading for the Staff.* London: W. Mitchell Stationer, Printer, Engraver and Bookbinder, 1862.
Lepage, Jean-Denis. *British Fortifications Through the Reign of Richard III: An Illustrated History,* Jefferson: McFarland, 2012.
Libal, Dobroslav. *Châteaux forts & fortifications en Europe du Ve au XIXe siècle.* Paris: Ars Mundi, 1993.
Lowry, Bernard, and Taylor Chris. *British Home Defenses 1940–45.* Oxford: Osprey, 2004.
Maurois, André. *Histoire d'Angleterre.* Paris: Fayard, 1937.
McInnes, C., and G.D. Sheffield. *Warfare in the Twentieth Century: Theory and Practice.* London: Unwin Hyman, 1988.
Mohr, A.H. *Vestingbouwkundige Termen.* Utrecht: Stichting Menno van Coehoorn-De Walburg Pers, 1983.
Moore, David. *A Handbook of Military Terms.* n.p.: Solent Papers & Victorian Forts Publications, 2011.
Morewood, Steve. *The British Defence of Egypt 1935–40.* Abington: Taylor & Francis, 2014.
Osborne, Mike. *Pillboxes in Britain and Ireland.* Stroud: Tempus, 2008.
Osborne, Mike. *Twentieth Century Defences in Britain.* Stroud: Tempus, 2003.
Paluzie de Lescazes, Carlos. *Castles of Europe.* Barcelona: EGC, 1982.
Parker, Geoffrey. *Warfare.* Cambridge: Cambridge University Press, 1995.
Penoyre, John, and Michael Ryan. *British Architecture.* London: Frederick Warne, 1951.
Phillips, Ellen. *The Enterprise of War.* New York: Time-Life Books, 1991.
Reid, Stuart, and Graham Turner. *Castles and Tower Houses of the Scottish Clans 1450–1650.* Oxford: Osprey, 2006.
Ropp, T. *War in the Modern World.* London: Duke University Press, 1959.

Ross, Stewart. *World War II Britain. History from Buildings*. London: Franklin Watts, 2006.
Rossi, Guido Alberto. *Grandi castelli visti dal cielo*. Venice: Atrium, 1978.
Ruddy, Austin. *British Anti-Invasion Defences 1940–1945*. Pulborough: Historic Military Press, 2003.
Sailhan, Pierre. *La fortification, histoire & dictionnaire*. Paris: Tallandier, 1991.
Saint John Parker, Michael. *Britain's Kings and Queens*. Andover: Pitkin Pictorials, 1990.
Saunders, Andrew. *Fortress Britain: Artillery Fortification in the British Isles and Ireland*. Liphook: Beaufort Publishing, 1989.
Stephenson, Charles. *Castles*. Lewes: Ivy Press, 2011.
Stephenson, Charles, and Steve Noon. *The Fortifications of Malta 1530–1945*. Oxford: Osprey, 2004.
Tardi, Jacques. *Loopgravenoorlog*. Paris: Casterman, 1993.
Treu, Herman, and Jaap Sneep. *Vesting Vier Eeuwen Vestingbow in Nederland*. Zutphen: Stichting Menno van Coehoorn, 1982.
Viollet-le-Duc, Eugène-Emmanuel. *Histoire d'une fortresse*. Paris: Berger-Levrault, 1978.
War Office. *Barrel Flame Traps, Flame Warfare*. Military Training Pamphlet No. 53, Part 1. London: War Office, July 1942.
Westermann, Georg. *Grosser Atlas zur Weltgeschichte*. Braunschweig: Druckerei und Kartographische Anstalt, 1956.
Wilkinson, Frederick. *Wapens en Wapenuitrustingen*. London: Hamlyn, 1980.

Index

Admiralty scaffolding 200, 201
air raid precautions 195
Albert Fort 160
Allan Williams turret 206
Ambleteuse 24
American Civil War 121, 122, 148
American War of Independence 77, 78
Anglican Church 3
Anglo-Dutch Wars 54
antitank obstacles 231–234
Atlantic Wall 242, 243
Augustus Fort 65
Ayr Citadel 53, 54

barbed wire 234, 235
Battle of Britain 194
Beausejour Fort 68, 72, 73
Berwick-upon-Tweed 32
Bison 205
Blériot, Louis 178
blind spot 180
blitz 196, 240
blockhouse 176, 177
Boer Wars 176–178
Borstal Fort 135
Bovisand Fort 131, 132
Brialmont, Henri 152
Bristol 47
Brockhurst Fort 127
bunker 192

Calshot Castle 16
camouflage 236, 237
Caponier 98, 101, 102, 116, 124, 129
Carisbrook Castle 33
Carnot Wall 115, 120, 129, 130
Catherine of Aragon 3
Channel Islands 21, 242, 243
Charleston 70
Charlotte Fort 82
Chatham Dockyard 105, 106, 135
Cherbourg 114, 123
Churchill, Winston 240; command bunker 240

Clarence Fort 107
Cockbourne 160, 161
Cockham Wood 57
coffer 145
Coles, Cowper, Philipps 148
Coles's Turret 148
colonialism 156, 157
concrete 145
Corgarff Castle 31
Cornet Castle 22
Cromwell, Oliver 36, 47
Cromwell, Thomas 7
Crownhill Fort 133, 134
Cumberland Fort 66, 67, 90
Cumberland Fort (USA) 73
Cumberland Lines 106

Dacoit fence 146, 147
Dad's Army 197
Darnet Fort 138
Deal Castle 10–13
decoy sites 208
De Gomme, Bernard 38, 39, 42, 47, 50, 51, 56–58
Delhi 173
disappearing gun 150
Donnington Castle 41, 42, 43
Dover quad pillbox 228
Dover Western Heights 100–105, 143, 144
Dürer, Albrecht 6
Dutch system 37, 38
Dymchurch Redoubt 99, 100

eared pillbox 225, 226
Eastbourne Redoubt 98, 99
Elizabeth I (Queen) 24, 25, 84
Elswick mounting 151
English Civil War 36, 37
Erie Fort 80, 81

Fareham Fort 139, 140
Fergusson, James 117
Feste 152
Fincastle Fort 83
fort 8
Franco, Franscisco 247
FW 3 Directorate 208–210
FW 3 pillboxes 208–223

Gazala Line 254, 246
George Fort 65, 66
Gibraltar 61, 62, 167, 247, 249

Halifax 164
Hampton Court 4
Harwich Redoubt 100
Haxo casemate 140, 144
Hexham, Henry 38
Hong Kong 249
Hoo Fort 137
Horse Sands Fort 142, 143
Horsted Fort 136
hull 23, 60, 61

Industrial Revolution 117, 156, 157
Ive, Paul 29

Jervois, William 117

Languard Fort 120, 121
Lawrence Fort 71, 72
Lee, Richard 8
Leicester 51
Lendy, Frederick 117
Ligonier Fort 75, 76
Lincolnshire pillbox 227
Littlehampton Redoubt 120
Liverpool 51, 52
London 49, 50, 153–155, 196, 201
Loudon Fort 74, 75
Louisbourg 71
Louisiana 68, 69
lozenge pillbox 224, 225

Madras 85, 86
Malta 167, 168–172, 247, 249
Manhattan Island 69
Martello towers 90–96, 162, 163
Maunsell Fort 238–240
moir pillbox 184
Montalembert, René 115, 123
Montresor, James 63
Morrison shelter 196
Muller, John 63
mushroom pillbox 206

261

New Amsterdam 69, 70
New England 67
New York 68, 69
Newark-on-Trent 45, 46
Niagara Fort 79
Norcon pillbox 228, 229
Nouvelle France 67

October Revolution 190
Operation Sealion 193–195
Organisation Todt 242
Oxford 42, 43

Palmerston, Henry John 120, 128
Palmerston Fort 128
Pasley, Charles William 116, 117
Pembroke Dock 119
Pendennis Castle 18, 33, 34
Perch Rock Fort 118
Pickett-Hamilton pillbox 203, 204
pimple 232
pirates 83, 84
Pitt Fort 76, 77, 108
PLUTO 130
Poilus 185
Portland Castle 18, 19
Portsmouth 20, 21, 50, 55
prepared battlefield 192
Purbrook Fort 138, 139

Quebec 165, 166
Queenscliff Fort 174

Raleigh, Walter 35
Réduit 126, 127
Regent Fort 8
retractable turret 148, 149
rifled artillery 121
roadblock 234, 235
Roanoke 34, 35
Rogers, John 8
Rommer, John 63
Royal Military Canal 108, 109, 198
Ruck pillbox 230

Saint Mawes Castle 17, 18
Sandgate Castle 14, 15
Sandown Battery 130
sconce 45
Sea Fort 140, 141
Shoreham Redoubt 129
shrapnel 144
Southsea Castle 19
spigot mortar emplacement 230, 231
Spitbank Fort 141
Stanton shelter 195
Star Castle 30
Steed, John 63
Suez Canal 157, 178

Tangier 39
Tenedos Fort 175
Tett turret 207
Ticonderoga Fort 78
Tilbury Fort 58, 59

timber fort 68, 75, 76
Tobruk 243, 244, 245
Torres Vedras 109–113
Tour Modèle 97
transitional fortification 6, 7
trench warfare 179–185
Turnbull mount 209
Twydall profile 145, 146

Upnor Castle 29, 30

Vauban, Sébastien Le Prestre 84
Victoria Line 168, 169
Von Hapschenperg, Stefan 7, 10

Wade, George 64, 65
Walmer Castle 14, 15
Ward, Robert 38
Wars of the Roses 3
William Fort 87, 88
William Henry Fort 73, 74
Windsor Castle 5
wolf pit 176
Woodland Fort 134
Worcester 48

Yarmouth Castle 21
Yaverland Battery 129, 130
Yorktown 80
Yule, Henry 117

Zulu War 175

www.ingramcontent.com/pod-product-compliance
Ingram Content Group UK Ltd.
Pitfield, Milton Keynes, MK11 3LW, UK
UKHW050538150426
5217IPUK00026B/1984

9 781476 689715